WAREHOUSE AND DISTRIBUTION AUTOMATION HANDBOOK

Other Books of Interest from McGraw-Hill

WAREHOUSE AND DISTRIBUTION AUTOMATION HANDBOOK

Nicholas D. Adams, P.E. Editor-in-Chief and Co-author

Terry W. Brown Co-author

Rowland V. D. Firth Co-author

Laura P. Misenheimer Co-author

McGraw-HILL

New York   San Francisco   Washington, D.C.   Auckland   Bogotá
Caracas   Lisbon   London   Madrid   Mexico City   Milan
Montreal   New Delhi   San Juan   Singapore
Sydney   Tokyo   Toronto

Library of Congress Cataloging-in-Publication Data

Warehouse and distribution automation handbook / Nicholas D. Adams ...
[et al.].
    p.   cm.
  Includes index.
  ISBN 0-07-000400-5
  1. Warehouses—Automation—Handbooks, manuals, etc.   I. Adams,
Nicholas D.
TS189.6.W37   1996
658.7'8—dc20   96-15423
                                                       CIP

McGraw-Hill

A Division of The McGraw-Hill Companies

1 2 3 4 5 6 7 8 9 0   DOC/DOC   9 0 1 0 9 8 7 6

ISBN 0-07-000400-5

*The sponsoring editor for this book was Robert Esposito, the editing supervisor was
Stephen M. Smith, and the production supervisor was Pamela A. Pelton. It was set in
Times Roman by Priscilla Beer of McGraw-Hill's Professional Book Group composi-
tion unit.*

Printed and bound by R. R. Donnelley & Sons Company.

McGraw-Hill books are available at special quantity discounts to use as premiums and
sales promotions, or for use in corporate training programs. For more information,
please write to the Director of Special Sales, McGraw-Hill, 11 West 19th Street, New
York, NY 10011. Or contact your local bookstore.

This book is printed on acid-free paper.

CONTENTS

PREFACE

The purpose of the *Warehouse and Distribution Automation Handbook* is to provide the information required to develop, organize, and implement an automation project for a warehouse or distribution center. This handbook provides a time-proven "template" on how to analyze, evaluate, design, and install a distribution automation project while utilizing the latest technologies to achieve the desired results. Throughout the chapters we explore industry trends and directions, technologies and their application, system design and implementation, material handling and mechanization, and many other subjects as they relate to the automation and mechanization of a distribution facility.

This book is divided into three parts. Part 1, Preparation, discusses industry trends and a project approach to warehouse and distribution automation. Part 2, Planning and Design, provides insight into the various considerations to be analyzed during a project, along with how to define requirements and evaluate systems. The use of simulation is reviewed to help analyze the complexities of an automation project. Finally, a method to quantify benefits and justify the project is discussed. Part 3, Implementation, focuses on a methodology and structure that ensures that the implementation is a success. Testing and training tips are also given on how to minimize the learning curve and achieve productive results.

A step-by-step approach is presented for improving the fairly complex distribution function. Such an approach is often overlooked or given a low priority by management, but if implemented successfully has the potential for significantly contributing to a company's bottom line. The project process gives all the considerations involved in an automation project, highlighting the cautionary review points for preventing and avoiding errors before, during, and after implementation.

Upon review of this book, a company can better make an informed decision on technology, software, and material handling as they apply to the business needs of that firm.

Nicholas D. Adams, P.E.
Editor-in-Chief

ACKNOWLEDGMENTS

The co-authors of this book and the contributors who helped them write it have participated in the automation of warehouse and distribution facilities throughout the world and draw upon numerous years of valuable distribution experience in many different industries.

Brief biographies of the co-authors follow:

Nicholas D. Adams, P.E., is a System Integrator with Worldwide Chain Store Systems (WCSS), a wholly owned subsidiary of IBM Corporation. He is an Industrial and Distribution Engineer who for the past 14 years has designed and installed numerous warehouse management systems and material handling and storage systems crossing over a variety of industries. He received a B.S. in Industrial Engineering from the New Jersey Institute of Technology and an M.B.A from the University of North Carolina at Charlotte. Mr. Adams is Vice-Chairman of the Material Handling Institute Association (MHIA) warehouse management system product section and the Industry of Industrial Engineering (IIE) society and recently received the Certificate of System Integration from IIE.

Terry W. Brown is a Systems Consultant with Worldwide Chain Store Systems (WCSS). He holds a B.S. in Mechanical Engineering from Purdue University. Mr. Brown has worked on numerous distribution and manufacturing automation projects, from design through implementation, as a system integrator and project manager over the past 11 years with both WCSS and IBM Corporation. He currently serves as a delegate to the Material Handling Institute Association (MHIA) warehouse management systems product section and is the WCSS representative to the Warehouse Education and Research Council (WERC).

Rowland V. D. Firth is an Industry Consultant with Worldwide Chain Store Systems (WCSS). He holds bachelor's and master's desgrees in Mechanical Engineering from Purdue University. His experience with major distribution centers has been worldwide and has encompassed various responsibilities from system integrator and project manager to the design and installation of warehouse management systems. He has been with WCSS since its inception; prior to that Mr. Firth was a Development Engineer with IBM Corporation for 18 years, during which time he received several patents in product development.

Laura P. Misenheimer is an Industry Specialist with Worldwide Chain Store Systems (WCSS). Her experience relative to distribution systems includes software design, systems integration, project management, technical proposal response generation, and product planning. Ms. Misenheimer holds a B.S. in Material Engineering from North Carolina State University. In addition to her distribution systems experience, she has authored numerous publications, and received a patent based on product development and manufacturing assignments within IBM.

The contributors to this book are as follows:

Jeanmarie Adams is an Advisory Engineer with IBM Corporation and is currently a team leader for an enterprise resource planning (ERP) implementation encompassing the planning, manufacturing, and distribution operations. She has performed simulation analyses for various manufacturing distribution facilities and is also APICS (American Production and Inventory Control Society) certified. Ms. Adams has a B.S. in Industrial Engineering from West Virginia University.

Dennis Castor is Manager of Marketing Communications for Teklogix. He has more than 10 years experience in the materials management and warehouse automation industries. Mr. Castor serves as chairperson of the Automatic Identification Manufacturers (AIM) USA's Radio Frequency Data Communications committee.

Richard C. Frye is Vice-President of Marketing for White Storage & Retrieval Systems, Inc., a major manufacturer of horizontal and vertical carousels and a subsidiary of Pinnacle Automation, Inc. Mr. Frye has been active in the materials handling industry for more than 25 years, having also served with companies in the hoisting equipment and fork lift truck businesses.

J. P. Hornak has been in the manufacturing industry for over 25 years, mostly with the IBM Corporation. While at IBM, he held many line management positions in engineering and manufacturing. In addition, he has worked in systems integration, professional services, sales, and marketing, all within the industrial sector. Mr. Hornak is presently responsible for managing the manufacturing and distribution markets for Symbol Technologies.

Lori Ives-Godwin is a System Integration and Installation Manager for Worldwide Chain Store Systems (WCSS) and has been involved in numerous large-scale warehouse management and distribution system installations. Ms. Ives-Godwin, formerly with Anderson Consulting, has been involved in software development since 1987 and has managed all phases of the installation project life cycle. She has a B.S. degree in Operations Research and Industrial Engineering from Cornell University.

Dave Kennedy is currently the Senior Director of Logistics for Barrie National, Inc., a pharmaceutical manufacturer. Prior to joining Barrie, Mr. Kennedy was employed for nearly 30 years with IBM Corporation and Worldwide Chain Store Systems (WCSS) in various operations and distribution management assignments. He is a graduate of the West Virginia Institute of Technology with a B.S. degree in Industrial Management.

Karen M. Longe is Marketing Manager for Zebra Technologies Corporation, a leading provider of bar code labeling solutions worldwide. She works with bar code manufacturers, distributors, and software integrators to assure that the development and dissemination of bar code printers and labels satisfy their market needs. Prior to joining Zebra, Ms. Longe was Manager of Program Planning and Development for the bar code program at the American Hospital Association, where she chaired the committee that developed the Health Industry Bar Code Provider Applications Standard. She co-authored the book *Bar Technology in Healthcare: A Tool for Enhancing Quality, Productivity, and Cost Management.*

Tom Myers has over 20 years experience in the field of automation, and has held positions in application engineering, system engineering, project management, sales, and marketing. He has held management positions at Procter & Gamble, General Electric, and Litton. He is currently Manager of Marketing for SI Handling Systems, a systems integrator and supplier of automated order selection systems. Mr. Myers has a B.S. in Electrical Engineering from Miami University and an M.B.A from Xavier University.

William Nordgren earned an M.S. in Computer Integrated Manufacturing from Brigham Young University. His expertise in simulation and simulation project management is well established. Mr. Nordgren was co-founder of ProModel Corporation and served as Vice-President until 1992. In 1993, he joined F&H Simulations B.V. as President of F&H Simulations, Inc. U.S.A. Through this union, Taylor II Simulation software was introduced into the U.S. market.

John Raab is Vice-President of Marketing at Rapistan Demag Corporation. He holds a B.A., M.A., and M.B.A., and has been employed by Rapistan Demag for the past 15 years. He has held positions in engineering, sales, and marketing in automated material handling companies for 20 years. Mr. Raab is currently Vice-President of the Conveyor

Equipment Manufacturers Association (CEMA) and is Rapistan Demag's delegate to the Material Handling Institute Association (MHIA).

Jane Rees is a System Integration and Installation Manager for Worldwide Chain Store Systems (WCSS) and was formerly with IBM Corporation. Ms. Rees has extensive experience with warehouse management systems and material handling automation projects. She has been involved in application development and project management since 1981 in a myriad of business areas including distribution, manufacturing logistics, accounting, and finance. She holds a B.S. in Computer Science from Louisiana State University and an M.B.A. from the University of North Carolina at Charlotte.

Susan Rider is Director of Marketing Services at Unarco Material Handling. Ms. Rider has over 15 years experience in the material handling field, has written various articles published in trade magazines, and has spoken at trade shows, including Promat, and the Warehouse Education and Research Council.

Milt Sedlak is a Senior Industrial Engineer with Worldwide Chain Store Systems (WCSS). Formerly of IBM Corporation, Mr. Sedlak has been in design and development as well as engineering for 27 years, during which time he has received several patents in the material handling and automated printing manufacturing areas. He is a graduate of the Rochester Institute of Technology, where he received a B.S. in Mechanical Engineering.

John G. Van Cleve is currently the Warehouse Market Segment Manager for Vocollect. He has been involved in the development and implemenation of voice-directed warehouse systems for numerous companies, including CVS, Lockheed, Freightliner, and Nabisco. He has spoken at ID Expo and other warehousing conferences about the application of voice within the warehouse. He holds a B.S. from Washington and Jefferson College and an M.B.A. from the Columbia Graduate School of Business.

Brenda Wrigley joined the Raymond Corporation in January 1990 as Marketing Communications Manager. In this capacity, she directs corporate advertising, public relations, and related communications programs for the corporation's Greene, N.Y., headquarters. Ms. Wrigley earned a Bachelor of Journalism degree from the University of Missouri at Columbia and will receive a master's degree in Public Relations from the S. I. Newhouse School of Public Communications at Syracuse University in 1996.

The co-authors would like to express gratitude to the following individuals for their creative input, editorial assistance, and research endeavors in the preparation of this book: Shannon Carney, Gina Graham, Steven Heyer, Gavin McIntyre, Dipak Patel, Michelle Romahn, Dick Shaffer, and Dale Yehle. The co-authors appreciate the sacrifice made by family members and the support by friends, without whom this book would not have been possible.

INTRODUCTION

With greater customer demand for products and services on a more timely basis at lower cost, distribution centers (DCs) have become the focal point in the distribution network. While companies use information systems to streamline their vital business support areas, such as Order Entry and Purchasing, many DCs have and continue to use a paper management or batch system. This significantly increases the potential for inaccurate inventory and shipment records, operational inefficiencies, and increased order processing time. Application software combined with the latest technologies in automated identification and information can add significantly in efficiency and control to warehouse operations by providing a highly traceable and real-time audit trail on all activity. These technologies include radio frequency terminals, bar codes, scanning devices, and distribution management software.

How does a company get involved in the evaluation of such technologies and their implementation into distribution operations? This book assists Information Systems, Operations, and Distribution departments in this process.

The procedure begins with the company's vision—recognizing where the company needs to be in two to five years to compete effectively, identifying customers' requirements, and defining the role distribution plays in achieving the desired results.

The next step is for the company to perform an analysis of its current operation efficiency, what trends are taking place within its industry, and where and how much it will benefit by using the latest technologies coupled with a warehouse automation package. For example, many companies are finding that a trend in distribution requirements is customers placing more frequent orders of reduced quantities with distribution centers. In effect, customers are reducing their inventory costs by keeping the inventory longer at the supplier's DC. As a result, the DC must be prepared to handle this strategic change to avoid significant distribution costs and service inefficiencies.

This book covers the strategic company plan, evaluation of automation packages, system justification, system design, requirements gathering, equipment analysis, testing, training, and implementation.

Chapters 5, 6, and 10 are uniquely focused on warehouse management system design and implementation.

Chapter 5 provides an in-depth analysis of the many functional details that should be considered when developing requirements for a warehouse or distribution system. This detailed level of analysis is not found in any other book to our knowledge. Having this information will maximize a company's return on investment when implementing a warehouse automation project, as it provides a vehicle by which system designers and distribution managers can easily and efficiently review and understand all of the processes and how best to re-engineer and simplify existing procedures.

Chapter 6 is intended to be a time-saver for engineers, in that it provides a comprehensive set of forms and examples for preparing the request for proposal. This working format also gives the reader a step-by-step guide on how to evaluate the various proposals and complete the vendor selection process.

Chapter 10 provides a focused discussion on the project management, tasks, and steps specifically required for a warehouse automation project. Books exist that discuss general project management, but this chapter provides the practical side of project management and the common errors and pitfalls directly associated with warehouse automation projects that can end up delaying the project and causing cost overruns.

Chapters 3, 7, 12, and 13 focus on design verification and user implementation of the system.

Chapter 3 provides a comprehensive review of the equipment available for a warehouse automation project and how these technologies apply to the project.

Chapter 7 discusses the value a simulation analysis plays in a distribution automation project, and explains the steps to perform an analysis. An example simulation project is included to illustrate the steps and assist engineers by providing an outline.

Chapters 12 and 13 can be grouped together, in that they provide insight when it is time to implement. Many companies have had unsuccessful distribution automation implementations as a result of poor training, preparation, and cut-over to the new system. These chapters provide successful strategies for the project team members to follow.

A more detailed description of the chapters in this book follows.

Chapter 1, Introduction: The general benefits of distribution automation and how technology plays a role in achieving these results are explained. The changing role of distribution is discussed as well as an analysis of industry trends. A few company success stories are included to highlight the potential benefit of automation.

Chapter 2, Automation Project Process Steps: A working summary of the steps necessary to successfully install a warehouse or distribution automation project is given. These steps are presented in reference form, providing companies with a road map to guide them through a lengthy and complex process.

Chapter 3, Technology Considerations: The reader is walked through the maze of distribution technologies that are available today. Explanations of how to effectively utilize these technologies, plus the advantages and disadvantages of each, are presented. Pictures of the various equipment are included to provide a visual reference.

Chapter 4, Material Handling Considerations: Every distribution center has some type of material handling equipment, from simple hand trucks to complex conveyor sortation systems. Material handling equipment considerations must be evaluated early in the automation process in order to achieve a properly integrated and effective system. A discussion of this important topic is provided, along with photographs of the various types of equipment and the advantages and disadvantages of each.

Chapter 5, Functional Characteristics: Numerous philosophies are available to management when designing and implementing a warehouse management system. Each functional operation of the warehouse must be considered separately, from receiving through shipping along with interfacing to other business applications and material handling subsystems. The various strategies that are commonly implemented are presented for each operation.

Chapter 6, Requirements Definition and System Evaluation: The company must clearly define the requirements of the distribution system. How to develop these requirements and create request for information and request for proposal packages are outlined and samples of each report and proposal are given. Processes for proposal evaluation for internal feasibility are also discussed with examples.

Chapter 7, Simulation: One of the most powerful tools to assist in the design and operation of distribution systems is computer simulation. Automated distribution systems are highly integrated, and consist of many interdependent functions and interactions. A simulation format is presented in this chapter to illustrate the importance of analyzing these interdependencies before implementation, so that omissions and errors can be identified and corrected quickly and inexpensively.

Chapter 8, Benefit Analysis and Justification: The existence of numerous system and equipment alternatives can make selecting the best ones difficult for any company. A benefit and cost analysis provides project team members and management with the tools necessary to make the best choice. A classic approach for performing a benefit analysis and cost justification for a warehouse automation project is explained in detail. Methods for calculating and identifying are discussed along with the forms required to complete the analysis. In addition, an example of an analysis is included to show the procedure.

Chapter 9, Design Phases: Design phases takes the proposed process and requirement concepts and translates and refines them into detailed specifications of deliverables. Commonly used methodologies for developing design specifications are discussed. Example external design specifications for the warehouse management system and material handling equipment are included.

Chapter 10, Project Management: Preparing a distribution center for the implementation of an automation/mechanization project directly affects the success and benefits of the project. This chapter discusses various techniques that can be used to minimize risks and prepares the distribution center and its personnel for the new way of doing business. A generic schedule is provided with the necessary project tasks to help guide those who are going through this process.

Chapter 11, Software Testing: Comprehensive testing of the warehouse application is a very important aspect of implementing a project to a successful end. The various testing phases and example test cases to be performed are described and discussions of the testing options available are presented.

Chapter 12, Training: If employees are not comfortable with the new system, attempts to get around the system are made. Accordingly, early involvement of personnel is critical. Various training methodologies are presented in detail and a timetable is given that emphasizes such early involvement. Tips for providing an efficient training session and effective documentation are given.

Chapter 13, System Conversion: A smooth and error-free implementation minimizes disruptions and risks to current operations. The various strategies available to help reduce risks are described, as well as tips based on actual experiences.

In summary, this book's purpose is to assist a company in utilizing technological advances in the distribution industry to gain a competitive edge by explaining all the steps required to successfully implement projects on time and provide a high rate of return.

ABBREVIATIONS AND ACRONYMS

ABC	Activity-based costing
AEM	Automated electrified monorail
AGVS	Automated guided-vehicle system
AIAG	Automtotive Industry Action Group
AIM	Automatic Identification Manufacturers
ANSI	American National Standards Institute
ASN	Advanced shipping notification
ASRS	Automated storage and retrieval system
BOL	Bill of lading
bps	Bits per second
CA	Collision avoidance
CAD	Computer-aided design
CASE	Computer-aided software engineering
CCD	Charge-coupled device
CD	Collision detection
CDC	Central distribution center
CEN	Comite European de Normalisation
CGA	Color graphics adapter
CIDX	Chemical Industry Data Exchange
CLM	Council of Logistics Management
COS	Corporation for Open Systems
CPM	Cartons per minute
CPM	Critical path method
CRT	Cathode ray tube
CSMA	Carrier sense multiple access
DASD	Disk access and storage device
DC	Distribution center
DOF	Depth of field
DOS	Decoded output scanner
DREF	Distribution Research and Education Foundation
EAN	European article numbering
EDI	Electronic data interchange
EGA	Enhanced graphics adapter
EIA	Electronics Industries Association
EIDX	Electronics Industry Data Exchange

FCC	Federal Communications Commission
FIFO	First in, first out
FPC	Flow process chart
GHz	Gigahertz
GUI	Graphical user interface
He-Ne	Helium-neon
ID	Identification
IPDS	Intelligent printer data streams
IRR	Internal rate of return
IS	Information systems
ISO	International Organization for Standardization
JIT	Just-in-time
LAN	Local-area network
LCD	Liquid-crystal display
LED	Light-emitting diode
LIFO	Last in, first out
LTL	Less than (truck) load
MH	Material handling
MHA	Material handling automation
MHIA	Material Handling Institute Association
MHS	Material handling system
MHz	Megahertz
MIS	Management information system
MRP	Manufacturing resource planning systems
MTM	Methods time measurement
NiCad	Nickel-cadmium
NPV	Net present value
OCR	Optical character recognition
ODBC	Open data base connectivity
OSF	Open Software Foundation
OSHA	Occupational Safety and Health Administration
OSI	Open system interconnection
PC	Personal computer
PERT	Project evaluation and review technique
PF&D	Personal, fatigue, and delay time
PLC	Programmable logic controller
PO	Purchase order
PRO	Progressive number
PTL	Pick-to-light
QC	Quality control
QR	Quick response
RA	Return authorization
RAM	Random access memory

RF	Radio frequency
RFDC	Radio frequency data communications
RFI	Request for information
RFID	Radio frequency identification
RFP	Request for proposal, request for price
RLC	Radio local-area-network controller
RMI	Rack Manufacturing Institute
ROI	Return on investment
ROM	Read-only memory
SKU	Stock keeping unit
SNC	System network controller
SOP	Standard operating procedure
SVGA	Super video graphics adapter
TCP-IP	Transmission control protocol–Internet protocol
UCC	Uniform Code Council
UHF-FM	Ultrahigh frequency–frequency modulation
UL	Underwriters Laboratories
UOM	Unit of measure
UPC	Universal product code
UPS	Uninterruptible power supply
VAN	Value-added network
VGA	Video graphics adapter
VICS	Voluntary Interindustry Communication Standard
WAN	Wide-area network
WCS	Warehouse control system
WERC	Warehouse Education and Research Council
WIP	Work in progress
WMS	Warehouse management system

PREPARATION

CHAPTER 1
INTRODUCTION

OVERVIEW

Successful application of the latest technologies to warehouse and distribution center (DC) activities enables a company to gain a competitive edge through their improved ability to deliver the right products, in the right quantity, at the right time. With customer demands increasing, the pressure is on warehouses and distribution centers to analyze present operations and implement new technologies. To achieve these ends, companies are utilizing distribution automation to improve productivity and order accuracy, reduce space requirements, increase volume capacity, obtain total inventory control and accountability, and ultimately increase customer service and market share. (See Fig. 1-1.)

In the past, many companies focusing on product development and manufacturing and marketing improvements have overlooked their warehouse operations. Survival in today's marketplace, however, now depends on prompt and accurate distribution, and those companies operating with batch systems and manual processes are finding it increasingly difficult to meet the changing customer demands. Warehouse automation integrates warehouse management systems, bar code technology, wireless terminals, scanners, material handling equipment, storage systems, operating procedures, and people into a single working unit that must effectively interface with all other business areas within the distribution business cycle. Distribution automation encompasses all the issues associated with warehouse automation plus its effective integration into the entire distribution business cycle (see Fig. 1-2).

It is extremely important to use a disciplined approach in developing a distribution or warehouse automation plan in order to maximize a company's return on the significant investment required. The remainder of this chapter emphasizes the link between distribution automation and a company's competitiveness, and the following chapters address the various phases and issues involved in an automation project.

CHANGING ROLES IN DISTRIBUTION

In a *Wall Street Journal* article entitled "The Economy's Power Shift," Peter Drucker describes the shift from a manufacturing economy to a distributor-retailer economy, emphasizing that the warehouse must become a "switching yard" instead of a "holding yard."[1] Transportation terminology is effectively used to visualize the speed required in today's supply pipeline, since the one who holds the inventory pays the carrying

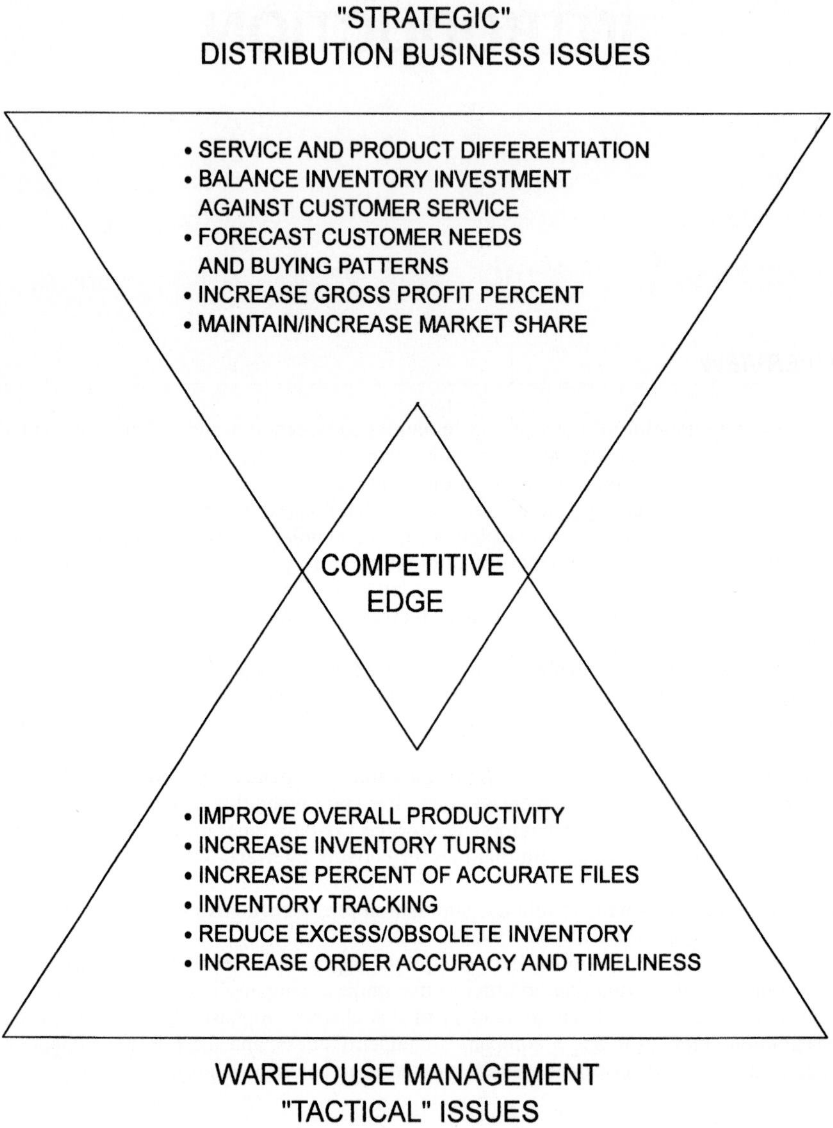

"STRATEGIC"
DISTRIBUTION BUSINESS ISSUES

- SERVICE AND PRODUCT DIFFERENTIATION
- BALANCE INVENTORY INVESTMENT
  AGAINST CUSTOMER SERVICE
- FORECAST CUSTOMER NEEDS
  AND BUYING PATTERNS
- INCREASE GROSS PROFIT PERCENT
- MAINTAIN/INCREASE MARKET SHARE

COMPETITIVE
EDGE

- IMPROVE OVERALL PRODUCTIVITY
- INCREASE INVENTORY TURNS
- INCREASE PERCENT OF ACCURATE FILES
- INVENTORY TRACKING
- REDUCE EXCESS/OBSOLETE INVENTORY
- INCREASE ORDER ACCURACY AND TIMELINESS

WAREHOUSE MANAGEMENT
"TACTICAL" ISSUES

FIGURE 1-1   Distribution automation benefits.

cost. This concept of switching is reflected in cross-docking, or product pass-through operations, where real-time system updates allow incoming inventory to transfer, or switch, directly to the outgoing docks, completely bypassing the storage locations. Figure 1-3 displays the shift occurring in the supply chain pipeline.

In an effort to reduce inventory costs while maintaining customer service, customers are placing increased demands on the distribution process. The requirements outlined

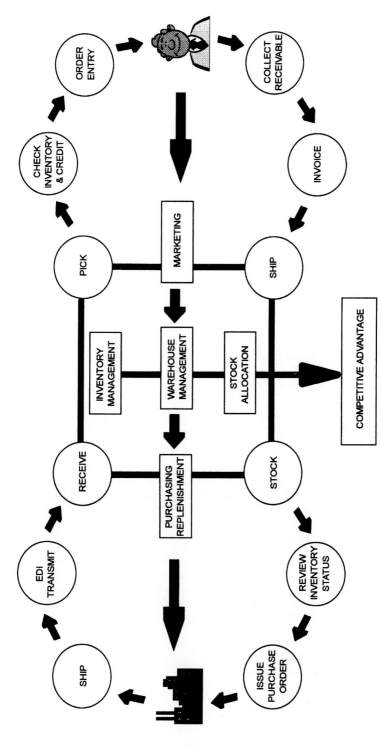

FIGURE 1-2 Distribution business cycle.

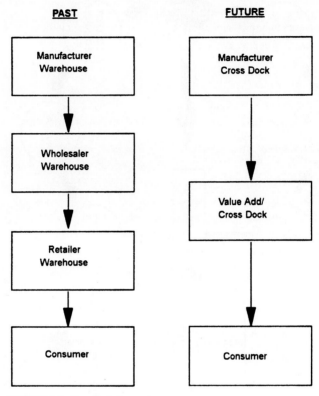

FIGURE 1-3   Supply chain warehouses.

below must be addressed to remain competitive in today's marketplace. It is important to note, however, that the operational impacts associated with these demands present a formidable task for a company that must also improve quality and control costs.

More Frequent Shipments and Smaller Quantities

Inventory is being forced back in the logistics process. Customers are less willing to carry the inventory when they can require the distributor to deliver smaller quantities more often. Reduced order sizes have put increased demands on warehouses to pick and ship partial cases rather than full cases. In order to provide this service, without increasing costs, the warehouse must select an efficient pick operation.

   Operational impacts:

- Increased receiving and shipping activity
- Increased labor due to number of picks
- Increased amount of equipment required
- Increased workload tracking
- Increased transportation planning

- Increased quality control activity
- More frequent inventory forecasting
- Stressed planning and purchasing activities

Increased Number of Products Stocked

Customers, i.e., retailers, are demanding their name on the products they sell, which results in carrying an increased amount of private labeling, in addition to carrying the manufacturer's labeled product. Customers also want smaller quantities in each replenishment order to avoid inventory costs. This leads to "broken cases" or smaller-quantity masterpacks, which become another product to track, process, and handle.
  Operational impacts:

- Increased inventory carrying costs
- More space required to stock stock-keeping units (SKUs)
- More shelving and flow racks required
- Decreased productivity per employee
- Increased administrative labor to handle additional paperwork

Reduced Cycle Times

Customers are demanding order arrivals to their location within a shorter period of time than previously from order placement. This requires warehouses to expedite orders through the system, with immediate allocation, picking, and shipping. Some recent analyses have indicated the need to reduce cycle time through the supply pipeline from 50 to 70 percent over the period from 1990 to the year 2000. This reduction will ripple all the way through the pipeline to the wholesaler warehouse and retail distribution center.
  Operational impacts:

- Increased labor requirements
- Increased need for mechanization
- Increased administration labor to track
- Increased need for cooperation among Inventory Control, Purchasing, and Distribution

Zero Tolerance for Errors

Customers are placing charge-backs into their contracts for items that are mis-shipped or not shipped to their specifications. As smaller quantities and more frequent shipments flow through at reduced cycle times, customer error correction is very costly and can result in significant charge-backs that impact the distributor's bottom line.
  Operational impacts:

- Increased quality control operations
- Negative customer relations
- Potential profit-margin reductions

Value-Added Services at the DC

Although it is recognized that the speed at which products flow through the distribution chain is important, it is also realized that somewhere, someone has to hold some inventory. This means that more and more inventory has to be in a ready condition where it can be moved quickly to the next link in the supply chain. That condition is, in many cases, no longer a full truck load, a full pallet, or a full case; rather, inventory has to be shelf-ready and possibly even prelabeled for a particular seller.

Customers are looking for the distributors to take over many of the supply chain functions that they once handled in-house. For example, distributor-provided services such as customer label and/or price-ticket application, product assortment kitting, private labeling and packaging, merchandise marking, and shelf-display building can help expedite activities at the customer's receiving end.

Operational impacts:

- Increased customer demands for the following services:

  - Quality control
  - Packing
  - Repack
  - Ticketing
  - Labeling
  - Kit assembly

- Increased labor
- Cycle time impacts
- Increased administrative labor to track work in process
- Increased number of products stocked

Requirement for Real-Time Information

Customers and suppliers are requiring information be transmitted electronically as soon as merchandise is received or shipments are made. Customers are increasingly requesting electronic data interchange (EDI) transactions such as an advance shipment notice (ASN), which requires real-time system updates and information transfer. Operating with reduced inventory and increased services requires a warehouse information tracking system that ensures customer orders are correct and complete before they are shipped. As shown in Fig. 1-4, the information is not only tracked within the warehouse but must also be visible throughout the entire supply chain.

Operational impacts:

- Value-added services tracking
- Interdependence of Inventory Control, Purchasing, Transportation, Manufacturing, Sales, and Distribution
- Faster response time to customer inquiries

Competition Based on Cycle Time and Quality

The days of being the low-cost provider are not gone, but value-conscious customers no longer consider that one factor alone. They also require speedy, predictable, and complete delivery without errors.

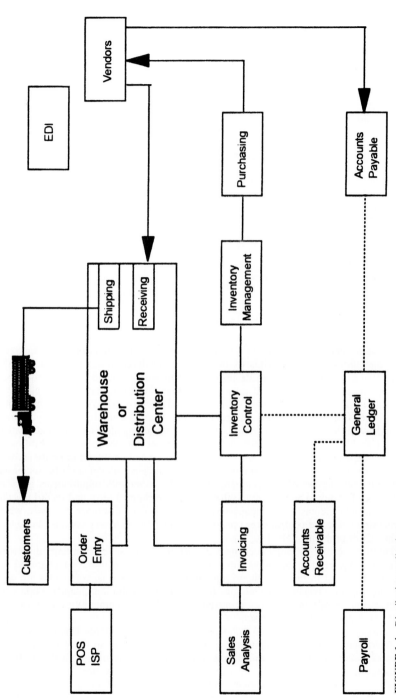

FIGURE 1-4 Distribution applications.

Operational impacts:

- Increased competitive pressures
- Increased customer demands
- Reduced supplier loyalty by customers
- Quality control and labor impacts
- Increased need for processes to improve cycle time and quality

Increased Labor and Facility Costs

This trend is not new but is exaggerated by increased urban land cost and the cost of hiring a skilled, technically adept work force. As inventory levels grow, expansion and potentially new facilities can be required.
Operational impacts:

- Employee turnover and/or training cost
- Increased space requirements
- Increased facility operating cost

TECHNOLOGY REACTIONS TO TRENDS

Until recent years, there have been minimal emphasis and attention given to warehouse operations. This was primarily due to the consideration and judgment that the changes in customer requirements just described were isolated events rather than growing and permanent trends. It can be easily seen, however, that a warehouse environment consisting of batch processes and manual, paper-intensive operations cannot support these emerging distribution requirements. Combine the increased cycle time demands with the additional operations required for value-added services and a rather difficult problem arises within the confines of present-day warehouse capabilities. A company dependent on paper-intensive operations to maintain record integrity is overwhelmed by a requirement for zero errors when combined with demands for increased shipments and number of products stocked. As the paperwork requirements grow to keep track of shipping and inventory, the potential for human error and record inaccuracies can only increase. A company trying to meet all its customer demands, while maintaining costs and control over operations, is faced with a very complex and challenging set of problems. (See Fig. 1-5.)

Had these seemingly isolated changes in customer requests been initially discerned as the growing trend in customer requirements, companies may have recognized the necessity to invest in their warehouse operations earlier. As it is, companies are only now looking for advanced technologies to help their distribution channels cope with the emerging trends in customer requirements and the accompanying increases in labor, equipment, and facility cost.

A survey conducted by the Warehouse Education and Research Council (WERC) and Ohio State University examined the fundamental changes in the way companies plan to conduct warehousing operations in the future.[2] One of the most significant trends to emerge is a high level of responsiveness in meeting more demanding customer requirements. Most respondents indicated that the use of advanced systems and distribution management approaches will continue to grow at a rapid pace. Their views include:

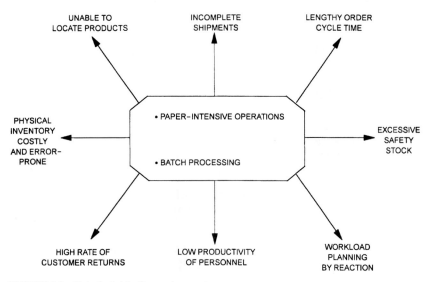

FIGURE 1-5   Today's distribution environment.

- Electronic data interchange
- Bar coding of shipments to meet customer specifications
- Warehouse automation
- Increased percentage of goods shipped using quick response (QR) or just in time (JIT) methods

Other studies confirm this shift in toward strategy warehouse automation. The August 1993 survey by the Distribution Research and Education Foundation (DREF) is displayed in Fig. 1-6. This depicts the degree of warehouse automation planned for implementation by the year 2000.[3]

Although the DREF report indicates a move toward technology implementation, a survey by KPMG Peat Marwick in *Distribution Center Management,* March 1994, confirms that companies have not progressed rapidly in automating their warehouses and installing high-technology systems. The survey also concluded that manufacturing, wholesale, retail, and service companies are increasing focus on logistics management as a way to reduce overall costs and improve customer service.[4] Figure 1-7 depicts how far companies have gone in implementing technology in their warehouses.

The various data from the different surveys all indicate a definite move toward increased technology in the warehouse and a reaffirmation that benefits can result. It is important not to become so enamored with being "state of the art" or "leading edge" that a company loses sight of the real goals of increased efficiency, accuracy, and flexibility in meeting customer demands. There is a process for installing technology successfully that starts with a complete examination and understanding of the company's operations. The series of questions that must be asked and the solutions that must be linked together into a cohesive automation strategy can become a complex puzzle for any distributor.

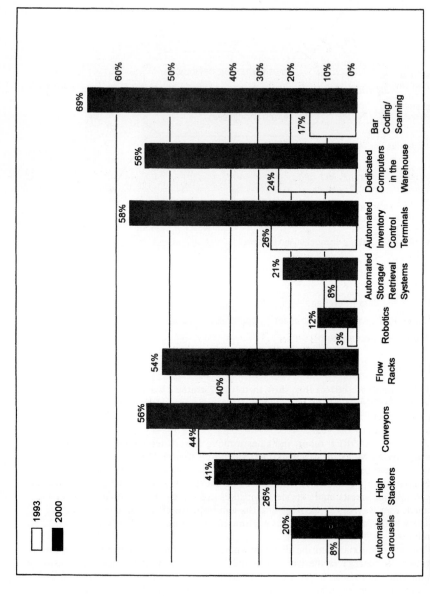

FIGURE 1-6 Warehouse automation plans. (*Source: Anderson, Arthur, Facing the Forces of Change 2000, Distribution Research and Education Foundation, Washington, D.C., 1993, p. 205.*)

	Complete	Partial	None
Bar coding	10%	49%	41%
Radio frequency	4%	18%	78%
EDI with vendors	1%	48%	51%
EDI with carriers	2%	36%	62%
EDI with customers	6%	51%	43%
Warehouse management	24%	52%	24%

FIGURE 1-7 Technology trends. (*Source: Distribution Center Management Newsletter, March 1994.*)

THE LOGISTICS PUZZLE

Successful warehouse automation is not just the application of the latest technology to existing warehouse operations. Maximizing the benefits derived from automating warehouse operations requires an examination of the complete logistics process and its ability to meet current and future customer demands competitively. Logistics is defined as the process of strategically managing the acquisition, movement, storage of materials, parts, finished inventory, and related information flows through an organization and its marketing channels to fulfill orders most cost effectively.[5] (Refer to Fig. 1-4.)

Examining the ability of the logistics process to meet current and future customer demands becomes quite a puzzle, considering all the questions that must be answered. There are five distinct pieces of this puzzle, as illustrated in Fig. 1-8.

Current Business and Product Attributes

A company must identify and clearly define the distribution processes that are driven by the current business and product attributes. While completing this exercise, a company identifies the inefficiencies in current processes that must be addressed. This process definition is also used to determine any deficiencies in meeting future product and service requirements.

Future Business and Product Attributes

By examining distribution trends, a company determines those that are critical to its future growth and competitiveness. Where current products may be pallets, future products may be individual cartons or pieces with unique labeling.

Consider a distribution center that is shipping pallet loads of their product to a retailer's distribution center. The retailer's DC then ships individual cartons along with other products to their retail stores. In an effort to reduce its inventory and double handling, the retailer now requires a supplier to ship carton quantities directly to their stores. This type of requirement changes the distribution center's functional processes. The picking requirement is no longer a fork truck picking pallets from a pallet rack, but is a carton label pick operation where a picker picks a carton, applies the appropriate shipping label, and then places the product on a conveyor belt for automatic sortation to the correct shipping lane. Since the shipment size to the individual stores is smaller than one large shipment to a DC, the carrier may have several destinations and therefore must be loaded accordingly.

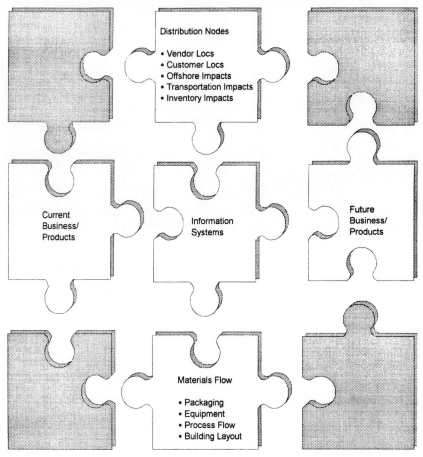

FIGURE 1-8 Logistics puzzle.

Distribution Nodes

Do the distribution nodes contained within the logistics pipeline, from the suppliers to the customers, still provide an efficient flow of goods? Does the market location change as a result of any future products or services? Continuing with the previous example, the DC is no longer shipping to the retailer's DC but rather directly to various stores located throughout the United States. If the company has multiple DCs to ship from, does it now make sense to service a particular region of stores from each of the DCs? This can change requirements for your suppliers, which now have to ship to multiple DCs instead of just one.

Are products going to be foreign-sourced or -marketed? The gradual lowering of trade barriers is leading to the creation of a global marketplace in which companies can now source suppliers and market on a multinational basis. The impact of this change can mean larger quantity receipts from new vendors to your DC. The DC must be able to handle receiving and storage of increased quantity receipts. On the output

side, foreign markets could mean new shipping processes with international paperwork. The extra time required to ship the product to and from overseas must be
accounted for in both the receiving and shipping processes.

Material Flow

The migration from current products to future products and services requires analysis
to determine the changes to the materials flow through the process. Changes caused by
the future products and services can have a significant impact on the optimum DC layout and the equipment required to manage material flow. For example, with smaller
quantities being requested from the customer, does pallet picking and shipping
become carton picking? With smaller pick quantities, does it now make sense to use
conveyors where fork trucks may have been previously used? How are the individual
cartons tracked as they flow through the process? Is the carton a shippable package? If
not, what needs to be done? Are the customers requiring special labeling on the carton? If so, at what point in the materials flow is the labeling operation added?

When analyzing material flow, it is important not to be overwhelmed by mechanization. With mechanization in warehouse operations, various types of material handling equipment are used to perform tasks. While mechanization is good, the decision
to implement mechanization must be thoroughly evaluated. For instance, floor-mount-
ed conveyors may create operational islands and constrict truck traffic through a warehouse. Will the planned mechanization limit flexibility in managing future products
and services?

Information Systems

With a strategy established for future products and services offerings and with the
impacts to the current distribution nodes and materials flow identified, it is time to
examine the capabilities of the current information systems in managing this future
environment.

How does a company control the data flow through the various processes? What are
the critical points in the process where product status must be captured? Can the data at
each operation best be captured by keyboard input, bar code readers, or radio frequency
(RF) terminals? What interfaces are required between the current systems and any new
systems to ensure data flow? Can the systems handle the expected peak volume?

While automation can be achieved in either a manual or mechanized operation,
information is critical to automation. Understanding the significance of information
flow and its lock-step relationship with the physical flow of merchandise determines
the successful modernization of a company's distribution center. Integrating application software and technology equipment with material handling equipment maximizes
the benefits achieved.

As the pieces in the logistics puzzle emphasize, companies must diligently examine
their operations and customer requirements prior to implementing an automated warehouse management system. This includes obtaining requirements prior to and after the
distribution activity as well as defining the relationship between mechanization and
information requirements. This is the time to reengineer processes, simplify operations, and then automate the DC. Too many companies attempt to force-fit a warehouse management system (WMS) into their current operations or simply mechanize
the current operations without the initial investment in process redesign. This significantly reduces the benefits that can be obtained from investment in automation.

Chapter 2 outlines the basic installation steps for a distribution automation project. A sound strategy and careful preparation are critical to a project's success.

SUCCESS STORIES

Successful warehouse automation can transform a distribution center into a highly efficient operation. Since almost every area of a company's operations is impacted, a warehouse automation project is a very major undertaking. But with careful planning and execution, warehouse automation can be a smooth and gratifying venture. Many stories about companies successfully implementing warehouse automation and mechanization projects, such as those summarized below, have been published in trade magazines and presented at conferences and seminars.

Kinney Drugs

Kinney Drugs, a retail drugstore in upstate New York, has recognized the importance that new technology has on warehousing performance.[6] Kinney Drugs automated their warehouse operations by implementing a warehouse management system and changes to their facility layout. This provided the company with the ability to plan, manage, and anticipate future requirements instead of reacting to problems as in the past. The solution included the implementation of radio frequency terminals, horizontal carousels, and a "fast pick" area.

As a result, service level has increased from 97 to 99 percent through increased store fill rates and greater pick accuracy. The layout improvements increased the number of SKUs stored by better utilizing existing space. The application of unit bar codes to track merchandise has improved accuracy and control. Due to the increased inventory accuracy, Kinney Drugs has eliminated two physical inventory counts per year, along with the associated shutdowns. Finally, increases in employee productivity have allowed the chain to expand the number of stores by 26 percent without increasing distribution personnel.

Results of the above actions were noted in the Kinney Drugs 1993 Annual Report as follows: "The automation, design and expansion of the distribution center carried out in 1991 and 1992 have resulted in improved inventory management, enhanced shipping capabilities and 10 percent increase in manpower efficiencies in that facility." The report also stated, "Improved distribution center shipping capabilities resulted in more in-time deliveries, better sell-through of seasonal and promotional products, less mark-downs, and higher gross profit margins."[7]

Williams-Sonoma

Williams-Sonoma, a leading mail-order and retail corporation selling upscale items for the home, has met and embraced the great challenge of the 1990s by managing its ever-increasing inventories more efficiently.[8] As the result of an automated distribution system utilizing bar codes and radio frequency data communications, the company has realized 99.7 percent inventory accuracy. Achieving that accuracy is astounding considering the size of its multiple inventory operations, with over 14,000 SKUs. They have also been rewarded with $400,000 in labor savings in one year while managing a 25 percent increase in receiving volume.

The automated distribution system is part of Williams-Sonoma's larger plan to revise its entire supply chain operation. The company believes the trend of the future is to have a tightly integrated supply chain that tracks where everything is, all the way from the supplier to the store. The warehousing environments view and distribute inventory information on-line. Personnel handle more products in less time with less chance of human error. Host systems are notified about product movement automatically. Picking and shipping operations are tracked step by step to ensure that orders are delivered correctly. By filling customer orders faster and more accurately, product throughput is improved and customers are delighted.

The strategic use of technology is of the utmost importance in the design of distribution systems in the 1990s. Williams-Sonoma has found that the technological advantages of bar codes and radio frequency communication combined with a total software distribution system give it the power to move smoothly from the 1990s into the next century.

Sharp Electronics Corporation

Sharp Electronics Corporation, a leading manufacturer and distributor of consumer electronic products and electronic components, ships from three central distribution centers and four regional warehouses throughout North America.

Sharp Electronics' distribution strategy includes a flexible warehouse management system that is designed to meet the requirements for all their distribution centers. Sharp's upfront investment in process analysis and reengineering led to a system that ensures similar distribution practices across all operations with a standard interface between the corporate host and the DCs. In addition, where unique requirements remain, the system is designed to enable the necessary customization.

To address Sharp's high volume of parcel shipment effectively, a mechanized package conveyor system was designed and fully integrated into the warehouse management system. This parcel system weighs, rates the freight, and prints and applies a shipping/packing list label automatically to the packages.

Within three years from the first installation, Sharp has installed the automation solution at six additional DCs. The benefits Sharp has realized from this distribution automation project include the following:

- A zero increase in distribution personnel four years running has been accomplished, while realizing double-digit growth in volume for each of the four years.
- The system provides regional allocation reports that measure accuracy of product placement in the regional distribution center to support regional sales. Adjustments to allocation have been made based on these reports; therefore, Sharp's less than load/truckload outbound transportation costs have continued to decrease as a percent of sales.
- Nationwide inventory accuracy of 100 percent and elimination of two physical inventories per year at the DCs with the implementation of cycle counting have been achieved.
- Duplicated functions such as shipment confirmation, warehouse receipt entry, adjustment entry, and entry of month-end book inventory from the distribution center have been eliminated.
- Printing, sorting, bursting, filing, and microfiching of 400,000 bills of lading annually have been eliminated, which under the old system was necessary to support small-package shipments.

Toyota

Toyota's 374,000-ft^2 parts DC combined storage and staging strategies with JIT, automation, and ergonomics to boost service and efficiency.[9] Toyota planned the facility with four key objectives in mind:

1. Improve accuracy of shipments, inventory records, and receipt processing.
2. Improve operating efficiency by automating movement of parts and personnel and eliminating nonproductive functions.
3. Use new material handling and information technologies, including automated materials handling, bar coding, distributive processing, and real-time paperless work processing.
4. Create a safe and pleasant work environment.

The glue that holds the entire system together is real-time information. Capabilities of the information system are the following:

- Real-time receipt, order processing, and inventory data
- In-process tracking of inbound and/or outbound parts and orders
- Radio frequency transmission of information throughout the warehouse
- Random storage location assignment by preferred zone
- Optimal location logic
- Bar code control for accuracy
- Paperless work processing
- Optimal work-path determination
- Full-time workload status reporting

These four success stories punctuate the benefits that are achievable through warehouse automation. Obviously, much thought and hard work are required to achieve such huge successes. The remaining chapters are provided to help guide the enterprising company on a distribution automation journey that can lead to its own personal and unique success story.

REFERENCES

1. Drucker, Peter, "The Economy's Power Shift," *The Wall Street Journal,* September 24, 1992.
2. LaLonde, Barnard J., and Delaney, Robert V., *Trends in Warehouse Costs, Management, and Strategy,* Warehousing Education and Research Council, Oakbrook, Ill., 1993, p. 9.
3. Anderson, Arthur, *Facing the Forces of Change 2000,* Distribution Research and Education Foundation, Washington, D.C., 1993, p. 205.
4. "Inventory Management—A Top Priority," *Distribution Center Management,* 29 (March):4, 1994.
5. Gattorna, John L., *The Gower Handbook of Logistics and Distribution Management,* 4th ed., Gower, London, 1990, p. 5.
6. Adams, Nick, "New Warehouse Automation Improves Customer Service at 'Kinney Drugs,'" *Industrial Engineering,* October 1993, p. 21.
7. Kinney Drugs Annual Report, 1993.
8. Denehy, Mike, and Broadnax, Junius, "From Catalogs to Retail Distribution, Inventory Accuracy is Enviable 99.7%," *Automatic I.D. News,* 10 (November): 20 (1994).
9. Kulwiec, Ray, "Toyota's New Part Center Targets Top Service," *Modern Materials Handling,* November 1992, p. 73.

CHAPTER 2
AUTOMATION PROJECT PROCESS STEPS

INTRODUCTION

Warehouse and distribution center automation can be a frightening and costly journey for the ill prepared. Properly done, it can revolutionize the way a company does business. Be aware, however, that it will affect almost every part of a company's business in some way. As such, a sound strategy and careful preparation are the keys to avoid being on an emotional roller coaster throughout the project. The rewards of success include increased productivity, increased customer satisfaction, and improved inventory accuracy. This book presents a methodology not only for applying and implementing automation, but also for devising new business processes to take maximum advantage of the benefits of automation.

OVERVIEW

This chapter describes the process for implementing a distribution center or warehouse automation system and forms a road map for the entire project. The process described herein has been used successfully in a number of distribution automation projects. Those readers familiar with concepts of business process reengineering and with activity-based costing (ABC) will recognize many similarities; the process presented here has arisen from the same industrial engineering origins as those techniques. Succeeding chapters will explore in more detail the various topics that must be considered when implementing a system.

The automation project process can be separated into three phases: preparing, defining, and implementing. The effort level required and time spent on each phase vary depending on a facility's current logistics process and existing automation level. It is recommended that before beginning a specific project, the entire project process as described in this book be reviewed and a customized strategy for automation be drawn up. Figure 2-1 shows the overview of the project process.

Phase 1: Preparation

This phase lays the groundwork for the automation of the project. The primary activities in this phase are to establish a set of goals and to form a team that will define and implement the business changes needed to accomplish these goals.

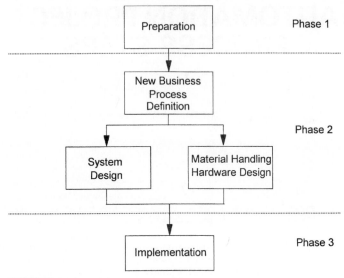

FIGURE 2-1   Automation process overview.

Phase 2: Definition

During phase 2, documents are written that describe in detail how the warehouse or distribution center will operate with automation. There are two major steps in this phase. First, define the new business processes for the facility. Second, specify the material handling hardware and system software required to support the new processes.

Phase 3: Implementation

The third and final phase of the automation project is the implementation phase. This phase typically constitutes the longest portion of the project cycle time due to vendor lead times for equipment and the time required to integrate and test all the new system components and make modifications to existing information systems.

PHASE 1: PREPARATION

This phase builds the foundation of the project. At this stage, the core of the project team is established. A set of project goals are defined that specify, in business terms, the desired results. Setting project goals early on will help the project team focus on the tasks required to attain those project goals. Project goals are also of great help later when making the inevitable decisions on what is critically important and what is merely a nicety. The scope of the project is also defined so that team members have a clear idea of what is to be included in their effort. Figure 2-2 shows the steps in this phase.

FIGURE 2-2   Phase 1—preparation.

1. Establish the Project Team

Before many key decisions are made, a formal project team is established with representatives from each of the main business areas that will be affected. In addition to their primary task of completing the automation project, these members are charged with ensuring that the project will satisfy the interests and needs of their business areas. Therefore, the team members should have the authority and knowledge to speak on behalf of the function they represent. At first, the team need only include members from those business areas most directly affected. As the project proceeds through the three phases, however, the team should be expected to grow by including members from peripheral business areas and people with specific knowledge critical to the project. Succeeding phases involve progressively finer levels of detail and require team members with corresponding knowledge and skills.

At a minimum, the initial team should include at least one full-time member from distribution and one from management information systems (MIS) departments. These members should be assigned to work full time on the automation project through its completion. Although this requirement for full-time assignment of these members may seem excessive, an automation project is almost always very extensive and there are many issues to resolve and details to be handled throughout the process. Frequently, the best team members are people who are viewed as critical to the day-to-day operations of the facility. The same knowledge of operations that makes them ideal team members also makes them "indispensable" to operations. Thus, it is very tempting to give them the automation project as an additional task to be handled, along with their normal assignment. This approach almost always results in a breakdown of the project process, daily operations, or both. The amount of effort a successful distribution automation project requires should not be underestimated.

The member from the distribution department should be familiar with the operation of the entire facility from receiving through shipping. Since no one person can be intimately familiar with the details of every operation, the distribution member must have the authority to call in other team members from various distribution departments as needed. These departmental representatives are authorized to speak for and make commitments on behalf of their departments.

The team member from the MIS department is responsible for ensuring that the automation project is compatible with other existing and planned business systems. For example, the purchasing system may require that vendor packing slip numbers are collected for each receipt. If so, the MIS team member must ensure that the new receiving processes record this information and that the warehouse management system (WMS) provides it to the purchasing system. In addition, this member is expected to ensure that the project is compatible with corporate MIS strategic goals. The MIS team member also needs the authority to call in additional MIS team members with detailed knowledge of various business systems as necessary.

A project manager should be appointed to lead the team. If the project is simple and small in scope, the project manager can be one of the above two members. On a project of any significant size, this is a separate job. The project manager performs the normal project management tasks such as coordinating team activities, tracking and managing schedules and status, and administering change control procedures. The project manager must also be able to request team members from other business areas as needed when project discussions get into topics that may affect those areas. Chapter 10 describes in more detail the work of the project manager.

One of the first tasks for the newly formed team is finding additional team members with expertise in WMS and material handling systems (MHS), which are the major components of distribution automation. The introduction of a WMS and possibly new material handling hardware into the distribution facility changes many business practices and work flows. The team will need to define these changes in detail during phase 2. Unless the team has members who are thoroughly versed in the many capabilities of warehouse management systems, including the latest material handling methods and state-of-the-art distribution practices, it will be very difficult for the team to determine what process changes can or should be made. In most cases, the existing warehouse or distribution center staff will not have people with the depth of knowledge required to provide these team members. This knowledge can only be acquired through direct involvement in the implementations of a variety of warehouse automation projects. These automation experts provide an added benefit by bringing an "outsider's" perspective to the group. Most successful automation project teams have a good balance of "inside" and "outside" members. The distribution and MIS members of the team are, of course, insiders. These are people who know all the details of the current operation and why things are done in certain ways. Insiders, however, tend to view things in terms of the current process and frequently have difficulty seeing other ways of accomplishing the same objectives. Outsiders, on the other hand, are people with no vested interest in defending the current status quo and can bring fresh ideas and outlooks to the table.

There are three potential sources of candidates for these team positions: technical representatives from prospective WMS and MHS suppliers, in-house experts from MIS and engineering departments, and outside consultants.

Many companies that have completed successful distribution automation projects have taken the approach of forming partnerships with their system suppliers early on in the automation process and have found that their vendors have become very valuable members of the project team. In order to use these sources of expertise, however, the team needs to select WMS and MHS suppliers much earlier in the project cycle than was considered standard practice in the past. There are good reasons for making the supplier selection while the team is being built. A conventional project team, in the early days of distribution automation, would spend many weeks drawing up a detailed set of system and hardware requirements, which were then sent out for quotes and the lowest bidder selected. Unfortunately, these detailed requirements documents were drawn up from scratch and did not match well with the capabilities of any available systems. This usually resulted in all potential vendors offering what were, essentially, total custom solutions to meet the requirements and at a very high cost. In addition, these requirement documents were written without knowledge of how to take maximum advantage of the capabilities of distribution automation technologies and usually reflected the current, essentially unchanged business process with the addition of bits and pieces of automation. The resulting completed automation projects generally failed to realize the full benefits of automation.

Settling on a WMS vendor prior to the start of phase 2 helps focus the business process discussions of that phase by having an established software base to use as a starting point. Most full-function WMS packages have many predefined common processes that will be usable without changes. This allows the team to limit its work to defining any modifications. Keep in mind that most system packages require some modification to fit any specific business exactly. The key goal in this phase is to select the system that provides the best starting point. Chapter 6 explains the details of evaluating warehouse management systems. There are, however, some key points to keep in mind. Does the system support the company's basic operating methodologies? For example, if the facility receives by container, does the vendor's package support container receiving? Does the package support the best practices of the company's specific industry? What is the vendor's reputation and technical competence? Attempt to find a supplier who will be a good partner in the automation project and who will provide good technical members for the project team.

Similar considerations also apply to selecting a MHS supplier during phase 1. The technical and engineering experts provided by this supplier are invaluable in providing insight into the capabilities and limitations of various hardware alternatives and generally are very knowledgeable in the best practices for various industries. They are able to advise the team on the latest, best methods for physically performing each type of operation. The key considerations in making this choice are to find an MHS engineering firm or MHS vendor with a broad range of capabilities and no obvious bias toward any one specific technology. In the case of selecting an MHS partner, it is also important to select one with experience either in the company's industry or one that is closely related.

A second source for the expert team members is the company's internal MIS and engineering organizations. If the company does not plan to use a system package but plans to write the automation system internally, an MIS architecture team should be formed for the automation project. One or more members of this architecture team then become the system members of the automation project team. This team will spend several weeks prior to the start of phase 2 becoming familiar with the features and capabilities of various WMS packages and becoming familiar with the best business practices in the industry. This helps the architecture team prepare to contribute to the business process discussions of phase 2. Similarly, many facilities have a strong and diversified internal industrial engineering group. If so, this group may be able to supply the MHS and industrial engineering expert for the team.

If the expert team members come from within the company, then the team should take care to find additional members to provide the "outsider" viewpoint. Business process consultants with distribution experience should be considered for, at least, the process design phase of the project.

A third potential source for the expert team members is a distribution automation consultant who can bring in experience with various projects. Select a consulting firm with "hands-on" knowledge of automation system design and implementation. It is essential to find a firm that has been involved in a number of successful projects from conception through completion.

Finally, a project sponsor on the executive level is required, one who is personally committed to the success of the project. This sponsor should frequently review the progress of the project with the project manager and ensure that milestones and timetables are being met. The sponsor ensures that all team members are available to work on the project and are not unduly distracted by day-to-day operations. The sponsor also ensures that the team receives cooperation from all of the business areas that are affected or need to provide input to the automation project. When required, the sponsor acts to resolve conflicts that the team cannot resolve on its own.

2. Define Project Goals

Consider first what needs to be accomplished. In order to determine the success criteria for the automation project, the business results expected must be defined in concrete terms. In addition, the project goals must consider, and blend with, the overall company goals for the future (i.e., how is the business expected to change in the next few years?). Several sources are considered when setting the goals for a distribution automation project, including logistics pipeline analysis, management input, and a customer survey. Using the information gathered from these sources, define a set of concrete project goals. Automation, in and of itself, is not a goal. Typical goals may be a minimum inventory accuracy measure, a maximum order cycle time, the number of shipments expected per day, and so on. Figure 2-3 shows a set of goals from a typical project.

> 1. Reduce labor costs to less than $0.35 per carton shipped.
> 2. Reduce inventory errors to less than 0.1 percent.
> 3. Ship 95 percent of orders within 24 hours of receipt.
> 4. Ship 97 percent of orders complete.
> 5. Reduce shipment errors to less than 0.3 percent.

FIGURE 2-3   Example of project goals.

Logistics Pipeline.   Examine the company's overall logistics pipeline. At the start of a distribution automation project, it is not clear what should be considered part of the project and subject to change and what will be beyond the project's scope. Initially, the complete logistics flow of the entire business is analyzed. Since the warehouse communicates upstream with manufacturing and vendors and downstream to customers, information is passed throughout the pipeline. Automation allows changes in warehouse operations that benefit other areas of the business. Conversely, changes in upstream or downstream operations may make automation easier to implement. For example, can the company's suppliers send shipment information via advanced ship notices to help speed receipts? Or, can the purchasing department use better information gathered in receiving and quality control departments to demand improved compliance from suppliers? Do customers, or stores, need better information on order status or shipment contents? If better inventory accuracy can be achieved, what effect will this have on inventory levels and on the frequency and size of receipts? Therefore, it is important to view the automation project as an integral part of the entire logistics pipeline in order to establish the true scope of the project and obtain meaningful goals.

Management Input.   Corporate management has strategic goals defined that need review to ascertain the future direction of the company. These strategic goals define and set the company's objectives and must be reflected in the automation project goals. For example, strategic goals to increase the customer base or increase sales volume require distribution systems that are capable of handling the increases.

   As with all large projects, management must agree with the completed list of project goals and support the initial project phases with the appropriate staffing and funding. Having top management's input and involvement guarantees the focus and attention a project of this magnitude requires.

Customer Survey.   Discuss with key customers their current and future needs. Determine what they like and dislike about the services they receive. A questionnaire is prepared to obtain consistent information including current service levels, quality problems, and most importantly, future requirements. For example, will they be mandating special bar coded labels? Will they be demanding reduced order cycle times?

Questions regarding the company's service to the customer:

1. Are order turnaround times (from order placement to promised shipment) reasonable?
2. Are you able to determine the status of your orders easily?
3. Are you informed when an order will not be shipped on the promised date?
4. Are shipments received on the expected dates?
5. Do you receive the correct items?
6. Are shipments received in the expected condition (paperwork, labeling, packaging, case pack quantity, etc.)?
7. Are shipments received in undamaged condition?
8. Is the backorder frequency acceptable?
9. When necessary, can you return items quickly and easily?

Questions regarding the customer's future plans:

1. Do you plan to change the quantities you order (more per carton, fewer per carton, or more mixed cartons)?
2. Do you plan to change how or where you will receive your shipments?
3. Do you plan to require special labeling of cartons or require special paperwork with receipts?
4. Do you have plans to implement EDI?

FIGURE 2-4   Example of a customer survey excerpt.

Figure 2-4 shows an excerpt from a typical customer survey questionnaire. It is important to understand the directions in which the company's customers are going in order to incorporate their requirements as part of the project goals. If the complete requirements of all customers are not known or understood, then the proposed warehouse automation project will not optimize the customer service potential of the company. Many project team members have made the mistake of not determining or anticipating future customer logistics requirements and have found themselves in "reaction" mode to satisfy unexpected requirements. Usually, this involves redesigning at a later point or even after implementation, where the impact to cost and schedule is significantly higher.

3. Define the Project Scope

Finally, define the scope of the project. Clearly delineate what is to be part of the distribution automation project and what is not. This helps prevent the project gradually growing to encompass more and more business areas, a frequently encountered phenomenon known as "scope creep." Such projects seem to take on a life of their own and are rarely ever completed. Will the project be confined to the traditional distribution center or warehouse functions or will it also include order processing and purchasing? Will the project include a new facility layout or even a new building? Typically, a distribution project is confined to those business processes and functions that are traditionally found within the "four walls" of a warehouse or distribution center. Functions such as order entry, purchasing, and accounts receivable and payable are generally considered to be outside of the scope of most distribution automation projects. That is not to say, however, that they should not or cannot be included, and many times they are. In those cases, though, the distribution automation project is frequently viewed as a large subproject within a company-wide business process reengineering effort.

PHASE 2: DEFINITION

Phase 2 encompasses the steps that define in detail how the distribution facility will operate in the new automated environment. The goals in this phase are first to design new business processes, "reengineering" if you will, for the facility and then design the required components of automation that will support and facilitate these new processes. Figure 2-5 illustrates the activities of phase 2.

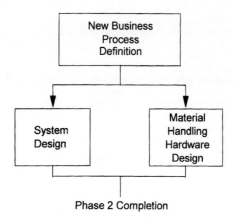

FIGURE 2-5   Phase 2—definition.

1. Define New Business Processes

In order to take maximum advantage of automation, the basic material and information flows of the facility, i.e., the business processes, must be redesigned. Adding the trappings of an automated system, such as RF terminals and bar codes, onto a set of unchanged processes is rarely successful, and will not realize the potential benefits of automation. Automating a poorly running process merely makes things go wrong faster. Automation should be viewed as a tool that allows better processes; it speeds up the movement of material and information moving from area to area and at the same time requires much less human intervention to support these flows. This in turn eliminates material sitting idly, waiting for transport or waiting for information to catch up before the next process can start. Figure 2-6 shows the steps in defining a new business process.

Review Current Processes.   After the team goals and project scope have been established in phase 1, current business processes within the project scope are reviewed. A simple process flow chart is developed for each functional process. This chart shows each physical step of the process and each decision that is made. Chapter 6 discusses in detail how to diagram a process. Along with this process diagram, all of the forms that are used in the process are collected. This step assists the team in learning and understanding each operation as well as the data that is collected and passed between operations. The team determines how materials and information are used in each area. Remember that each time material moves, an invisible flow of information occurs simultaneously. For example, when a pallet is moved, information about the pallet

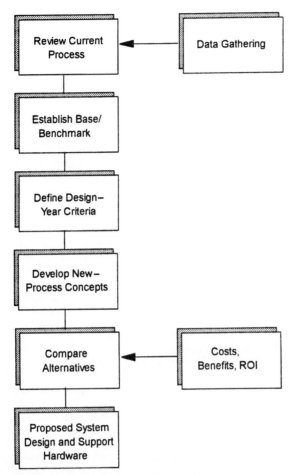

FIGURE 2-6 New business process definition.

changes: where it is, when it was last moved, who moved it, why it was moved, etc. The team determines which information is important, which information is not really needed, which material processes provide added value, and which processes are merely moving material without apparent added value. For each operation consider the question, "Would I pay for this operation if I were a customer?"

This review of current processes should not be overly concerned with documenting every nuance of the existing process. The goal here is to understand what a process does, why it is done, and to whom it is important that it be done.

Data Gathering.   Data gathering adds to the current process definitions and forms the input for the establishing of benchmarks. This involves obtaining, collecting, and documenting data regarding current operations. The production volumes for each type of process are determined, both average and peak. Frequently, this data is only known on an aggregate facility-wide basis. The team must break down this data to the level of each process. Any questions that arise concerning the processes are researched and answered. This data is used later to help determine unit costs of processes. Later, during the hardware design step of the project, this data is used to size material handling and data-processing hardware.

Establish Baselines and Benchmarks.   In order to compare alternatives to the current way of doing business and define the savings a proposed system can generate, the costs of the current processes in terms of time and materials are established as a baseline. The most accurate method of determining the time expended in current operations is to develop time estimates for each step in the process. If the company has time standards already in place, this becomes a very easy step. Then, the time per unit, multiplied by a key volume measurement for the operation, determines the labor effort. A verification of this procedure is obtained by multiplying this time standard by current volumes, summarizing to a facility-wide level, and comparing the resulting calculated labor to actual labor expended. In the absence of existing time standards, reversing this verification method can also be used to infer estimated time standards by dividing actual labor by current volumes. Once again, the goal here is not to analyze the cost of every last step; the goal is to determine the overall cost of each current process.

In addition to determining a baseline for the current processes, try to determine, if possible, corresponding benchmarks for analogous processes that competitors and related industries use. Do not get too involved in understanding the details of these other processes; the other companies will have different situations and corporate goals. It would be a mistake to copy their processes directly and attempt to apply them in a different environment. Instead, this benchmarking effort is used as a means of setting target levels that the new processes must exceed.

With baselines and benchmarks established, a very precise analysis is conducted to determine the true benefit of an alternative or a change in a process.

Define Design-Year Criteria.   To evaluate alternatives and determine requirements for the automation project properly, assumptions are made in regard to volume and product line growth rates. The automation project should be designed based on volume growth rates forecasted for the next three to five years in order to allow growth within the new system without requiring major changes. These projected growth rates were discussed and identified during the phase 1 step of defining project goals.

Now, having the baseline benchmarks and design-year criteria established, the labor requirements calculated in the previous step are projected into the future to establish the baseline case that can be expected if no changes are made to current business processes. The resulting future labor requirements are then converted into dollars by multiplying them by the projection of labor rates. These labor requirements and dollars become the base case (current process) and are used to compare against process alternatives.

Brainstorm New-Process Concepts.   The new processes for each area are now devised to take advantage of the material handling, information handling, and management capabilities of distribution automation technologies. The project goals and benchmarks that were established earlier are used as guidelines and targets for the new-process concepts. For example, if one of the goals is to increase the capability to ship less than full case ("broken case") orders efficiently, then alternative processes are developed that improve broken-case picking. Alternative methods for performing each process are brainstormed by the project team and a list is created. This continues with the remainder of the goals and business processes until a comprehensive concept list is formed. When brainstorming new processes, key process component considerations include technology and material handling equipment and systems, which are explained in Chaps. 3, 4, and 5. Since the next step will be a detailed analysis of the leading alternatives for each process, the list is then winnowed down to the top two or three most viable alternatives for each process.

Compare Alternatives.   The leading alternatives are compared against each other and against the current process baseline to select the best new process. Before this can be done, however, each of the leading candidate processes is analyzed in detail. First, a proposed process flow chart for each alternative is constructed. This flow chart is similar to the ones that were designed for the current process. The processes predefined in the selected WMS can be used as a starting point. If the WMS fits well with the proposed alternative, it is used as is. Do not be afraid, however, to propose changes if they are needed to support a better process. Wherever possible, the WMS is allowed to handle the information flow and use distribution automation technologies such as radio frequency terminals and bar coding to eliminate paperwork and data entry. In most cases, automation enables major simplification of many business processes. Simulation is another tool to assist in the analysis of the alternatives. Chapter 7 presents a simulation strategy and provides an automation project example.

Once alternative processes have been diagrammed, the labor costs for each alternative are estimated. When a step is the same as the current process, the current labor standard for that step is used. If a step is new, however, there are two options. For many common steps, the labor time can be found in an industrial engineering reference book. A second option for determining labor is to create a simple mock-up of the proposed operation and time several different people performing the process. This need not be elaborate. Fairly good estimates can be done using tables of the correct height and empty boxes of the right size.

Next, a cost estimate is made of new or additional equipment that will be required to support the alternative process. The costs should include any installation of any equipment required by the process and any increased maintenance costs to support the equipment. Factors such as inventory relocation, cables, power outlets, sprinkler systems, and lighting systems must be considered and, if applicable, included in the cost estimate. Chapter 6 provides a framework for a request for information and a method to evaluate and choose vendors.

In addition to any quantitative savings that will result, a list of qualitative benefits is identified for each alternative. These include benefits such as improved picking accuracy and increased space utilization. If possible, the team should try to estimate the value of these benefits to the business or to the company's customers.

Finally, the alternative processes are compared against each other and against the baseline. Using the cost estimates, the return on investment is calculated, the qualitative benefits are evaluated, and the optimum alternative selected. Chapter 8 explains the evaluation process in detail.

New-Process Selection.   After analyzing and selecting the best alternative processes in each business area, the overall new-process flow is documented in a comprehensive new business process report. This report is reviewed by all team members and then by the management of each affected business area before proceeding with the design steps of phase 2. The goals of the review are to ensure consistency from business area to business area and to ensure that the managers of areas that will be affected are familiar with and agree with the planned direction of the automation project. This is a crucial milestone to the success of an automation project; obtaining the "buy-in" from all the managers at this point establishes the proposed new processes as the plan of record and helps avoid unnecessary conflict and redesign later. In order to obtain this buy-in, the management reviews are face-to-face meetings with the managers and their key staff, where the team members walk through and explain the new processes in detail. If any problems are uncovered, and they almost always are, they are resolved, and any needed changes are then reflected in the report. Above all, avoid simply sending out copies of the report to the managers and asking for their comments. Inevitably, the reviews will be cursory or details will be misunderstood, and late in the project a major oversight will be uncovered.

2. Detail Design

Once the new processes have been documented and reviewed, the activities in phase 2 split into two tracks that can proceed more or less concurrently. The customization to the warehouse management system and the design of the material handling system and other hardware needed to support the new processes must be done simultaneously because the design of each influences the other. The details of the two designs, however, require different areas of expertise and typically involve different subgroups of the project team.

System Design.   Once the new business processes have been defined, the team defines how the WMS will support these processes. This means defining how the system will interact with the warehouse personnel, with other corporate systems, and with the facility's material handling systems. These aspects of the WMS are termed *system externals.* System externals include things such as information display screens, data input screens, reports, and interfaces. These system externals are designed and documented in detail. This ensures that both the project team and the WMS supplier understand exactly what is to be delivered. Chapter 9 explains what constitutes a system external design and describes a process for doing such a design. System details that are not directly visible to distribution operations personnel such as the programs and databases to support this external behavior are *system internals* and will be designed during phase 3. If the WMS will be written in-house by MIS or if a custom WMS is being written by a vendor, the system externals of the entire WMS are specified. If a WMS package has been selected, many of the externals are already specified and can be used as is. This does not, however, completely eliminate the need for specifying system externals. It would be very unusual for a WMS package to fit the new processes perfectly. In almost every case, some degree of modification is needed to fit the special requirements of the operation, but the team need only specify the changes from the basic package.

Material Handling and Hardware Detailed Design.   While the WMS external design is proceeding, a similar design is done for the MHS and any other hardware needed to support the new business processes. Where material handling equipment is required, the volume input and output requirements of the system are defined and concept drawings made. Finally, a technology equipment plan is formalized. The exact number of terminals, radio frequency terminals, scanners, bar code printers, and the like are defined and their location determined. Any necessary facility changes are designed, including electrical power needs, data cabling, air conditioning, sprinkler systems, lighting, emergency backup electrical power, and compressed air supply. Specifications are written to document formally the requirements for the various equipment suppliers and the conveyor control software supplier. Material handling equipment specifications are written to contain the new-process flows, preliminary equipment drawings, other electrical and/or mechanical specifications, and any interface file requirements to the WMS. These designs are a very important step in that they communicate the new-process design to others who will estimate the costs and schedules required. These estimates are critical to the overall project implementation budget and schedule. The designs are clearly written in order to avoid misinterpretations that could lead to future redesigns and subsequent cost and delivery overruns.

Equipment and Vendor Selection. A conventional bid process is used to select the vendors for items that are going to be purchased and such items for which there is no existing partnership with a supplier, e.g., bar code scanners. The following are the steps in this process.

A. *Prepare bid packages.* Sets of identical bid packages, containing the appropriate specifications, are prepared for each type of equipment. The purpose of standard bid packages is to allow different suppliers to submit proposals against the same requirements and allow the company to compare the proposals equally and select the best vendor. Additionally, the criteria that will be used to evaluate the vendors objectively are defined and documented by the team.

B. *Select and qualify vendors.* A list of vendors to be contacted for qualification is prepared in cooperation with the company's purchasing department. There are other sources of vendor lists including technical magazines, professional societies, and trade shows. To ensure a qualified vendor is selected, a discussion not only with the sales representative but with a design engineer and a project manager is helpful. The comprehensiveness of this qualification process is dependent on the investment amount and customization required for the equipment or software.

C. *Bid process.* The selected vendors are sent the appropriate bid packages and are all given a fixed time period to review the package and develop their proposals. During this time, vendors are able to contact a designated project team member to obtain further clarification on any questions. If any changes to the initial specifications develop, all vendors are updated with the new information.

D. *Final design.* Vendor proposals are evaluated based on price, delivery, ability to meet requirements, and on other criteria previously identified. If the selected vendors have worked with other clients on similar projects, they may propose new ideas or slightly different ways to accomplish the same tasks. These new ideas received in the vendor's proposal are considered and evaluated. If accepted, they are integrated into the final design. Vendor selection is completed and presented to the management team for concurrence and then submitted to the purchasing department.

E. *Award contracts.* Upon final agreement of the contract price, delivery, and schedule, purchase orders are placed with the appropriate vendors.

3. Phase 2 Completion

At the end of this phase, all the components of the automation project have been designed in detail and assigned to the appropriate providers, either vendors or internal departments or interdepartmental groups, for completion.

PHASE 3: IMPLEMENTATION

The third and final phase of the warehouse automation project is the implementation phase. This is where all of the processes, software, hardware, material handling, and

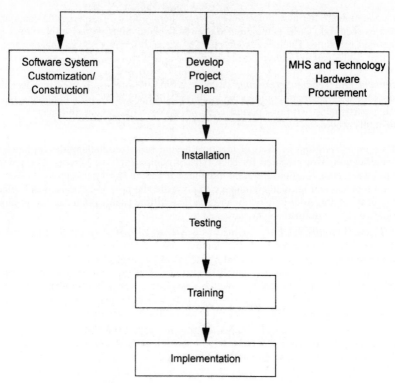

FIGURE 2-7   Phase 3—implementation.

technology designed during phase 2 come together. Figure 2-7 depicts the steps for the implementation phase of the project. This phase typically constitutes the longest portion of the project cycle time due to vendor lead times for equipment and the time required to integrate and test all the system components as well as the existing information systems. Phase 3 is also often the most difficult phase to manage due to the many complex and interrelated events that occur in what is frequently a very tight timetable. In addition, any deficiencies in the execution of the steps of phase 2 will not become evident until phase 3 is well under way. For example, suppose the entire business process design step of phase 2 was skipped and software and hardware designs were started immediately based on the current processes. The project could proceed quite smoothly and everyone concerned would have the impression that the project was on track. Not until sometime late in phase 3 would it then become obvious that automation not only does not save money but that it slows down the whole operation. Therefore, be sure that phase 2 has been successfully completed before getting very far into phase 3.

1. Project Management

Develop a Project Plan.   At this point, the various suppliers and providers of the components of the automation project have begun working on their portions of the project. There are still many details to be planned and executed before the project is implemented. Chapter 10 will describe how to plan and manage the project. However, in brief, a project schedule that defines all the activities, milestones, and responsibilities is developed and used to communicate, manage, and track the project. This project plan includes all the activities and responsibilities of everyone involved, including suppliers, internal departments, and project team members. Milestone dates obtained from the providers are included in the project plan to ensure that all tasks occur in the proper sequence. As with any project, it is impossible to plan with too much detail. Do not forget to include the following in the plan and specify the who, where, when, what, and how of each.

Testing. Even though everyone has agreed on the designs prepared during phase 2, plan to thoroughly test the system with production data to ensure what is delivered is what was expected and that it works correctly.

Training. When distribution automation is installed, many familiar processes will change. Plan to train the entire staff so that they know how to use the system and know how their jobs will be affected.

Staffing. Although automation may eliminate some positions, other new positions will be created. Do not forget to plan for these, e.g., computer system operators and a help desk.

Procedure Documentation. Plan to document all of the new business processes carefully so that new hires can be trained easily. Typical documents include:

- Procedures manuals
- System users guide
- Training manuals
- Technical reference manuals
- Maintenance manuals

Material Handling Installation. Plan in detail for the installation of any material handling changes. Where will equipment be stored before it is installed? How will operations be maintained as the equipment is installed?

Installation of Technology Equipment. Plan where each terminal, printer, etc., will go. Do not forget tables or stands to hold the equipment. Include tasks in the project plan to procure these ancillary items and install them.

Wiring and Cabling. Computers, terminals, material handling equipment, etc., will require power and cabling. Tasks for their installation must be included in the project plan.

Bar Code Labeling of Storage Locations and Mapping into WMS. If the new processes include bar coded locations, a naming sequence for the locations must be defined, labels must be printed, a label must be attached to each storage location, and the location must be entered into the WMS. A typical distribution center may have thousands of locations. Developing the location map and ordering and installing the labels can be a significant effort. Plan this well ahead of time.

Initial Loading of Inventory Data into WMS. This can also be a significant effort. Alternative strategies include downloading information from the current inventory system or entering a physical count of each location into the new WMS. Downloading requires less effort, but may be less accurate. The team needs to examine the risks and tradeoffs of the various ways of doing the inventory load and prepare a strategy. Once a decision has been made, analyze how many people will be required, what equipment will be needed, and how long the inventory load will take, and include the preparation and execution of this step in the project plan.

Disaster Recovery Plans and Procedures. Plan for power failures and natural disasters. How will backup data be stored and where? What is the plan for an extended power outage? Consider uninterruptible-power supplies (UPSs) for vital equipment such as computers and terminals to protect against lightning and brief power interruptions.

Postimplementation User Support. Even with good training, workers will have many questions about the system or about their terminal. Plan for a help desk to answer questions.

Phased-In Implementation or Single Cut-Over. It can be less disruptive to day-to-day operations to implement the new processes in phases rather than all at once. A typical phased approach would install inbound processes—receiving, quality control, put away, and inventory control—in the first phase. Outbound processes—order release, picking, and shipping—would then be implemented in a second phase. Plan how the cut-over will take place and what automation components will be needed for each phase.

Employee Buy-In. Employee support of the automation project is critical to its success. If the employees that use the new system want to make it work, small problems will be accepted in stride. If the employees view the system as a threat, reasons will be found to say that "the system does not work." There will be problems with the automation system, especially immediately after start-up. These problems, relevant or not, will lend credence to the employee complaints and they will find a receptive audience with anyone in corporate management who opposes automation. The result is that ways will be found to go around the system and control of inventory will be jeopardized. Employee buy-in does not just happen; one must plan to obtain it. Be sure to keep employees well informed as the project progresses. Let everyone know how the new system will help make their job easier.

Customer Notification. Well before the final cut-over, plan to review the company's distribution automation strategy and implementation plans with key customers. This will help to avoid any unexpected incompatibilities. Include in the project plan tasks to inform all customers about the project and what changes they can expect after cut-over. Prepare them for the possibility of some delays in deliveries immediately after implementation.

Cut-Over to the New Automation System. When the new system is first turned on, problems are inevitable. Plan to start slowly so that problems can be caught and fixed as they are found. Schedule practice runs before starting real production. Do not plan on a heavy workload the first day. Start with a few small shipments and have more "observers" than workers. During the first day of using the new system, all activities should be planned in detail, down to the hour.

Status Reviews. The project manager should carefully track progress against the implementation plan. Regular status meeting are held to keep all team members informed. Any changes to the schedule are documented and integrated into the new project schedule and distributed to the team members. Keep the "big picture" in mind. By their very nature, automation projects tend to be complex and involve a large part of the company. It is almost inevitable that the project will be perceived as being in trouble at some time. Plan in advance on how to respond.

Change Control. As the implementation progresses, unforeseen issues can force changes in the designs or in the master plan. *Plan to change the plan.* Have a change-control procedure and follow it. The project budget should have a "change-control bucket" as a line item to allow time and funds for necessary changes to the project.

2. Software System Customization and Construction

During this step, the providers of the various automation systems are taking the finished external designs from phase 2 and producing the finished products for delivery. The first step that takes place is the conversion of the external designs into what are called *internal designs*. These internal designs describe programs to be written, what each program is to do and the logic it will follow, what database files are needed, and so forth. As the internal designs are written, parts of the external designs may be found to be unclear or inconsistent. The change-control procedure is then used to correct the external designs. The internal designs are assigned to programmers to be converted into programs. As each program is completed, it is tested to ensure that it conforms to the internal design. This is known as *unit testing*. After several individual programs are completed and unit tested, the group of programs needed to support a process is tested to ensure that they conform to the intended external design. This is usually called *functional testing*. Each software supplier provides the project manager with a schedule of these activities and advises him or her of their progress and completion so that the project team may be assured that things are progressing on schedule.

Many times, software suppliers ask their customers to review and approve the internal designs. It may be an informative activity for the team to review the internal designs, if only to verify that the designs have been completed on schedule. The review, however, should not be construed as an authorization for the supplier to substitute the internal designs as the plan of record. The external designs remain as the agreement of what is to be delivered. If in the process of writing the internal designs, the supplier finds a needed change or clarification to the external design, the change-control procedure is used to make the change. If the external designs are not kept up to date, it will later be very difficult for the project team members, who are not computer science professionals, to use the internal designs and test the software to determine if it meets the agreed requirements.

3. MHS and Technology-Hardware Procurement

In a fashion similar to that for the software, the suppliers of the material handling equipment and technology hardware are using the conceptual designs to prepare detailed engineering drawings. The suppliers then present final detail design and installation drawings to the team for their approval. Prior to approval, engineers from the facility or maintenance group and from the industrial engineering department review the drawings for completeness and identify any tasks required for implementation. This usually includes such items as current-equipment relocation, ceiling supports, power, and compressed air installation.

4. Installation

Installation includes the actual placement of hardware, equipment, and software in the facility. This is performed either by the selected vendors, the project team, or other internal departments. During the project plan development, the installation phase is carefully considered. The methods used to install equipment into the facility vary, depending on the amount of new equipment to be installed, production volumes during the installation, and the space available for the installation. Since many times the warehouse normally must continue to receive and ship product throughout the implementation phase, it can be advantageous to install the system in phases. For example, the new line for returns or a particular company division can be installed first in a vacated area. Then, once that portion of the installation is up and functioning, the old, replaced equipment can be removed and another phase installed in the newly vacated space. This reduces the risk of any disruption to customer service levels.

5. Testing

Verify Installation.   It is necessary to verify that the hardware and equipment that have been installed meet the agreed-to specifications. The best time to perform this activity is during and right after installation. These tasks include verifying the model numbers, quantities, and location of the equipment. In the case of material handling equipment, such as conveyors and carousels, verify that the operational characteristics meet the specifications. This includes the operation of sensors, internal logic, safety features, and throughput capabilities.

Interface Testing.   Since various departments and/or vendors are involved with inter-faces among automation systems, a separate testing session is conducted to ensure that proper communication and data transfer occur among all systems involved. A set of test cases is developed by a project team member that mimics daily operations and unique daily situations and confirms any restrictions that must be met. The expected results are documented and compared to the actual results when the testing occurs. Any deviations are addressed and software is fixed and retested.

Software and Hardware Integration Testing.   This step involves verification that all system components perform together and meet the required system specifications. Test cases are developed that define the scenarios required to verify system capabilities and throughput. All too often, this step is not given the attention it requires. Allowing the proper time during this step ensures that complete testing is performed and prevents frustrating problems during cut-over and "going live" with the system. A key success factor to a successful automation implementation is the acceptance of the system by the employees. Find and correct problems with a small select employee group before involving the full staff. The new system reliability will erode quickly in the minds of all employees if many serious problems are identified after cut-over. The employee buy-in, critical to the project's success, will then be much more difficult to obtain. In addition, fixing problems in a test environment is much easier than fixing problems in a live environment since there is no risk of disrupting ongoing operations. The time invested early on with testing provides significant payback. Chapter 11 describes the different testing phases and provides a framework to develop a test strategy.

6. Training

Employee Software and Hardware Training.   Training sessions are conducted so employees can experiment with the system and learn their jobs under the new system. At the completion of the training session, the employees must be able to:

- Operate the new equipment
- Understand the new processes
- Perform their job assignments using the new system
- Understand how to handle various situations
- Be familiar with safety procedures

Two helpful documents during training are the *User Guide* and *Operating Procedures*. The *User Guide* contains the computer system screens for the supporting software and instructions on their usage. The *Operating Procedures* explains the steps required to complete a job. Using these documents during the training sessions allows the employees to become familiar with the contents of the documents so that they can be used as future references. Many studies have confirmed that training is a critical success factor. Chapter 12 explains and provides training strategies.

Maintenance Training.   Now that the facility is becoming automated, maintenance and preventive maintenance become more significant. A formal training session with the maintenance department on the equipment and the associated maintenance manuals is conducted by the vendor prior to acceptance and sign-off.

Having a maintenance representative available during the vendor hardware and equipment installation step assists in the understanding and functioning of the new system and minimizes the time required for the department to become proficient with the new equipment.

7. Implementation

Chapter 13 describes the adventures associated with implementation. If properly prepared, the project will be a success.

Initial Loading of Inventory Data into WMS.   The current inventory and where it is located are loaded into the WMS just prior to implementation. The accuracy with which this task is done is critical to the early success of the WMS. If workers cannot find inventory because the WMS was not correctly loaded, they may begin to mistrust the WMS. This will lead to "working around" the system, which will compound inventory errors. The accuracy of the inventory load should be double-checked prior to cut-over.

Cut-Over Implementation.   This is the time when all of the team's hard work and efforts pay off. The cut-over tasks and schedule have been designed to minimize customer service impacts. The project team and all technical personnel from all major vendors are present the first day the new system is in production. Any problems that arise are addressed and corrected immediately. In addition, changes to the documentation are noted and updated as soon as possible. As each day progresses, the number of problems decline, employees become familiar with the system and operating procedures, and warehouse supervisors begin managing the system directly. The project team's and vendors' support continues but the number of team members needing to be present is reduced until the warehouse workers can operate on their own. Thought should be given to the amount of support, especially vendor support, that the project will require until it reaches a steady state.

Final Acceptance.   Now that the system is successfully in production, the project team finally accepts the new system as complete. This means the vendor contracts have been fulfilled and final payment can be made. By clearly establishing the project scope in the first phase of the project, a completion point can be established and identified.

Follow-Up Evaluation.   After the automation system reaches a steady-state point, one last task for the project team is to review the actual results of the system compared to the initial goals that were established. A final report documenting attainment or nonattainment of the goals is created and presented to management.

SUMMARY

The foundation of a successful warehouse automation project is a clear understanding of what needs to be done and how to do it. Part 1 has covered the main steps, described what must be accomplished in each step, and laid out a road map for proceeding from start to finish. All of the details needed for a specific project will not be known at the beginning of the project. Each step, however, is structured to determine and specify the details for the succeeding steps. Part 2, Planning and Design (Chaps. 3 through 9), covers the various considerations that enter phase 2 of the project in which the functions of the automated facility and its supporting systems are defined. These chapters are dedicated to the components of this project process and the technology that can be brought to bear on the distribution and warehouse environment.

PLANNING AND DESIGN

CHAPTER 3
TECHNOLOGY CONSIDERATIONS

INTRODUCTION

With today's business environment demanding increased accuracy, timeliness, service, and availability of products; reduced order cycle times; improved productivity; and decreased processing costs, warehouse management systems (WMSs) are being designed to utilize modern technology to assist businesses with the required parallel flow of material and associated data. Additionally, with more and more companies opting for a combination of mechanized and conventional approaches rather than large-scale, highly automated material handling systems, a wide gamut of technology offerings, which can be integrated to maximize the benefits of a WMS implementation, is available.

It is important for a WMS strategic direction to be established based on relevant technology information, since the proper specification of technology for a proposed system is vital to ensure optimum use of inherent WMS functionality, adequate system and end-user performance, a meaningful return on investment (ROI), and system scalability. Modern warehouse management systems are being implemented with the following technologies:

- Bar coding
- Automated data collection (i.e., scanners, batch terminals, radio frequency devices, etc.)
- Computer systems
- Printing
- Electronic data interchange

OVERVIEW

The objective of this chapter is to describe the technologies currently utilized with an automated WMS. An understanding of each of these technologies and their relationship to a WMS is essential for the specification of hardware and software requirements and/or the evaluation of third-party vendor WMS or technology vendor proposals. Information for each technology detailed in this chapter encompasses the following:

1. Components required to implement
2. Examples of utilization with a WMS
3. Associated advantages and disadvantages

BAR CODE TECHNOLOGY

Bar code technology, which assimilates the Morse code by utilizing a series of varying width parallel bars and spaces rather than dots and dashes to represent alphanumeric characters, is a printed form of data communications. When a light beam is moved across a bar code, it measures the length in terms of time of each highly reflective white space and nonreflective dark bar, and it converts the spaces and bars into a binary digit (bit) representation. (Bits that can have values of 0 or 1 are strung together in groups of eight to form bytes. Each byte represents one character.) As schematically depicted in Fig. 3-1, the scanner converts the pattern of wide and narrow spaces and bars into an electronic signal that is decoded according to the bar code symbology it represents. It is important to be cognizant of the fact that the scanner interprets the signal and bar code format and subsequently transmits the interpreted bar code data to a data collection device.

Normally a bar code symbol is comprised of the following elements:

- *Leading quiet zone.* A clear space without markings that is positioned ahead of the start character of the bar code symbol. It provides a reference point for the scanner, and it defines the relative reflectiveness of a space. To ensure reliable interpretation, it is recommended that the quiet zone be greater than 10 times the minimum bar width.

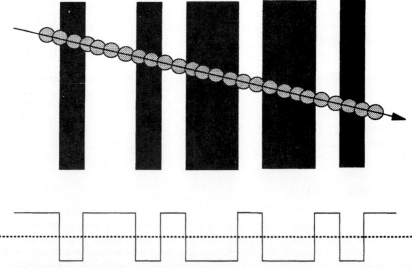

FIGURE 3-1 Bar code decoding methodology. As a light beam is moved across the bar code symbol, an electronic signal that delineates the reflectivity of the spaces and bars is created. The electronic signal is then decoded according to the bar code symbology it represents.

- *Start character.* A special bar code character consisting of a pattern of bars and spaces located immediately following the first quiet zone. The start character yields reading instructions including reading direction, element widths, and, in certain scenarios, symbol definition.

- *Data characters.* A group of bars and spaces that, in accordance with the bar code symbology used, represents a number, letter, ASCII character, symbol, and so forth.

- *Check character.* A character embedded within a sequence of data, the value of which is utilized to perform a mathematical check to ensure the accuracy of data interpretation.

- *Stop character.* A special bar code character consisting of a pattern of bars and spaces positioned just prior to the trailing quiet zone. The stop character indicates the conclusion of the symbol information.

- *Trailing quiet zone.* A clear space without markings that is located immediately after the stop character. The recommendations and limitations specified for leading quiet zones also apply to trailing quiet zones.

Figure 3-2 illustrates the standard bar code elements.

FIGURE 3-2 Standard bar code elements. A module or X dimension is the narrowest nominal bar or space in a bar code symbol. To ensure reliable interpretation, it is recommended that the leading and trailing quiet zones be greater than 10 times the minimum bar width, and that the symbol height be a minimum of 15 percent of the symbol length.

A set of rules for encoding information in a bar code symbol and/or an explicit set of characters used to depict and convey information define a bar code symbology. Based on the specific rules established, bar code symbologies are further categorized as continuous or discrete. A continuous bar code symbology utilizes both the spaces and bars to define the characters. Conversely, a discrete bar code symbology begins and ends each character with a bar, and permits an intercharacter space (the space between the last bar of one character and the first bar of the next character) that is not part of the code to be included between each character. Since discrete bar codes only use bars to represent characters and allow loosely toleranced intercharacter gaps, they tend to be less dense then continuous bar codes. This difference in density results in discrete bar code symbologies requiring more space than continuous bar code symbologies to depict the same information. It is important to note, however, that the relative data security of bar code symbologies is not influenced by whether they are discrete or continuous.

Although many different types of bar codes exist, the most commonly employed bar code symbologies within the distribution field today are as follows.

Code 3 of 9.   The Code 3 of 9 bar code is an uppercase alphanumeric, binary encoded, discrete, self-checking, and variable-length bar code that utilizes three wide and six narrow elements to represent each character. A Code 3 of 9 bar code can be scanned bidirectionally because of its use of the same character to symbolize start and stop as well as its placement of this character before the most significant character and following the least significant character. Because of its ability to encode alpha and numeric characters, it is often used in the warehouse for location and unit identification and tracking, by the Department of Defense, and by the automotive industries. An example of a Code 3 of 9 bar code is shown in Fig. 3-3.

FIGURE 3-3   Code 3 of 9 bar code.

Interleaved Code 2 of 5.   The Interleaved code 2 of 5 (I 2 of 5) bar code is a continuous, number-only bar code that employs two wide and three narrow elements to define each character. The characters, which are grouped into pairs with the first character being represented by bars and the second being depicted by spaces, are said to be interleaved. Since interleaving permits the characters to overlap, it produces a higher density, more compact bar code. One drawback of the interleaved bar code is that its compressed format increases the likelihood of misinterpretation. Unique start and stop elements allow the I 2 of 5 bar code to be scanned bidirectionally. Applications of the I 2 of 5 bar code include pick labels, general warehousing scenarios in which label space is at a premium, and product or container identification. An example of an Interleaved 2 of 5 bar code is illustrated in Fig. 3-4.

FIGURE 3-4   Interleaved code 2 of 5 bar code.

Code 128.   The Code 128 bar code is a continuous, self-checking, variable-length, high-density alphanumeric bar code in which each character is represented by three bars and three spaces that span eleven modules of width. (A module or X dimension is the narrowest nominal bar or space in the bar code symbol.) The use of three different character subsets, start characters, and shift codes in Code 128 bar codes provides for the encoding of any characters from the ASCII 128-character set. (In contrast to the Code 128 bar code, which permits all 105 characters on a standard keyboard to be encoded, Code 3 of 9 and I 2 of 5 bar codes only have the capability to represent 43 and 9 characters, respectively.) Two additional features of the Code 128 symbology are its data message management codes that allow two or more messages to be decoded and transmitted at once and its ability to encode pairs of numeric digits in place of alphanumeric characters. The inherent data set complexity associated with the Code 128 symbology results in Code 128 bar codes of greater length than Code 3 of 9 or I 2 of 5 bar codes. The Code 128 bar code is commonly used for inventory control and retail container marking (i.e., shipping labels). An example of a Code 128 bar code is depicted in Fig. 3-5.

123456789

FIGURE 3-5   Code 128 bar code.

Universal Product Code.   The Universal Product Code (UPC) Version A symbol is a fixed-length, continuous, 12-digit numeric bar code. A UPC includes a six-digit manufacturer number (digit 1 is the product category, while digits 2 through 6 are the vendor's identification number) assigned by the Uniform Code Council (UCC), a five-digit product code designated by the manufacturer, and a one-digit check digit. Each UPC character is created through a combination of two bars and two spaces that populate a total of seven modules. UPCs are used by all industry segments to identify products. An example of a UPC Version A bar code is illustrated in Fig. 3-6.

0 1234   56789   5

FIGURE 3-6   UPC Version A bar code.

UPC Version E is a six-digit variation of the UPC symbology that is used for number system "0" manufacturer identification codes. (A number system character of 0 is assigned by the UCC to standard retail and nonretail items.) Normally, UPC Version E is utilized to label small items.

The European Article Numbering (EAN) system is a superset of UPC, and is the international standard bar code for retail food packages. A thirteen-digit version, EAN-13, and a eight-digit version, EAN-8, which are similar in construction to UPC Versions A and E, are currently available. Each country using the EAN system has a numbering association established to assign codes to manufacturers and maintain the numbering data base.

A summary of the different bar code symbologies previously discussed along with examples of their utilization in an automated warehouse is detailed in Fig. 3-7.

The evolution of bar code technology initially resulted in a myriad of bar code symbologies with no commonly accepted standards. However, as individual industries recognized the need to communicate information effectively through the use of bar coding within or across industries and to their respective customers, they began standardization efforts. Presently, industry standards for bar coding are issued by many organizations, including:

- Automatic Identification Manufacturers (AIM)
- American National Standards Institute (ANSI)

Bar code symbology	Warehouse utilization examples	Advantage to symbology use
Code 3 of 9	Product and rack labeling	Alphanumeric capability
Interleaved 2 of 5	Pick labels and product or container ID	Excellent for numeric coding in limited space
Code 128	Product and rack labeling	Alphanumeric capability
UPC	Item labeling	Required for conformance to retail identification standards

FIGURE 3-7   Automated warehouse bar code symbology utilization.

- Automotive Industry Action Group (AIAG)
- Comite Europeen De Normalisation (CEN)
- Electronics Industries Association (EIA)
- Uniform Code Council (UCC)

Although bar codes are presently viewed as an industry standard for product identification as well as a gateway to efficient use of data collection systems, it is essential that the benefits and limitations of bar code technology be examined prior to implementation. A summary of the advantages and disadvantages associated with bar codes is detailed in Fig. 3-8.

Bar coding, like any other modern technology, mandates an initial investment. At a minimum, the implementation of bar coding requires the following:

1. *A bar code formatting computer software application* that permits bar code formats to be produced and maintained and creates associated print instructions is needed. A computer to execute the application software is also necessary.

Advantages	Disadvantages
Improved data collection accuracy through bar code scanning (typical error rate of 1 in 1 to 2 million scans) rather than keystroke data entry (typical error rate of 1 in 300 keystrokes)	Technology investment including printing equipment and supplies and/or purchase of preprinted bar code stock, labor costs for the application of labels, equipment costs for terminals and scanning devices, and maintenance required
Support of real-time rapid communications	Resource required to monitor or verify bar code print quality
Improved sharing of data between suppliers and customers	Inherent susceptibility of bar code labels to damage, which can result in read errors
Elimination of the need to rekey data	Increased dependency on automated data collection
Availability of built-in check digits to eliminate errors, misreads, and character substitutions	
Increased productivity resulting from scanning	

FIGURE 3-8   Advantages and disadvantages of bar code technology implementation.

2. *A printer* (i.e., dot matrix, thermal, laser) that is capable of satisfying the published print standards for each bar code symbology to be printed is required. Appropriate printing supplies (i.e., ribbons, labels, and so forth) also must be available. (See the section entitled "Printing" in this chapter.)

3. *A symbol verification device* that is used to inspect bar code quality and verify that the symbol conforms with the requirements of a specification or standard is mandatory.

Bar Coding Alternatives

In addition to traditional linear bar coding, the following alternatives are also available:

Radio Frequency Identification (RFID). Requirements such as the ability to read data that is located in a hostile environment and/or is not optically visible to a system and the capability to read data from and write data to a medium have spurred the invention of RF tags and RFID technology. RFID systems utilize transponder technology, which involves the encasement of a small radio transmitter, antenna, integrated circuit chip, and receiver in a RF tag attached to the item to be tracked. The RF tags are categorized as active (internal battery) or passive (powered by the polling reader). When a RF tag receives a transmitted radio signal, it responds by sending back information stored in the tag. The range and speed of RF tag data transmission are regulated by the antenna size. Data contained on the tags can be preprogrammed during fabrication, or it can be altered or updated based on the radio signal received.

Currently, RFID systems are used to identify packages and pallets in warehouses. Additionally, in applications where a harsh environment (i.e., an oily or dirty surface limits effective application of a bar code) is standard and the cost is justified, RF tags are an excellent alternative to standard bar coding. A summary of the advantages and disadvantages associated with RFID is included in Fig. 3-9.

Two-Dimensional Bar Codes. In order to increase the information density in bar code symbols, two-dimensional bar codes have been introduced. A two-dimensional bar code is created by decreasing the standard bar height and stacking the symbols (set apart by separator bars to eliminate interference) on top of each other. Depending on the particular bar code symbology used, two-dimensional bar codes, like linear bar codes, are categorized as discrete or continuous. Two commonly utilized two-dimen-

Advantages	Disadvantages
Excellent for use in harsh environments or locations that are not optically visible	Ten to fifty times more expensive than one-dimensional bar codes
Ability to encode more information than a one-dimensional bar code	Requires a specific RF tag and transponder device that are not widely available
Enhanced security due to difficulty in counterfeiting tags	
Capability to be a read/write system	
Capacity to be read and written to from a distance	

FIGURE 3-9 Advantages and disadvantages of RFID.

FIGURE 3-10   Example of a PDF417 two-dimensional bar code. (*Courtesy of Symbol Technologies, Inc.*)

sional bar coding structures are Codablock Multi-line Code (MLC) and PDF417. See Fig. 3-10.

Since two-dimensional bar codes allow a significant number of characters to be stored in a minimum amount of space, they are often used to specify item and shipment information on product and/or shipping container labels. Encoding detailed information in the two-dimensional bar code permits data to be readily available in a paper-based data file.

A disadvantage of two-dimensional bar codes is that, with the exception of PDF417 bar codes, they require special two-dimensional charge-coupled device (CCD) scanners or video cameras rather than hand-held laser scanners to be interpreted. (A degree of skewing normally occurs due to the laser beam's inability to intersect all of the cells.) Because the availability of CCD scanners is currently restricted and current applications must be modified to utilize the information made available by the denser format, the use of two-dimensional bar coding is probably going to remain somewhat limited.

Optical Character Recognition (OCR).   OCR is a two-dimensional technology utilizing uniquely styled fonts that are readable by both humans and machines. Although OCR provides a significantly improved error rate (1 in 10,000 characters entered) over that of manual data entry (1 in 300 keystrokes), it does have several drawbacks, including:

- Low initial read rate due to the need for precise document orientation
- Sensitivity to scanning motion, especially for an extended line of data
- Difficulty reading information from moving objects
- Higher error and substitution rates than bar codes

OCR is most widely used to enter pages of information into a computer system (i.e., word processing) via an optical page reader.

Manual Data Entry.   The practice of entering data by typing information at a keyboard is referred to as manual data entry. Needless to say, the information collected as a result of manual data entry is not as accurate (a typical error rate is 1 in every 300 keystrokes) and is gathered at a much slower rate than with bar code technology.

AUTOMATED DATA COLLECTION

Automated data collection is divided into three equipment classifications, namely bar code reading, data collection, and data transmission. Bar code reading technology deals with the scanning and interpreting of various codes including bar codes and RFID tags. Data collection technology relates to terminals, how they are used to collect, receive, and process data, and their transmission mode such as batch or on-line. Data transmission technology includes terminal connectivity issues such as hard-wired or wireless local area networks (LANs). When these three technologies are properly selected and integrated, paired with a WMS, and installed in a warehouse environment, the result is a warehouse that operates at efficiencies unheard of even a few years ago.

Bar Code Reading Technology

In today's warehouse environment, bar code reading devices, which are normally referred to as scanners, are utilized to extricate the data that is optically encoded in a bar code symbol and to decode the extracted data so it may be appropriately used. The two major components of a bar code reader are the scanner and the decoder.

Scanner. A bar code scanner illuminates a bar code symbol with light energy and measures reflectance across the symbol. Spaces (white bars) are viewed as domains of high reflectance, while bars are sensed as areas of low reflectance. This measurement of reflectance or reading of the bar code symbol is accomplished through the use of an electro-optical system that is composed of a light source, transmitting lens, receiving lens, photodetector, amplifier, and waveshaper. In order to discern the bar code, the transmitting lens focuses the light source (i.e., either an infrared light source in the 900-nanometer (nm) wavelength range or visible-red light source in the 600- to 700-nm wavelength range) onto the symbol. Light reflected from the bars and spaces of the symbol is collected by the receiving lens and the photodectector. The relative reflectance of the light is then quantified and converted into an electrical signal by the photodectector. Next, an amplifier within the scanner modulates the electrical signal produced by the photodetector and transmits the signal to a waveshaper. In turn, the waveshaper converts the signal into a time-varying binary digital signal that is ready to be translated by the decoder.

Decoder. A decoder utilizes a series of logical steps to interpret the binary digital signal representing the length in terms of time of each highly reflective white space and nonreflective dark bar of a scanned bar code symbol and to translate the signal into meaningful data. The steps performed by a typical decoder include:

1. Determining the width of each bar and space
2. Confirming that valid quiet zones are located at both ends of the symbol, thereby ensuring the entire bar code is read
3. Comparing the number of widths contained in the bar code to the number of widths for the particular symbology
4. Verifying that the number of element widths observed meets the encoding rules for the symbology
5. Mapping the interpreted bar and space pattern to a table of stored values to finalize the translation of the data within the symbol

Additionally, if the scanner is designed to perform autodiscrimination, the decoder must also automatically determine the symbology of the bar code scanned while stepping through the translation process. Because the introduction of autodiscrimination increases the opportunity for misreads, it is important to limit the number of selected symbol options to the number to be encountered during normal warehouse operations.

The decoder's output is a binary signal that is ready for transmission to a data collection device such as a computer or terminal.

Classifications of Bar Code Readers

Bar code reading devices are classified by their physical presentation (fixed or handheld) to the bar code symbol and the dynamics of their scanning beam (fixed or moving). In addition to the bar code reading devices that can be classified through the aforementioned criteria, there is also a classification of scanning devices that use CCD technology to create an image of the bar code to be decoded. A summary of the different classifications of bar code readers is illustrated in Fig. 3-11.

| | Dynamics of scanning beams | |
Physical presentation	Fixed beam	Moving beam
Fixed mount	Noncontact	Noncontact
Hand-held	Contact	Noncontact
	Noncontact	

FIGURE 3-11   Bar code reader classifications.

Prior to proceeding with written and pictorial descriptions of bar code readers from each of the classifications, scanning terminology (as illustrated in Fig. 3-12) that is commonly used to express the optical limitations of scanning systems needs to be defined. *Optical throw* is the minimum distance between the bar code symbol and scanner that can result in a successful read, while *operating range* is the maximum

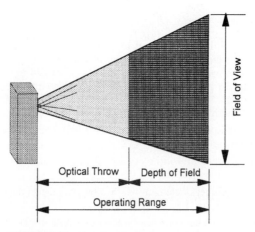

FIGURE 3-12   Scanning terminology.

distance permitted between the scanner and symbol. *Depth of field* (DoF) is the range between the maximum and minimum distance that a bar code reader is capable of reading symbols of a specified module or X dimension. Lastly, *field of view* is the linear dimension defining the maximum length of a bar code that can be read in one scan.

Hand-Held, Fixed Beam, Contact Scanners

A wand scanner, often referred to as a light pen, is a hand-held, fixed beam, contact bar code scanning device that is excellent for scanning symbols on smooth, flat, rigid surfaces. The scanning motion of the wand is provided by the operator steadily and smoothly moving the wand tip across the bar code symbol at a speed of 3 to 30 in/s. Because wand scanning is a manual operation that is somewhat sensitive to scanning velocities, it is suggested that the scanning of long bar code symbols be avoided.

As shown in Fig. 3-13, a wand scanner is composed of a long, cylindrical housing (similar to a pen) and a light source (visible light-emitting diode in the 670- to 700-nm

FIGURE 3-13   Wand scanner. (*Courtesy of Symbol Technologies, Inc.*)

range) from which light is released through a sophisticated tip assembly. To read a bar code symbol properly, the wand's scanner tip must either be in contact with the bar code or no more than 0.05 in above the bar code.

Two crucial elements to consider with wand scanners are the aperture size required and the degree of tilt allowed during scanning. For optimum reading conditions, the

Advantages	Disadvantages
Inexpensive to purchase and operate	Smooth, flat, rigid surfaces are required for bar code application and scanning
Established and well-accepted technology	Operator training is required due to wand's sensitivity to scanning velocities
Durable and compact design	Difficult to read long bar code symbols
Minimal power usage	Lower first-pass read rate than experienced with moving beam scanners
	Symbol degradation experienced due to abrasion caused by contact with the wand tip
	Periodic replacement of the wand tip is required

FIGURE 3-14   Advantages and disadvantages of wand scanners.

aperture width is to be approximately the same size as the module or X dimension of the symbol to be scanned. If the aperture is too large, more than one contiguous bar or space may be present in the scanner's effective field of vision. On the other hand, if the aperture is too small, bar code imperfections such as printing defects can result in the bar code's bar and space pattern being misinterpreted. Additionally, bar code readability is also dependent on the optics in a wand scanner being utilized within the degree of tilt for which they have been designed. Normally, the tilt range permitted is $\pm 45°$; however, there are scanners with tilt specifications of $\pm 25°$.

Wand scanners are best suited for warehouse operations, such as returns and packing, that are performed in workstation environments. The advantages and disadvantages associated with wand scanners are listed in Fig. 3-14.

Hand-Held, Fixed Beam, Noncontact Scanners

A hand-held, fixed beam, noncontact scanner is typically a pistol-shaped device that employs a solid-state laser diode to permit bar code symbols to be read. As is the case with wand scanners, the operator is responsible for manually providing the scanning motion. In contrast with a wand scanner, however, the operator is able to scan bar code symbols on pliable or irregular surfaces, through glass or laminates, and at greater working distances. If an infrared light source that is not visible by humans is used in these scanners, a visible alignment beam is included to assist the operator in accurately positioning the scanner with respect to the symbol.

Like wand scanners, hand-held, fixed beam, noncontact scanners are best suited for warehouse operations, such as returns and packing, that are accomplished in a workstation environment. The advantages and disadvantages of hand-held, fixed beam, noncontact scanners are described in Fig. 3-15.

Fixed-Mount, Fixed Beam, Noncontact Scanners

A fixed-mount, fixed beam, noncontact scanner utilizes a light-emitting diode (LED) or incandescent light source to read bar code symbols moving past its scanning aper-

Advantages	Disadvantages
Capability to scan bar code symbols on pliable or irregular surfaces or through glass or laminates	Operator training is required due to scanner's sensitivity to scanning velocities
Inexpensive to purchase and operate	Lower first-pass read rate than experienced with moving beam scanners
Durable and compact design	Difficult to read long bar code symbols
Minimal power usage	Old technology that is being displaced by hand-held, moving beam scanners
Greater scanning depth of field than wand scanners	
A visible alignment beam is often provided to facilitate aiming of the scanning beam	

FIGURE 3-15   Advantages and disadvantages of hand-held, fixed beam, noncontact scanners.

ture. Because of the static nature of these systems, fixed-mount, fixed beam scanners have only one opportunity to interpret and decode a passing symbol. With only a single read opportunity per symbol, it is crucial that the bar code symbols exhibit good print quality, are within the limited depth of field for the scanner, and are properly positioned on the object being moved in front of the scanner.

Fixed-mount, fixed beam, noncontact scanners are utilized on a limited basis in the warehouse environment for conveyor applications including carton sortation and truck loading. To improve the accuracy of sortation, warehouse conveyor systems utilizing fixed-mount, fixed beam, noncontact scanners can employ additional photosensors to signal the proximity of bar coded units, gap belts to ensure proper spacing and timing of the bar coded units relative to the scanner, and recirculation or exception handling routes to facilitate a second scan opportunity. The advantages and disadvantages associated with fixed-mount, fixed beam, noncontact scanners are summarized in Fig. 3-16.

Advantages	Disadvantages
Less expensive than moving beam scanners	Limited depth of field
Unlike moving beam scanners, no moving parts are required	Single bar code read opportunity
	Very sensitive to poor bar code print quality especially if interfaced to mechanization
	Sensitive to the speed of the bar coded object as it passes through the scan field
	Proper location of bar coded object relative to the scanner is crucial

FIGURE 3-16   Advantages and disadvantages of fixed-mount, fixed beam, noncontact scanners.

Hand-Held, Moving Beam Scanners

A hand-held, moving beam scanner, as shown in Fig. 3-17, uses an internal electro-mechanical system, involving mirrors that are polygonal or oscillating in nature, to emit a helium-neon (He-Ne) laser or laser diode light source to scan bar code symbols at a typical scan rate of 40 scans per second. He-Ne lasers utilize a helium-neon plas-

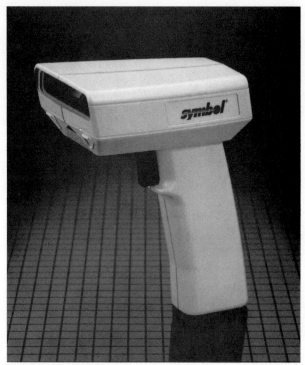

FIGURE 3-17   Hand-held, moving beam scanner. (*Courtesy of Symbol Technologies, Inc.*)

ma tube to produce a small, powerful, monochromatic (one-color), coherent (wavelengths are in phase) visible light beam. Although the technology employed by the He-Ne lasers is proven and effective, the substantial amount of battery power required to operate them is resulting in their replacement with lower-power-consuming, equally effective laser diodes. Currently, two laser diode devices, infrared and visible-red, are available. The latter of the two, visible-red, is in more demand mostly due to its light source being discernible by the human eye and its capability to read bar codes produced from a wider variety of inks. (See the section entitled "Printing" in this chapter for more information.)

Hand-held, moving beam scanners, with both long- and short-range reading capability, are often required for warehouse applications. It is common to find warehousing requirements both to read a 40-mil X dimension bar code symbol at distances ranging from 5 or 6 in to 3 ft or, conversely, at distances ranging from 6 to 12 ft. In either case,

Advantages	Disadvantages
Higher first-pass read rate than experienced with fixed beam scanners	Power consumption for the older technology He-Ne type is high
Incorporation of an aiming beam to facilitate aiming of the scanning beam	More expensive than hand-held, fixed beam scanners
Improved flexibility through mobility provides greater depth of field and operating range as compared to fixed beam scanners	Less durable than hand-held, fixed beam scanners
Utilization of multiple scans to read bar code symbols	Protective encasements may be required in industrial environments
Ease of use	

FIGURE 3-18   Advantages and disadvantages of hand-held, moving beam scanners.

the hand-held, moving beam scanners are attached to a portable data collection device or fixed terminal. Extra-long-range scanners with their extended depth of focus are typically attached to a truck mount terminal and are employed for warehouse operations such as put-away, full pallet picking, and replenishment. Long-range scanners, which are normally coupled with hand-held terminals, are often used for warehouse operations such as cycle count and carton or piece picking where the user is more mobile. Short- or standard-range scanners are utilized for warehouse operations, such as returns and packing, that are performed in a workstation environment.

To facilitate the use of the hand-held, moving beam scanners at greater distances, many scanners have been equipped with "aiming beams" that enhance the visibility of the laser light. A practical implementation of an aiming beam is a two-position scanner trigger that allows an *aim* and *scan* operation. When the trigger is moved to the first position, the angular scan amplitude is decreased, thereby causing the scan line to be significantly shorter and therefore more visible. Once the scanner beam has been properly aimed, the trigger is then pulled to the second position. In this second position, the angular scan amplitude is restored to its standard mode, and the symbol is scanned. The advantages and disadvantages of hand-held, moving beam scanners are described in Fig. 3-18.

Fixed-Mount, Moving Beam Scanners

Fixed-mount, moving beam scanners (see Fig. 3-19), which are often used in conjunction with material handling systems to perform automatic sortation, employ a He-Ne or laser diode light source to read bar code symbols moving through their field of view. These scanners utilize an internal scanning mechanism that oscillates a light beam at rates ranging from 40 to 1000 scans per second to create a scanning pattern. In order for the bar code symbol to be correctly read, the light beam must intersect all the bars and spaces in the same scan path.

In contrast with the single read opportunity associated with a fixed-mount, fixed beam scanner, moving beam scanner systems are designed to allow a minimum of four or five scans of each passing symbol. The increased number of read opportunities characteristic of moving beam systems results from controlling system variables, including scanning line length, scanning rate, bar height, symbol orientation, and con-

FIGURE 3-19   Fixed-mount, moving beam scanner. (*Courtesy of Metrologic, Inc.*)

veyor speed. An additional method utilized to ensure more reads is the introduction of complex scanning patterns or omnidirectional scanning that permits more of a symbol area to be interpreted at the same time and decreases misreads caused by poor symbol orientation, placement, and/or print quality. The raster pattern, which is the most cost effective and commonly used pattern, is created through the addition of mechanical motion in a direction perpendicular to the original scan line.

When a conveyor system is designed that includes a fixed-mount, moving beam scanner, the relative positioning of the scanner and the symbol are of paramount importance. The two standard symbol orientations considered for these applications are normally referred to as *picket fence* and *ladder*. With the picket fence orientation, the bars are vertically aligned and the direction of motion of the symbol through the scan field is perpendicular to the bars. Conversely, with a ladder orientation, the bars are horizontally aligned and the direction of motion of the symbol through the scan field is parallel to the bars. A summary of typical combinations of scanner and symbol positioning is depicted in Fig. 3-20. It is important to note that most of the standards for conveyor systems are written for fixed beam scanners and therefore specify the use of the picket fence orientation and side-mounted scanners. In actuality, however, moving beam scanners read labels in the ladder orientation more effectively.

Due to the fact that fixed-mount, moving beam scanners are required to operate at high scanning speeds, the number of symbologies to be interpreted is normally kept at a minimum. This limitation is especially practical if the fixed-mount, moving beam scanner utilizes a voting algorithm during the decoding process. Basically, a voting algorithm dictates that for a given symbol there must be one to three subsequent reads that agree with a previous read for the bar code information to be passed to a data col-

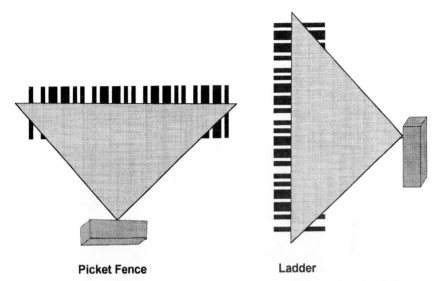

Picket Fence   **Ladder**

FIGURE 3-20 Scanner and symbol positioning norms. With the picket fence orientation, the bars are vertically aligned, and the direction of motion of the symbol through the scan field is perpendicular to the bars. Conversely, with a ladder orientation, the bars are horizontally aligned, and the direction of motion of the symbol through the scan field is parallel to the bars. If multiple bar codes are required for a single unit, a combination of orientations is used to represent the bar coded information.

lection device. The advantages and disadvantages associated with fixed-mount, moving beam scanners are listed in Fig. 3-21.

Charge-Coupled Device Scanners

A charge-coupled device (CCD) scanner, which is often referred to as an image scanner, uses a xenon-gas, incandescent, or laser diode light source to flood-illuminate,

Advantages	Disadvantages
Higher first-pass read rate than experienced with fixed beam scanners	Sensitive to poor bar code print quality, if interfaced to mechanization
Utilization of multiple scans to read bar code symbols	Sensitive to proper symbol orientation for accurate bar code interpretation (with the exception of advanced scanning patterns)
Greater scanning depth of field than fixed beam scanners	More expensive than fixed beam scanners
Greater operating range than fixed beam scanners	
Availability of multiple scanner patterns to enhance bar code read rates for varied orientation and location	

FIGURE 3-21 Advantages and disadvantages of fixed-mount, moving beam scanners.

produce, and interpret an image of a bar code symbol. CCD scanners, which operate in a mode similar to a conventional photographic camera, focus the captured image of the reflected symbol onto the CCD's linear array of multiple adjacent photodiodes. Reflected spaces cause the photodiodes to be saturated, while the reflections of bars exhibit a limited amount of photodiode saturation. An internal microprocessor is then employed to interpret the photodiode saturation levels, create an electronic signal, and decode the bar code symbol. A CCD scanner is depicted in Fig. 3-22.

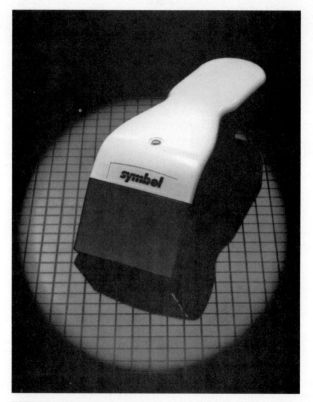

FIGURE 3-22 CCD scanner. (*Courtesy of Symbol Technologies, Inc.*)

Since CCD scanners must capture the entire bar code symbol image, they tend to have a limited depth of field (the working range is from contact to 7 in) and field of view (limited by the number of photodiodes in the CCD's array). Even with the above limitations, it is important to note that CCD technology provides a significant advantage in the interpretation of high-density one- and two-dimensional bar code symbols. Currently, CCD scanners are procurable in fixed-mount or hand-held configurations. The advantages and disadvantages of charge-coupled scanners are further delineated in Fig. 3-23.

Advantages	Disadvantages
Ease of use	Limited depth of field
Durable and compact design	Limited field of view
Capability to interpret high-density one- and two-dimensional bar codes	Difficult to read long bar code symbols

FIGURE 3-23   Advantages and disadvantages of CCD scanners.

Based on their capabilities and limitations, CCD scanners are best suited for warehouse operations, such as returns, packing, or parcel shipping, which are performed in a workstation environment.

Data Collection Technology

Data collection technology addresses computer terminals and their ability to collect, receive, and process data. After a bar code has been interpreted by a scanner, the derived information is transmitted to a data collection terminal. The terminal, which can be stationary or portable in nature, transfers the received data in either a batch or on-line mode to the host computer for processing.

Wedges

Wedges are devices that provide an interface between input devices and a host computer or terminal. A serial wedge utilizes an RS232 signal to transmit scanned or other input data between a host computer and a terminal, while a keyboard wedge (discussed in more detail below) presents the same data to a terminal as a series of keystrokes. Two conditions that mandate the use of a wedge in conjunction with a terminal are as follows:

- When two or more different scanning devices are required at a single decoding station
- When an auxiliary port is needed to allow for the integration of scales, printers, or other serial input/output devices

A keyboard wedge, which is normally installed between a keyboard and a cathode ray tube terminal or personal computer, is a device designed to permit data to be entered via a keyboard or scanner without any manual switching between devices. Usually, a keyboard wedge receives a scanner's bar code symbol information in an undecoded or decoded form, and presents decoded information to a computer as if it had been entered with standard keystrokes. In an effort to reduce the number of keystrokes futher required by a user, wedges are frequently configured with options that allow ASCII control characters such as automatic return (enter) or tab to be included with the data to be transmitted from the wedge to the data collection device or terminal. A keyboard wedge is shown in Fig. 3-24. The advantages and disadvantages of wedges are described in Fig. 3-25.

Decoded Output Scanner

A decoded output scanner (DOS) is a wireless hand-held, moving beam, noncontact bar code scanning device that has the capability to perform both the decoding and

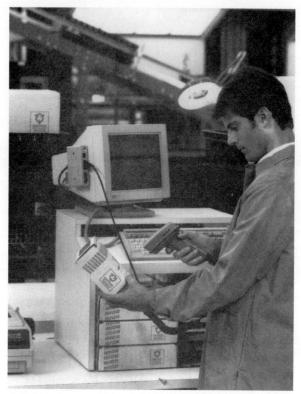

FIGURE 3-24   Keyboard wedge. (*Courtesy of Symbol Technologies, Inc.*)

wedging functions. Utilizing the same laser diode light source technology employed for hand-held, moving beam scanners, DOSs integrate bar code symbol scanning, bar code data decoding, transmission of data via radio frequency communications, and connection to their host (i.e., personal computer, computer terminal, and so forth). Because of their use of unlicensed low-power radio frequency transmission instead of

Advantages	Disadvantages
Capability to allow data to be entered via a keyboard or scanner without manually switching devices	Implementation requires a CRT terminal
Reduction of keystrokes required by user if interfaced and programmed for an external input device	Portability of scanning activity is limited to the length of the scanner cord
Provides integration of data collection devices with standard or existing CRT terminals	Additional cabling is required to attach the keyboard wedge
Less expensive than an RF terminal	

FIGURE 3-25   Advantages and disadvantages of keyboard wedges.

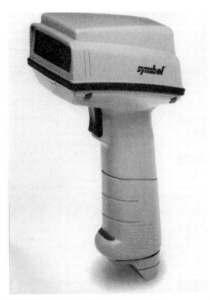

FIGURE 3-26 DOS scanner. (*Courtesy of Symbol Technologies, Inc.*)

wires to communicate, DOSs provide operators with the flexibility of unrestrained movement up to 50 ft from the host computer. A DOS scanner is pictured in Fig. 3-26.

DOSs with their advantage of wireless scanning are appropriate for warehouse operations, such as returns and packing, that are normally performed in a workstation environment. Additional advantages as well as disadvantages inherent to DOS scanners are detailed in Fig. 3-27.

Cathode Ray Tube (CRT) Terminals

A cathode ray tube terminal is a television-like device that presents data in a visual form by permitting a user to enter information via a keyboard and by utilizing a vacuum tube in which a beam of electrons is moved to draw lines and to form characters or symbols on a luminescent screen. Communications between a CRT terminal and its host computer are normally accomplished through hard-wiring, although to reduce cabling costs wireless local area networks are becoming more common.

Because today's users are demanding and implementing more applications that include graphics and graphical user interfaces (GUIs), resolution requirements must be considered during the selection of a CRT terminal. Resolution, which is expressed as dots per inch, improves as the dot size decreases. Accordingly, augmented resolution, as well as color array options, result in higher prices and increased requirements for processing power and memory. Four commonly encountered resolution categories, listed in increasing resolution capability, are as follows:

- Color graphics adapter (CGA)
- Enhanced graphics adapter (EGA)

Advantages	Disadvantages
Flexibility of unrestrained operator or scanner movement within an increased operating zone relative to its attached terminal or host computer	More expensive than a standard keyboard and external device
Incorporation of decoding and wedging functions in one device	More susceptible to being lost or misplaced
Utilization of the hand-held, moving beam scanning technology	Protective encasements may be required in industrial environments

FIGURE 3-27 Advantages and disadvantages of DOS scanners.

Advantages	Disadvantages
Fairly inexpensive to purchase and operate	Limited mobility due to the stationary nature of device
Established and well-accepted technology	Increased space requirement to facilitate the device's footprint
Availability of information in a full-screen format	External device (i.e., scanner) attachment requires the use of a keyboard wedge
Ability to attach a full-sized keyboard	Industrial usage may require specially designed terminals or protective encasements
Improved response speed due to hard-wired attachment to the host computer	

FIGURE 3-28 Advantages and disadvantages of CRT terminals.

- Video graphics adapter (VGA)
- Super video graphics adapter (SVGA)

Based on their configuration and usage, CRT terminals are typically grouped into two categories: nonprogrammable terminals and intelligent workstations. Equipped with a display and keyboard and directly wired to a computer, nonprogrammable terminals only provide the capability to access and interact with the computer system to which they are attached. Conversely, an intelligent workstation, such as a personal computer, has computational capability and can be programmed to perform user-determined functions. When an intelligent workstation is connected to a remote host system, it can be configured to emulate a direct attach terminal, upload and download files, and independently execute local programs. An additional feature of intelligent workstations and nonprogrammable terminals is that most of them allow for the direct attachment of a printer.

A nonprogrammable terminal or intelligent workstation in combination with a keyboard wedge is often used in warehouse operations such as label receiving, returns, label and paper picking, or packing. In these types of workstation operations, the user requires limited mobility of the scanning activity and access to a maximum amount of information. The advantages and disadvantages associated with CRT terminals are described in Fig. 3-28.

Portable Data Collection Terminals

Portable data collection terminals are microprocessor-based, hand-held terminals that allow for the entering of data through the movement of the terminal to a data source. Basic elements common to all portable terminals include a keyboard, scanner (integrated or attached), display, printer port, memory, microprocessor, and battery.

Data capture into a portable terminal can be accomplished through a reduced-function keyboard, scanner, or, as discussed later in this chapter, voice. Generally, the keyboard is used to enter data or system options that cannot be bar coded. Terminal keyboards exhibiting some combination of numeric, alphanumeric, and special characters come in two standard configurations: horizontal (similar to a typewriter or CRT terminal keyboard) and vertical (similar to a calculator). Areas to consider when evaluating terminal keyboards include whether or not the keyboard contains all the characters that can be scanned during normal warehouse operations, the amount of multiple key entry (shifting) that may be needed to allow necessary characters to be entered, the number of available standard function keys, and the proper key size selection relative to the warehouse operating environment (i.e., the need to use gloves in a refrigerated warehouse environment can hamper an operator's ability to enter information accurately on a keyboard with inadequate key spacing).

The second mode of data input to a portable terminal is a bar code scanner or RFID reader. Scanners or RFID readers can either be attached via a cord as shown in Fig. 3-29 or integrated with a portable data collection terminal as illustrated in Fig. 3-30. A wide variety of scanners including wand, hand-held, and CCD scanners can be connected to today's portable data collection terminals.

FIGURE 3-29 Portable terminal with a peripheral scanner. (*Courtesy of Teklogix, Inc.*)

A visual presentation of the data is supplied by the portable terminal's display. Two types of display methods, light-emitting diode (LED) or liquid crystal display (LCD), are typically utilized in portable terminals. LEDs are displays in which characters are formed from a dot matrix of semiconductor chips that release visible or infrared light when activated. An LED's ability to be viewed in dimly lit environments and perform in adverse operating conditions (i.e., refrigerated environments) is considered to be advantageous. Disadvantages of LEDs include their inability to be read in bright light (i.e., sunlight) and their high energy usage. LCD displays contain a grid that is polarized to alter the properties of the liquid crystal medium to appear opaque or transparent and thus form alphanumeric characters using reflected or background light. In contrast to LEDs, LCDs are easily read in brightly lit areas, but are hard to discern in dim

FIGURE 3-30   Portable terminal with an integrated scanner. (*Courtesy of Symbol Technologies, Inc.*)

lighting and are affected by environmental conditions (i.e., a cold environment induces the liquid to freeze, while high temperatures cause the displays to cease to function). To permit portable terminals with LCD displays to be used in a variety of lighting scenarios, backlighting is often employed.

Portable terminal display size is normally specified in lines and characters. A wide gamut of line and character combinations including 2 by 16, 8 by 16, and 16 by 21 is available today. To allow for adequate viewing of the data, many terminals are equipped with options to vary the display size through the selection of alternative fonts and/or to scroll the display.

Due to the increasing need to print labels and documents on demand, portable data collection terminals are outfitted with serial RS232 printer ports. These printer ports allow for the attachment of portable label printers. Since the printer is consider to be directly attached to the host computer, print instructions are issued via the host computer rather than the portable terminal.

Ordinarily, the memory of a portable data collection terminal includes two components: read only memory (ROM) and random access memory (RAM). The ROM, which is generally programmed by the terminal vendor, contains the emulation information along with the terminal intelligence and instruction set. In most cases, ROM size is determined by the terminal vendor and is in the range of 128 to 256 kbyte. A significant benefit of ROM is its ability to retain information coded into its memory even when its power source is removed. In contrast with the ROM portion of the terminal's memory, the RAM, which includes the buffer memory, data memory, and any executable programs to be performed by the terminal, does not retain its information if its power source is unavailable. Most terminal vendors offer RAM sizes in the range of 128 kbyte to 1 Mbyte and allow the customer to select the amount of RAM required to meet application requirements.

The utilization of ROM and RAM in portable data collection terminals is controlled by internal microprocessors. By definition, microprocessors are microchips containing electronic integrated circuits that are capable of receiving and executing programmed instructions. With today's portable data collection terminals including microprocessors rated at 8, 16, or 32 bits, it is important to note that the greater processing capability provided with the 32-bit technology usually results in more functionality and better performance, power, and speed (i.e., higher-capacity processors allow more terminal functionality such as graphical or color displays rather than monochrome character–based formats). This additional power can also be used to support advanced features including multiple sessions with the operating system and variable font capability.

All portable data collection terminals consume power and therefore require an energy source such as a battery. Even though three types of batteries, namely rechargeable nickel-cadmium (NiCad), single-use alkaline, and single-use lithium, are presently available, the cost associated with replacing and disposing of batteries has caused rechargeable batteries to be the most popular option. Although rechargeable batteries do provide the advantage of reuse, it is important to note that the maximum effective life of a battery is obtained by completely discharging the battery prior to recharging it. This is based on the fact that the effective life of a battery is equivalent to the recharge time. For example, if a battery is continually recharged after four hours of use when its rated life is eight hours, its effective life becomes four hours. Battery reconditioners or "smart" chargers that allow batteries to accept their rated or original charge may be utilized to overcome battery recharge and effective life problems.

Batch versus On-Line, Real-Time Operations

Portable data collection terminals can be operated in two modes: batch and on-line. For batch operations, the terminals (acting as electronic clipboards) are downloaded instructions that are used to direct the collection of data to be uploaded to the computer at a later time. Typically, batch terminals are placed in a cradle that is attached to a host computer via a cable (i.e., RS232, RS422) connection. Programs, instructions, and data are then downloaded to the batch terminal's memory. Following the displayed instructions, the user performs the required operations and the terminal stores the collected data. Once the user has completed all required activities, the batch terminal is placed back in the cradle, and the data is uploaded to the host computer. The final data processing step involves the updating of the host system's data base. A hand-held batch terminal is shown in Fig. 3-31.

Even though the RAM of batch terminals can be appropriately specified to permit downloaded programs to be executed and data to be collected, these devices are not designed to perform data base access and editing functions. Because of their store and forward technology, batch terminals are truly best suited for warehouse operations, such as flow rack eaches (piece) picking and truck deliveries to stores, where direct

FIGURE 3-31   Hand-held batch terminal. (*Courtesy of Symbol Technologies, Inc.*)

access to an on-line data base is not necessarily practical due to response time limitations. The advantages and disadvantages of batch terminals are depicted in Fig. 3-32.

In the case of on-line operations, the portable data collection terminals can be connected to a local data base in real time utilizing radio frequency data communications (RFDC). This real-time transmission allows the host computer to verify the accuracy of all incoming information against its data base, as well as transmit instructions or corrections to the operator without the need for a cradle. On-line, real-time operations and the resulting data are the premise on which most state-of-the-art warehouse operations are based.

Radio Frequency Data Communication

Radio frequency data-communications technology (commonly referred to as RF), which was developed in the late 1970s and first implemented in the early 1980s, is credited with being the first true alternative to batch data collection systems. Through the use of radio bands for communication within the electromagnetic spectrum, an RF system permits the interactive exchange of information, including on-line inquiries

Advantages	Disadvantages
Improved data collection accuracy through bar code scanning rather than keystroke data entry	Not capable of performing on-line, real-time data base access and editing functions
Less expensive than radio frequency systems	Application software may need to be written to permit data files to be transmitted between the terminal and host computer
	Application programming at the terminal level is required
	Operator directions are limited to down-loaded programs and information
	Inability to handle exception situations

FIGURE 3-32   Advantages and disadvantages of batch terminals.

and record and file updates, between a host computer and a user operating a portable or mobile RF terminal. As shown in Fig. 3-33, the components typically defining an RF system encompass the host computer, the system network controller, the radio local area network controller, and the RF terminal.

The host computer, which is designed to centralize information management activities, execute software applications, and organize and integrate data for use in other areas, is ultimately responsible for driving the RF system. Based on the software application being run, the host computer issues operator or terminal specific instructions and tasks for the RF system to communicate, responds to RF terminal inquiries, and manages data collected through RF terminal operations. This seamless mode of communication is the result of the host computer viewing terminals on the RF network as directly attached.

Hard-wired to the host computer through an Ethernet, token ring, or RS232 connection is the system network controller (SNC). The SNC, which is also referred to as a *base station,* is responsible for communication protocols and data transmission speeds between the host computer and RF terminals. In addition to regulating communications and transmission speeds, some SNCs also provide emulation protocol. The incorporation of emulation capability within the SNC allows for the transmission of screen changes rather than retransmission of the entire screen and for the storing of multiple screens within the memory of the SNC and RF terminal. Another advantage of SNC emulation is that a terminal session can be interrupted (i.e., loss of power) without causing disruption to the system. When the terminal is available again, the SNC is capable of resending the stored screen to the terminal.

Directly attached to the SNC via an RS232, RS485, or Ethernet connection is the radio local area network controller (RLC). An RLC typically consists of a control unit, a serial cable for connection to the host computer and/or SNC, an antenna, and an antenna cable. Mounted in a warehouse location in close proximity to the RF terminals being utilized, the RLC directs the communications between the SNC and specified RF terminals. Although many factors influence the number of RLCs required for a facility, most RF vendors infer that multiple RLCs may be necessary for larger facilities with high transaction rates or sites using spread-spectrum technology (refer to the section entitled "Narrow Band versus Spread Spectrum" later in this chapter).

The two most common protocols used by the SNC-RLC combination to retrieve data from the RF terminals are polling and contention. In a polling system, the SNC via the RLC transmits appropriate information and surveys each RF terminal for data

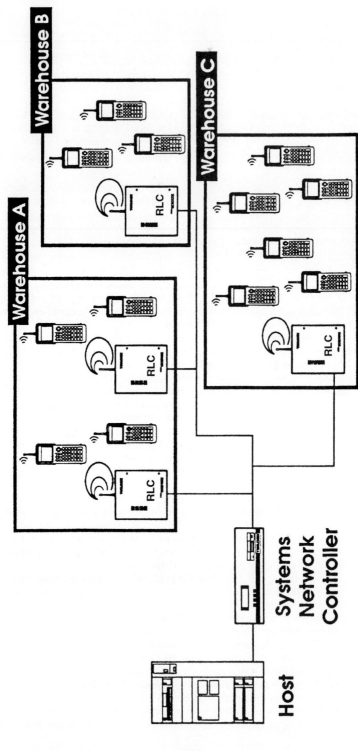

FIGURE 3-33 RF system configuration. (*Courtesy of Teklogix, Inc.*)

or messages in a specific sequence. Due to the fact that the polling method can introduce inefficiencies (i.e., polling of terminals without data to transmit, thereby resulting in longer response times for terminals with information to send), it is best suited for an RF system with a high transaction rate and a small number of terminals.

In contrast to a polling system, a contention system permits individual terminals to initiate transmissions and vie for RF channel usage. The premise of a pure contention protocol is that terminals transmit data or information whenever they are inclined. If no other terminals are sending data at the same time, the information normally is communicated successfully. However, if terminals are transmitting in unison, a collision results, and no information is effectively relayed. To minimize collisions and allow contention to be productively utilized, a number of alternative contention protocols have been developed. Carrier sense multiple access (CSMA), one of the more popular alternative contention protocols, is commonly employed for RF terminal communications within warehouse systems. In a CSMA system, RF terminals ensure that a selected RF channel is clear prior to initiating a transmission. If a selected channel is busy, the RF terminal delays a random number of milliseconds and then checks the channel again. With CSMA, it is presumed that collisions only occur if units start transmitting at the same time. Based on its benefits and limitations, a contention system is more practical for an RF system with many terminals and a low transaction rate.

Regardless of the data-communication protocol used, RF system operations can be complicated by multipathing. The roots of multipathing can be traced to the fact that RF transmissions emanate from a terminal's antenna in all directions. This multidirectional broadcasting results in the same transmitted signal, which is absorbed or reflected by various materials within the warehouse, arriving at the receiver through multiple paths at different times. Fortunately, many systems are equipped with timing mechanisms that alleviate multipathing problems.

The RF terminal is used to enter and access information from the host computer via the RF network. In addition to containing all the basic elements described for portable data collection terminals, RF terminals also include a radio transceiver that utilizes ultrahigh-frequency–frequency-modulation (UHF-FM) radio technology, in either the narrow-band or spread-spectrum range, to transmit and receive information. Four portable RF terminal form factors, namely rectangular hand-held, pistol-grip, wearable (back-of-hand), and vehicle-mount, are commonly available today.

Rectangular Hand-Held. Where data once had to be carried (on paper) to fixed terminals for entry and retrieved from computer output reports or displays, hand-held terminals have permitted the movement of data collection and availability to the point of activity. Because of the mobility they offer to the user, rectangular hand-held terminals are widely employed in warehouse operations, such as detailed receiving and picking (especially carton or eaches picking that require man-aboard pickers and/or picking onto charts) that do not require a lift or pallet truck to access materials. An example of a rectangular hand-held terminal is illustrated in Fig. 3-34.

Pistol-Grip. By integrating a laser scanner and allowing point-and-shoot capability for bar code reading, the pistol-grip RF terminal provides an alternative to the rectangular hand-held terminal. Due to the fact that its ergonomic design also frees up one of the operator's hands, the pistol-grip RF terminal can be used for warehouse operations such as receiving, picking, and shipping. An example of a pistol-grip RF terminal is depicted in Fig. 3-35.

Wearable (Back-of-Hand). Where terminal direction, hands-free scanning, and limited keyboard input are required, wearable RF terminals are a viable option. Mounted on

FIGURE 3-34   Rectangular hand-held RF terminal. (*Courtesy of Teklogix, Inc.*)

FIGURE 3-35   Pistol-grip RF terminal. (*Courtesy of Teklogix, Inc.*)

the back of the operator's hand, the wearable RF terminal integrates a display and keyboard with a finger-activated bar code scanner to create a hands-free operating environment. The display-keyboard-scanner apparatus is connected to a lightweight RF terminal, which is secured to the operator's body. Warehouse operations most likely to benefit from a wearable RF terminal include piece picking, sorting activities, or high-volume receiving areas. An example of a wearable RF terminal is shown in Fig. 3-36.

FIGURE 3-36 Wearable RF terminal. (*Courtesy of Symbol Technologies, Inc.*)

Vehicle-Mount. In addition to being designed to withstand the shock and vibration associated with warehouse vehicles (i.e., lift and pallet trucks, man-aboard cranes, and so forth) they are attached to, vehicle-mount RF terminals also provide keyboard, display, and input/output options. In contrast with the previously discussed hand-held RF terminals, vehicle-mount RF terminals utilize a larger display and keyboard and are powered by the vehicle's power source. Vehicle-mount RF terminals are best suited for full pallet warehouse operations such as put-away, full pallet picking, replenishment, and shipping. An example of a vehicle-mount RF terminal is illustrated in Fig. 3-37.

When the capability to perform emulation is not contained in the system network controller, it is normally resident in the terminal itself. The advantages of terminal emulation center around the ability to program a terminal to operate as a hybrid by executing other transactions in a batch mode and sending the collected data in a real-time mode and the possibility of being more conducive to GUI environments in the future. Disadvantages inherent to terminal emulation include the system requirement to transmit the entire RF screen rather than some subset, such as changes only, and the need to reinitiate and resend screens from the host computer if a terminal session is interrupted. It is important to note that the listed disadvantages may be minimized,

FIGURE 3-37   Vehicle-mount RF terminal. (*Courtesy of Teklogix, Inc.*)

respectively, through the use of the of the higher RF data transmission speeds associated with spread-spectrum technology and the employment of session mirroring at the
system network controller level.

Since warehouse conditions vary significantly, RF terminals have been designed to
withstand environmental considerations including windblown rain, humidity, dust,
shock, and vibration. In addition, the use of RF terminals in temperature extremes
such as refrigerated areas is permissible if the terminals are equipped with special
low- and high-temperature options (i.e., an optional freezer keyboard for a vehicle-
mount RF unit). If environmental conditions dictate the need, intrinsically safe RF terminals are also available.

Narrow Band versus Spread Spectrum

An RF system in a warehouse environment can transmit and receive data over one of
two Federal Communications Commission (FCC)–defined bandwidths, namely *narrow
band* or *spread spectrum.* In order to select the appropriate radio band and supporting
technology, it is vital that an understanding of the strengths and weaknesses of each be
examined in relation to the facility in which the RF system is to be implemented.

Narrow-band RF terminals are designed to broadcast on an FCC-assigned frequency between 450 and 470 megahertz (MHz). Utilizing one frequency for transmission
and one for reception, narrow-band RF terminals normally operate at 2 watts (W) of
power. Theoretically, a 2-W radio transmitting within the narrow-band frequency is
capable of covering 1 square mile; however, due to environmental and physical condi-

tions that may exist, the effective range is normally more limited. (See the section entitled "Physical Site Survey" later in this chapter for more information.)

Within the 450- to 470-MHz frequency band, narrow-band RF terminals transmit at speeds ranging from 1200 to 9600 bits per second. Although the fastest data transfer rate is typically most desirable, there are many factors that the influence the actual data transmission rate achieved. For instance, the slowest communications speed within the RF system (system network controller, radio local area network controller, and terminals) controls the speed of the overall system. Additionally, if RF terminals are expected to communicate over large distances and there is a possibility that noise interference can be introduced, it is prudent to use slower transmission speeds to reduce data errors and system retransmissions.

Because narrow-band devices broadcasting on the same frequency in the same location are known to create communication problems, narrow-band RF systems utilize a specific frequency that is protected by a mandatory FCC license. This license secures the frequency for the user, and provides a means for reporting interference problems to the FCC. Once the application is completed, the FCC licensing process, including approval and frequency assignment, takes approximately two weeks. A summary of the advantages and disadvantages associated with narrow-band RF systems is depicted in Fig. 3-38.

Although originally developed by the United States military during World War II to provide increased throughput while resisting interference, spread-spectrum technology has also proven to be a viable option for RF data communications in the warehouse environment. Approved for use in 1990, spread-spectrum RF terminals transmit in the 902- to 928-MHz radio frequency band without requiring an FCC license. Offering less coverage than traditional narrow-band systems, spread-spectrum RF systems operate at 1 W of power or less with a effective range of 1500 ft.

Spread-spectrum RF systems communicate through direct sequencing or frequency hopping. In the direct sequencing mode, the RF signal is broadcasted over a wide range of frequencies or channels and the data is, therefore, equally spread across the entire band. Due to the number of frequencies available, interference is not normally encountered. Direct sequencing modulation is usually selected for applications that require very high data rates. When the frequency hopping modulation technique is employed, the terminal(s) and transceiver are permitted to move from frequency to frequency in a random manner and then transmit. If interference is found on a particu-

Advantages	Disadvantages
Facilitates on-line, real-time operations	FCC license required prior to RF system implementation
Maximizes the benefits of implementing automatic data collection	More expensive than batch terminals
Transmission frequency is protected from interference by a mandatory FCC license	Lower data transmission rates than experienced with spread-spectrum RF technology
Higher transmit power equates to better coverage than is possible with a spread-spectrum RF system	
Less expensive to implement than a spread-spectrum RF system	

FIGURE 3-38 Advantages and disadvantages of narrow-band RF systems.

Advantages	Disadvantages
Facilitates on-line, real-time operations	Sensitive to transmission interference from co-located or adjacent systems due to the absence of FCC licensing requirements in the spread-spectrum RF bandwidth
Maximizes the benefits of implementing automatic data collection	Saturation over time of the 902- to 928-MHz spread-spectrum bandwidth through the proliferation of wireless equipment
Higher data transmission rates than experienced with narrow-band RF technology	Lower transmit power equates to less coverage than is possible with a narrow-band RF system
An FCC license is not required to operate a spread-spectrum RF system	More sensitive to multipathing than a narrow-band RF system
	More expensive to implement than a narrow-band RF system if greater coverage is required

FIGURE 3-39   Advantages and disadvantages of spread-spectrum RF systems.

lar channel, the terminal(s) and transceiver can select an alternative channel on which to retransmit. Frequency hopping is the most practical solution if maximized interference rejection and coverage are the goal.

One of the primary benefits of a spread-spectrum RF system is its high-data-capacity throughput of over 100,000 bits per second. This high throughput is the result of the spread spectrum's 26-MHz bandwidth that allows the system to disseminate the data over the entire band. It is important to note, however, that although spread-spectrum systems are capable of transmitting a significant amount of data at high speeds, their low transmission power limits the coverage provided by each RLC. Also, due to the fact that spread-spectrum systems do not transmit well through walls, additional RLCs may be needed to provide adequate coverage.

A disadvantage of spread-spectrum systems is the potential problem of RF signal interference. Since FCC licensing is not required and the frequencies are open to appropriate devices, the spread-spectrum bandwidth may be saturated over time by the widespread use of wireless equipment such as LANs, security systems, and vehicle location systems. Although direct sequencing and frequency hopping may minimize potential interference, these modulation techniques cannot overcome higher-power transmissions that effectively eliminate the spread-spectrum RF signals. The advantages and disadvantages of spread-spectrum RF systems are listed in Fig. 3-39.

RF System Response Time

The performance of an RF system is measured in response time. By definition, response time is the time required to transmit information from a RF terminal to the host computer and back to the RF terminal. Since response time can vary between subseconds and several seconds, and thereby greatly impact warehouse productivity, it is important to evaluate the factors affecting response time thoroughly. These factors include:

Data Transmission Speed.   This is the speed rated in bits per second that data is transmitted through the RF network. The typical data transmission speeds for narrow-band and spread-spectrum systems are 9600 and greater than 100,000 bits/s, respectively.

Number of Radio Channels. Because radio channels are travel paths for transmitted data, higher data throughput can be achieved by using multiple channels.

Quantity of Data Transmitted. The number of transactions, amount of data in each transaction (as determined by the location of the emulation), and number of RF terminals influence the amount of RF support that is required to obtain an adequate response time.

System Links. The communication capability between the host computer and the SNC as well as the SNC and the RF terminals must be established at a level that prevents bottlenecks during periods of high data transmission. Maximum processing power in each component is the key to compensating for changes in transaction flow.

Host Computer Response Time. It is important to be aware of the fact that host computer processing time is normally the most significant portion of an RF system's response time. (See Fig. 3-40.) The host system and any associated applications must be tuned to minimize host computer response time.

Being cognizant of the factors affecting response time, current transaction rates, and future system requirements can provide a significant advantage in the specification of an RF system that can meet increased throughput rates and sustain acceptable system response times.

Physical Site Survey

Once the initial RF system specifications are established, it is wise to request that the selected RF vendor perform a physical site survey. During the site survey the following facility characteristics are examined with respect to their impact on system performance.

Number of Sites and Facility Layout. If the RF system is to be implemented at multiple sites, it must be determined whether one large system or several smaller systems allow for the best coverage. Additionally, the specific facility layout including size (i.e., square footage) and shape (i.e., square, rectangular, etc.) impacts the type of RF system required as well as the placement of the RF equipment.

Building Composition. Due to the fact that materials, such as fire walls and steel, reflect RF signals differently, it is crucial that building composition be analyzed to assess its effect on a potential RF system.

Inventory Material and Storage Environment. As is the case with building composition, the inventory material (i.e., corrugated paper) and the storage environment (i.e., metal shelving) are examined to determine how they influence the RF signals and system.

Existing Equipment. The effects of existing equipment (computer systems, communication devices, etc.) on the operation of the potential RF system are studied to eliminate possible problems.

Physical Environment. If the RF system is to be used in a unique environment, such as a freezer or outdoors, RF terminals designed for these special environments are specified, and RF system requirements are modified to ensure adequate coverage.

Geographic Location. The geographic location of a facility is evaluated to determine the effects of interference from other facilities' systems and/or equipment on the RF system to be installed.

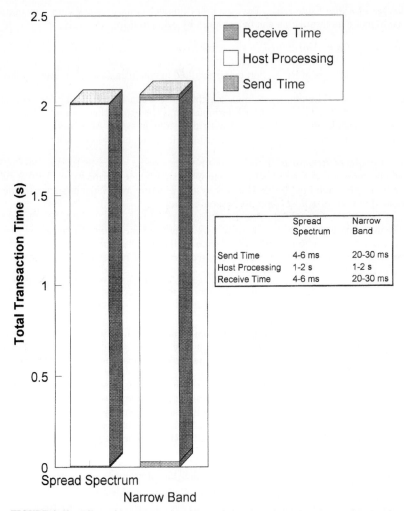

FIGURE 3-40   Effect of host processing time on RF response time. Host processing time is dependent on host sizing and application processing. Send and receive times are based on a 20-byte message being transmitted to and from the terminal. Message length is dependent on the type of emulation being used (i.e., terminal-based or controller-based). Spread-spectrum data is based on a 60 kbits/s transmission speed, while narrow-band data is based on a 9.6 kbytes/s transmission speed.

Application.   System performance is optimized and acceptable coverage is verified through the testing of components of the potential RF system. It is best to conduct site-survey testing in a live environment using a worst-case scenario (a significant level of inventory in storage, material handling equipment in operation, and so forth).

   Information typically supplied by an RF vendor at the conclusion of the physical site survey includes a preliminary RF system design specifying the suggested placement of all the RF equipment, a summary of the building characteristics, a synopsis of the survey results, and a list of recommended equipment and associated power requirements. Even more significant than the survey results is the "guaranteed coverage" promised by most RF vendors. Guaranteed coverage basically equates to the RF vendor being responsible for ensuring that all areas specified for coverage during the survey actually meet the system specifications.

Voice Recognition

Since the introduction of computers, users have sought more natural and easier ways to interact with software productively. In pursuit of this goal, two major steps, graphical user interfaces and programming shells, have allowed computers to be more accessible to users with little technical expertise. One innovation that also has been eagerly anticipated since the early 1970s has been the development of voice-driven computers that permit users to do away with keyboards. Benefits cited for voice recognition systems in the warehouse environment include users' being able to collect data or be given instructions while keeping their hands and eyes free to perform a task, minimized training requirements, and the creation of a very natural and user-friendly system.

A voice recognition system is composed of a personal computer (PC) acting as an SNC and portable data collection terminal. Typically, the PC, which is hard-wired to the host computer via a token ring, Ethernet, or RS232, RS485 connection, contains software to accommodate audio communications and permit images of words to be stored. Several components including a headset with earphones (for audio direction) and a microphone (for audio data collection), a radio transceiver unit, a microprocessor-equipped terminal unit, and a battery are combined to create the voice recognition portable data collection terminal pictured in Fig. 3-41. Depending on the selected configuration, voice recognition systems can operate in either a batch or real-time mode. Real-time communications are achieved through RF signal transmissions between the terminal and the PC.

Conceptually, all speech recognizers work in the same manner. Sound enters the system through a microphone attached to the headset of the voice recognition portable data collection terminal. Much as a movie camera represents an event by using a sequence of snapshots, the speech recognizer depicts the entering sound as a sequence of audio "snapshots." To interpret the speech, the terminal compares the snapshot sequence it is hearing with stored sequences that define particular words. The matching of the incoming snapshot sequence to a stored image permits a particular word to be recognized.

Voice recognition systems are available in two varieties, namely speaker-dependent and speaker-independent. The more commonly available speaker-dependent recognizers require individual users to prerecord their voiceprints, so that the system can create an image of each word for later comparisons. Because of its use of individual user voiceprints, speaker-dependent systems provide for a larger vocabulary and more accurate recognition. In contrast with speaker-dependent systems, speaker-independent systems utilize recognition templates containing previously recorded images to recognize words spoken by any system user. Unfortunately, the diversity required to enable the templates to be effective has limited the breadth of the speaker-independent vocabulary to digits and a few words.

Another distinguishing characteristic of speech recognizers is whether they are discrete or continuous. By definition, discrete systems require a brief pause after each word is spoken. Conversely, continuous systems are capable of recognizing a naturally spoken sequence of words with no mandatory pauses. Continuous systems are considered to be the standard in today's market.

FIGURE 3-41 Voice recognition portable data collection terminal. (*Courtesy of Vocollect, Inc.*)

Warehouse operations most likely to benefit from a voice recognition system include receiving, quality control, and picking (especially eaches). In these areas, productivity can be improved and paper can be eliminated through the implementation of a voice recognition system that allows the users to keep their hands and eyes free to perform required tasks. Advantages and disadvantages associated with voice recognition systems are described in Fig. 3-42.

Data Transmission Technology

Data transmission technology addresses communication issues such as lines, interfaces, and protocols that permit connectivity between computer systems and other associated devices to be achieved. One segment of this technology that has recently received a significant amount of emphasis and is considered to be important for warehouse computer system communication is local area networks.

Communications Interfaces and Protocols

Traditionally, communications between a computer and its peripheral devices (SNCs, printers, terminals, and so forth) has occurred through the use of communications lines

Advantages	Disadvantages
Hands-free operation permits users to keep their hands and eyes free to perform tasks	Input method is prone to error due to incorrect operator data (saying the wrong number or word) and computer misinterpretation of spoken information
Training requirements are minimized	Without applicable noise-reduction devices, noise in industrial environments limits the use of voice input and output
Voice communication creates a very natural and user-friendly system	Lack of visual output can lead to misunderstood instructions
High rate of operator acceptance of technology	With RF voice transmission, a different frequency is required for each operator
Operators are not required to "read" directions	System operation requires additional (as compared to RF terminals) specialized devices to be worn
	Safety issues associated with the wearing of headphones and/or noise-reduction devices may arise

FIGURE 3-42   Advantages and disadvantages of voice recognition systems.

(i.e., cables) equipped with the appropriate electrical interface and following a defined communications protocol. The term *interface,* in this context, refers to the connector configuration, required electrical voltage levels, and connector pin function necessary to allow for the attachment of the communications line. Interfaces most commonly required to provide communications among systems and devices in a warehouse environment are serial connections such as RS232, RS422, RS485, token ring, and Ethernet.

The final component needed to permit successful communications is a communications protocol. By definition, a communications protocol is the formal set of rules controlling the format, timing, sequencing, and error control of exchanged messages in a data network. Protocols typically utilized with the aforementioned serial connections include the following:

Asynchronous.   This protocol is a low-speed, low-cost communications method that transmits individual bytes of information at a rate of one bit at a time with no fixed relationship between bytes. Start and stop bits are utilized to signify the beginning and end of a particular transmission. RS232, RS422, and RS485 serial connections can use asynchronous communications.

Synchronous.   With this communications protocol, special characters that are employed preceding the transmitted information synchronize the receiver and allow many bytes of information to be sent in a single block. The capability to transmit blocks of characters rather than one bit at a time permits synchronous communications to be more efficient than asynchronous communications. The synchronous communications protocol can be utilized with RS232, RS422, and RS485 serial connections.

Token Ring.   In a token ring network, packets of information are passed from one station to another along an electrical ring. When a particular station is ready to transmit, it takes possession of the token frame, sends its data, and then frees the token to indicate that the ring is again inactive and able to carry information from another source. The version of the token ring network selected determines whether the information on the ring flows at a rate of 4×10^6 or 16×10^6 bits/s.

Ethernet. With the Ethernet protocol, all systems and devices are attached as stations on a common cable or bus. Because all the systems and devices on the network are configured with equal access capability, the carrier sense multiple access/collision detection (CSMA/CD) contention protocol is used to regulate transmissions and ensure that communications are adequately accomplished. On an Ethernet network, information is transferred at a rate of 10×10^6 bits/s.

Local Area Networks

Through the proper integration of cabling, interfaces, communications protocols, and selected devices, local area networks (LANs) can be formed. By definition, a LAN is a data network located within the user's facility in which serial transmission is used for direct data communications among network stations. Two of the most popular networks used today are Ethernet and token ring. The type of LAN specified for a particular application is most often dependent on the computer system implemented. Examples of available LANs are listed in Fig. 3-43.

Wireless LANs

Wireless LANs operate in the same manner as standard (wired) LANs with the exception that the communications between devices (workstations, terminals, and so forth) are provided by RF transmissions in the 915-MHz, 1.9-gigahertz (GHz), 2.4-GHz, or 5.8-GHz bandwidths instead of through serial cables. To facilitate RF communications, devices on a wireless LAN are equipped with radio transceivers that utilize a contention system to vie for RF channel usage. Contention between devices attempting to access another LAN medium is resolved through the carrier sense multiple access/collision avoidance (CSMA/CA) protocol. The CA mechanism within the CSMA/CA protocol requires acknowledgment of transmissions to ensure data integrity.

Vendor	Network name	Capacity (Mbits/s)	Media	Protocol
AT&T	StarLAN	10.0	Twisted pair	CSMA/CD*
Gateway	G-Net	1.0	Coaxial cable	CSMA/CD/CA†
IBM	Token ring	4.0 to 16.0	Twisted pair or fiber optic	Token
Novell	Netware/S	0.6	Twisted pair	Polling
Xerox	Ethernet	10.0	Coaxial cable	CSMA/CD

*CSMA/CD: Carrier sense multiple access/collision detect.
†CSMA/CD/CA: Carrier sense multiple access/collision detect/collision avoidance.

FIGURE 3-43 Examples of available LANs.

With today's technology, wireless LANs provide an economical way to facilitate computer and device communications in an office environment; however, they are not practical if a device is considered to be mobile and communications coverage is to be provided over a large area.

COMPUTER SYSTEMS

With all the information available through hardware and software vendors, publications, Information Systems departments, and consultants, the process of selecting a computer system including the platform, operating system, and data base can seem overwhelming. In an attempt to dispel erroneous claims and ease the selection process, this section is designed to provide the information (available technologies, common terminology, standard definitions, and so forth) necessary to allow objective system comparisons and evaluations to be made. Additionally, because the computer system chosen is to be used to run a WMS instead of support processing requirements for enterprise-wide applications, an emphasis on the selection of a proven computer technology with growth potential as well as an appropriate perspective relevant to leading-edge technologies is also presented.

Classifications of Computers

Computer systems are commonly classified with respect to attributes such as physical size, computing or processing power, storage capability (memory), and connectivity. Based on logical divisions within these attributes, the following computer system classifications listed in decreasing size and complexity have been developed.

- *Mainframe computer.* A large, centralized, multiprocessing and multiprogramming computer that provides end-user computational and data base services for the entire company or enterprise and performs network control functions such as interconnection, communications, and support for locally attached and remote devices as well as additional computers.
- *Midrange or minicomputer.* An intermediate-size computer with less processing power, storage capacity, and speed than a mainframe computer, but capable of performing the same kinds of services. Minicomputers are often connected (locally or remotely) to a mainframe to execute ancillary operations.
- *Microcomputer.* A small, integrated computer containing a microprocessor chip, memory chips, and control circuitry capable of handling input, output, processing, and storage functions. Although primarily designed for stand-alone operation, microcomputers can be utilized as workstations in terminal emulation mode, LAN servers, or LAN clients.

When evaluating potential systems, it is important to note that recent computing technology trends toward increasing the computing power and storage capability of microcomputers and decreasing the size of minicomputers are resulting in less of a distinction between the minicomputer and microcomputer classifications.

Computing Methods

In today's business environment, the computing and information needs of a company are typically met through the use of centralized computing, distributed computing, or a

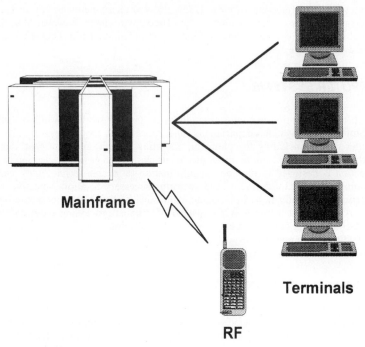

Mainframe

Terminals

RF

FIGURE 3-44 Central mainframe configuration.

combination. Centralized computing, which is traditionally performed by a mainframe computer (although midrange computers are becoming more prevalent), involves the containment of all data processing on a single unit. As depicted in Fig. 3-44, a central mainframe configuration is designed to fulfill a company's computing requirements by sending and receiving data and instructions to and from locally or remotely attached input and output devices.

Benefits commonly cited for a centralized computing configuration executing a WMS include the consolidation of Information Systems personnel in one location, the capability to perform maintenance and implement system upgrades more easily, and the inherent security. Limitations attributed to this configuration consist of vulnerability to computer downtime and performance issues related to remote communications. Additional advantages and disadvantages associated with centralized computing are summarized in Fig. 3-45.

In contrast to centralized computing, distributed computing involves the physical dispersal of computing resources to additional processes and/or sites. With a distributed computing configuration, a secondary computing system (i.e., minicomputer) locally performs the processing, storage, control functions, and input/output functions, and communicates pertinent data to a primary or host computer (i.e., mainframe) via an established network. An example of a distributed midrange computing system configuration is illustrated in Fig. 3-46.

Improved system performance and reduced response time due to factors such as local attachment to the system, consolidated communications, and independence from the host computer and data base are justifications frequently stipulated for implement-

Advantages	Disadvantages
Consolidation of Information Systems personnel in one location	Required hardware and software for a mainframe is more expensive
Easier to perform system maintenance and upgrades	Scheduling of job runs is controlled by the Information Systems department
Provides for centralized system backup, disaster recovery, and security	Priority jobs and demand interrupts cause frequent aborts and the need for reruns on routine jobs
Greater computing and processing power are available with a mainframe system	Operations are vulnerable to system downtime
	Performance issues related to communications are common

FIGURE 3-45   Advantages and disadvantages of centralized computing systems.

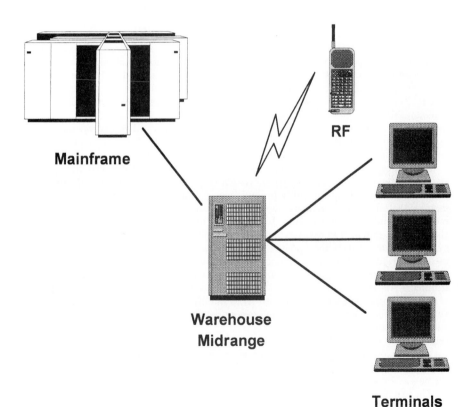

FIGURE 3-46   Distributed midrange configuration.

Advantages	Disadvantages
Local attachment of terminals normally equates to improved system performance (i.e., reduced response time).	More difficult to manage multiple computers
Communications to the host computer are consolidated prior to transmission	Security issues arise with remote users
System can run independent of the host computer and data base	Processing power of distributed computer is typically less than a mainframe
Distributed processing is less expensive than mainframe computing	Interfaces are required for host computer applications

FIGURE 3-47   Advantages and disadvantages of distributed computing systems.

ing a WMS on a distributed computing system. Issues to be considered include the ramifications associated with managing multiple computers and the security aspects introduced with remote (i.e., host computer or other systems) users. A more complete listing of the advantages and disadvantages identified with distributed computing systems is depicted in Fig. 3-47.

Two additional distributed computing system configurations are the client/server and LAN server. By definition, client/server computing is a cooperative processing architecture in which a single application is partitioned among multiple devices that work together to complete a given task. Applications are considered to be *clients,* which are permitted to access authorized services. Conversely, *servers* contain shareable resources (i.e., the data base) and provide one or more services (data base access, communication, etc.). As shown in Fig. 3-48, a distributed client/server configuration incorporates client units (a microcomputer, terminal, etc.) and a LAN server unit (i.e., typically a minicomputer or microcomputer), which affords facilities to the clients and communicates with a host computer (mainframe, minicomputer, etc.).

Benefits typically associated with a WMS implemented on a distributed client/server computing system include improved system response time due to the utilization of localized data, enhanced capability to implement graphical user interfaces (GUIs), and increased efficiency of warehouse personnel as a result of better functional division and more timely information. Liabilities with a WMS distributed client/server system consist of the requirement to manage multiple computers and data integrity issues such as synchronization of data and updates. Additional advantages and disadvantages attributed to distributed client/server computer systems are delineated in Fig. 3-49.

An alternative to the previously mentioned client/server computing system is a distributed LAN server computing system configuration. The premise behind a LAN server system is that users employ networked intelligent terminals or workstations that share a common processor and data storage device (i.e., the server). Furthermore, since any device on a LAN can be accessed by any other, all devices on a LAN are regarded as extensions of each other. As illustrated in Fig. 3-50, a typical distributed LAN server configuration includes microcomputer workstations and a minicomputer or microcomputer LAN server responsible for providing data processing and accessing the host computer.

The benefits and liabilities cited for distributed client/server computing systems are also applicable to LAN server systems. A summary listing additional advantages and disadvantages of distributed LAN server computing systems is presented in Fig. 3-51.

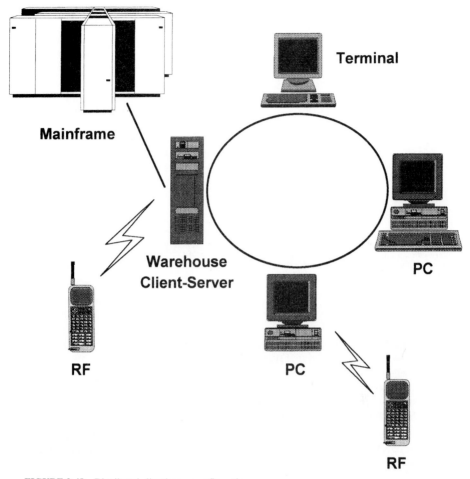

FIGURE 3-48 Distributed client/server configuration.

Types of Data Bases

Another computing system technology consideration that is often debated and discussed during the evaluation of system and application options is the type (and sometimes brand) of data base to be implemented. The term *data base* refers to a systemized collection of data that can be accessed and operated upon by a processing system or application programs to provide a functional service. Although many types of data bases exist, the four most common are defined below.

- *Relational.* A data base that is organized and accessed in accordance with the associations, dependencies, or links between data items. Within a relational data base, the data items are depicted as tables in which rows signify entities and columns symbolize attributes. Relationships among these tables are represented by data values that are matched at the time of access.

Advantages	Disadvantages
More localized data is available with a server implementation	Data integrity and security issues relating to synchronization of data and updates are possible
Distributed processing is less expensive than mainframe computing	
	More difficult to manage multiple computers
GUIs are more easily implemented	More extensive training is required to operate and maintain system
Use of existing personal computers to provide processing reduces server processing power requirements and creates a more cost-effective system	Interfaces are required for host-computer applications
Improved personnel efficiency due to better functional division and more timely information	

FIGURE 3-49   Advantages and disadvantages of distributed client/server systems.

- *Hierarchical.*   A data base that is organized in a treelike format such that each record or segment has only one owner. The owners define how the records or segments are related and establish the access paths to the data contained in the data base.
- *Network.*   A data base that is organized with respect to the ownership of records. With the exception of the root record, all other records can have many owners and access paths.
- *Object-oriented.*   A data base organized in terms of objects and classes of objects. Typically, object-oriented programming languages are used to access the data and the functionality provided by the objects.

Figure 3-52 includes an abbreviated list of some commercially available data bases in terms of vendor, platform, and type. The type of data base selected is most often dependent on the application chosen.

Open versus Proprietary Systems

The most controversial term in the world of computing today is *open systems.* By definition, the open systems concept provides a set of standards that permits portability, interconnection and operability, data access, and connectivity across a variety of platforms. In other words, open systems standards theoretically relay the principles necessary to allow for the interconnection of systems complying with the same standards. These open systems standards, which evolved based on the difficulties envisioned with the interconnection of proprietary systems, are published by many groups including the International Organization for Standardization (ISO), American National Standards Institute (ANSI), Open Software Foundation (OSF), and Corporation for Open Systems International (COS).

The open systems interconnection (OSI) architecture designates a format for the development of standards for the interconnection of computer systems. As illustrated in Fig. 3-53, the network functions are segmented into seven layers. Each layer is to

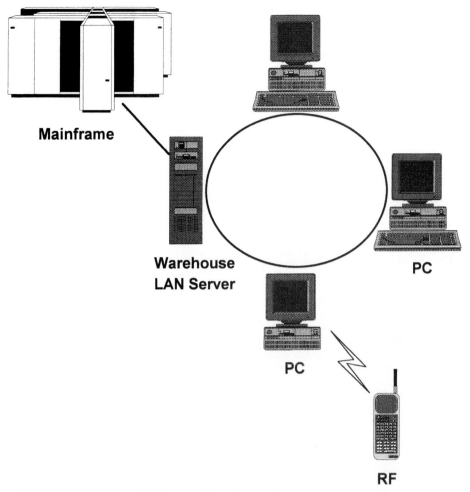

FIGURE 3-50   Distributed LAN server configuration.

contain the necessary data processing and communication functions that meet the open systems standards and support different applications. For example, in a true open system, an application residing in its specified layer can request information from any installed open data base and thereby cause the appropriate intermediate layers to react and allow for data access and retrieval. (In actuality, however, most "open" applications are developed with the capability to access particular data bases. If an alternative data base is to be used, new routines may need to be created to permit data base access.)

Before proceeding any further, it is important to be cognizant of the fact that the open systems concept is a goal that has not been achieved. In today's world, a platform, operating system, data base, or application is considered to be open if it complies with any of the well-known specifications that have been issued to date. In other

Advantages	Disadvantages
Ease of use	More difficult to manage multiple computers
Established and well-accepted technology	Processing capability is more limited on microcomputers
GUIs are more easily implemented	Data integrity issues relating to synchronization of data and updates are possible
Utilization of resource sharing that is more efficient and economical than separate data processing and storage capabilities at individual workstations is beneficial	Interfaces are required for host computer applications
Wide variety of commercially available applications can be obtained for these systems	Network line speed is limited due to serial transmissions
Inherent networking capability	

FIGURE 3-51   Advantages and disadvantages of LAN server computing.

words, a system that is capable of communicating with the Transmission Control Protocol/Internet Protocol (TCP/IP) or a data base that complies with Open Data Base Connectivity (ODBC) (i.e., any standard query language can be utilized to access the data base's flat tables) are both "open," because each meets typical requirements included in a specification for open systems.

In contrast with the open systems concept, proprietary systems are based on a confidential fully integrated architecture and/or operating system. If a system is truly proprietary, there is no external support for interconnection available (with the exception of other data bases and systems by the same vendor). Unfortunately, the term *proprietary* is almost as controversial as *open,* since all open systems touted today are to some extent proprietary. If, for example, an operating system cannot be moved from one vendor's hardware to another's hardware, it is technically considered to be proprietary.

PRINTING

Most of today's state-of-the-art warehouse management systems strive to achieve the goals of creating a paperless warehouse environment and operating as a true on-line, real-time system. Even with any imposed printing limitations and the advent of

Computing system type	Vendor	Platform	Operating system	Data base type
Centralized	IBM DB/2	Mainframe	MVS	Relational
Distributed or client/server	IBM DB/2	AS/400	OS/400	Relational
Distributed or client/server	IBM DB/2	RISC/6000	AIX	Relational
Distributed or client/server	Informix	Open systems	UNIX	Relational
Distributed or client/server	Ingres	Open systems	UNIX	Relational
Distributed or client/server	Oracle	Open systems	UNIX	Relational
Distributed or client/server	Sybase	Open systems	UNIX	Relational
Client/server or LAN server	IBM DB/2	Personal computer	OS/2	Relational

FIGURE 3-52   Abbreviated list of commercially available data bases.

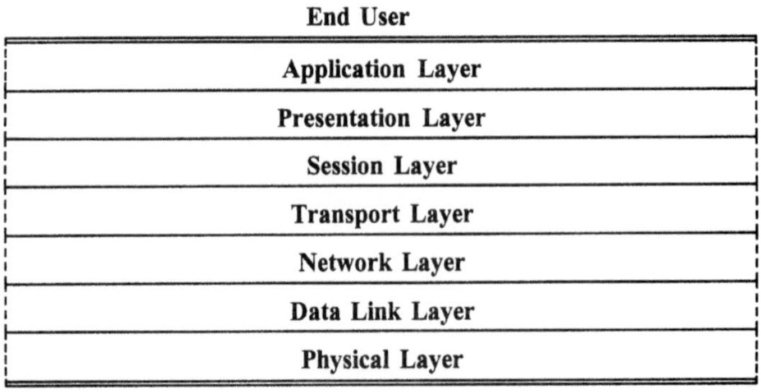

End User

Application Layer
Presentation Layer
Session Layer
Transport Layer
Network Layer
Data Link Layer
Physical Layer

FIGURE 3-53   Open systems interconnection architecture.

advanced shipping notifications (ASNs) and electronic data interchange (EDI), documents such as bar coded pick lists, packing lists, bills of lading, and selected reports are commonly printed to accommodate warehouse operations and meet customer requirements. Fortunately, the print technologies currently available in the warehouse can often be integrated with a WMS to provide the capability to print the aforementioned documents.

One emerging warehouse technology that merits the utilization of the latest print technologies, however, is bar code labels. Uses for bar code labels in the warehouse environment entail unit identification (license plate concept), location identification, customer compliance labeling, shipping labels, and so forth. In addition to supporting automated data collection and real-time communications, bar code labels can also be rendered as a high-speed replacement for RF technology (i.e., label picking). Due to the widespread employment of bar coding with most warehouse management systems, the information in this section concentrates on the significant print technologies used in the creation of bar code labels.

Printing of Bar Code Symbols

Although bar codes can be printed with almost any printing technology, the differences in the quality and durability of the symbols resulting from various technologies affect how well the system using them performs. Bar code quality is measured with respect to the following criteria:

- Conformance of the bar code symbol to symbology specifications, including
  - Correct bar and space patterns
  - Check digits that are present and properly calculated
  - Bar and, for some codes, space dimensions within specifications
- Incorporation of adequate print contrast for defined scanning wavelengths
- Presence of adequate quiet zones
- Absence of voids, edge effects, and other print artifacts
- Ability of a user to decipher any human-readable data (provided as a backup in case of bar code or scanner failure, or for non–bar coded portions of the system) contained in the bar code

Equally as important as quality is the durability of a bar code. Durability relates to how well both the image and background (as well as the face stock and adhesive for bar code labels) endure the expected conditions of storage and use. Stresses such as contact scanning, shipping, handling and other sources of abrasion, temperature extremes, ultraviolet light, moisture, and chemicals typically encountered by bar codes in a warehouse environment must be factored into expected durability. Additionally, to ensure adequate readability throughout a bar code's usage, it is crucial that the bars do not fade and the spaces do not darken.

Based on a warehouse application's specific requirements for bar code quality and durability, the following factors deserve consideration during the selection of the appropriate print technology:

- *Bar code symbology.* The space available on the object to be marked, the amount and type of information to be encoded, customer labeling requirements, the relative importance of a high first-read rate and low probability of misreads assist in the specification of an optimum bar code symbology and character density. Character density determines the minimum bar and space width and thereby defines or at least constrains the printing technology.

- *Scanning technology.* Because bar codes can be scanned in either visible or infrared light, the use of some types of direct thermal materials and dye-based inks that produce a visible black image, but considerably less contrast at infrared wavelengths, is limited. To avoid readability issues, special direct thermal coatings and carbon-loaded inks may be preferable.

- *Quantity, variability, and lead time.* The quantity of labels needed at once, the degree of variability in the information on them, and the lead time between when that information first becomes available and when it must be printed on labels govern whether on-site or on-demand printing is required and the choice of print technology.

- *Life.* The required life of the bar code under the expected temperature, humidity, and other aging factors may restrict the selection of certain direct thermal labeling materials. Abrasion resulting from repeated contact scanning or other sources and/or exposure to solvents or other chemicals can limit the choices among direct marking inks and thermal transfer materials.

Off-Site versus On-Site Bar Code Printing

Once the bar code symbology and scanning technology have been selected, a decision on whether to procure bar code labels from an off-site vendor or to produce them on-site must be made. Off-site printing is one method of obtaining bulk quantities of low-cost preprinted bar code labels that meet established quality standards. Nearly any standard commercial printing process including offset, rotogravure, and flexographic printing can be used. Normally, a film master or computer data file is provided to the printing house, or if the code is part of the product packaging, the bar code can be incorporated into the artwork. Compensation for product shape, ink bleed, and so forth can be designed into the film masters, thereby allowing for more reliable scanning.

Warehouse operations targeted for the use of preprinted bar code labels include high-volume or sequential applications, and picking into preprinted boxes or totes. For example, sequential unique bar code labels that contain no information other than the bar code number can be applied to received units and serve as a license plate for the identification of the unit. Another example is the use of preprinted labels to identify totes or boxes of picked items. (Both of the preprinted bar code label examples cited assume that some level of WMS is implemented to utilize the unit identification num-

Advantages	Disadvantages
Excellent way to obtain bulk quantities of low-cost preprinted bar code labels	Timing of order relative to availability of bar code labels
Utilization of vendor expertise in bar code creation	Not advantageous for the printing of small quantities of different bar code labels
Control and responsibility for complying with established standards is transferred to vendor	Storage for preprinted labels
Available print technologies employed in off-site printing processes often result in very high quality bar codes	Variable data is more difficult to support
Incorporation of bar code in product packaging	

FIGURE 3-54   Advantages and disadvantages of off-site bar code printing.

ber.) The advantages and disadvantages associated with off-site bar code label printing are detailed in Fig. 3-54.

On-site printing is used when there is variable data on the label, a need for fast turnaround, or a requirement for small total quantities of each bar code. Typically, on-site label printing is categorized into two forms, namely, direct marking and label production. Direct marking is used on production lines to apply printed symbols to corrugated cartons. Label production, which can be accomplished in either on-demand or batch mode, pertains to the printing of labels on media such as pressure-sensitive labels, tickets, or tags. A typical on-demand application involves the printing of shipping labels for complete orders on the dock. Conversely, a typical batch application can encompass the advance printing of 5000 pick labels for order picking.

The fact that some level of applications development and programming is required to allow on-site printing systems, such as a direct marking head on a production line, a label printer-applicator positioned along a conveyor, a demand printer located on the receiving or shipping dock, or a line printer situated in a back room printing rolls or stacks of pick labels for the next shift to operate, has spurred the availability of label-preparation software. This software, which permits the developer to concentrate on label layout rather than programming, is designed to provide printing device instructions about which bar code symbology to print, where to print it, and the data to include. In addition to creating simple labels, complex label designs involving multiple bar codes, text, and even graphics symbols, which are increasingly required to meet industry standards for shipping and regulatory labels, can also be addressed through the use of commercially available label-preparation software.

The ease of developing a bar code label varies widely among printers. Document and line printers have commands for accepting rows of dots in addition to text. These commands allow installed label-preparation software to image bar code symbols as well as any graphics or text not native to the printer and to transfer the imaged data as dot rows for printing. Although the above process is viable with most information sources including a host computer, local PC, terminal, scale, or some combination and a wide variety of printers, image generation can be slow on older hardware, and serial and parallel ports can present a significant bottleneck in printing large batches of labels.

Conversely, specialized label and line printers equipped with bar code controllers have additional commands for specifying bar codes by symbology and data rather than as rows of dots. This allows the printer to image bar code symbols at a faster rate as well as incorporate data from other information sources with previously loaded bar code and graphics commands to print a finished label. In addition, the better bar code command sets are text oriented so labels can be defined and manipulated in any application or language and thereby pass through protocol conversion transparently. Through the utilization of specialized label and line printers, a developer has the flexibility to design labels by hand and embed the commands in an application or employ label preparation software to create the label and output commands to drive the printer or to upload label data to another application.

On-Site Bar Code Print Technologies

As stated earlier, on-site bar code print technologies are typically grouped into two categories, direct marking and label production. The following section delineates the print technologies used to satisfy on-site bar code printing requirements and the advantages and disadvantages that differentiate them.

Direct Marking. Direct marking is a print technology involving the direct application of text, graphics, and bar codes to warehouse packaging media such as corrugated boxes. The two major technologies used for direct marking of bar codes are ink jet printing and laser etching. When these print technologies are implemented to accommodate the direct marking of cartons located on a conveyor, certain material handling parameters, such as how close the product can be brought to the marking station, how repeatably it can be positioned, and how rapidly and smoothly it is moving, must be optimized to allow for adequate bar code print quality.

Ink Jet Printing. Ink jet printing, which is the most common direct marking process, is often used on high-speed production lines for coding products and cartons with bar codes, human-readable data, and lot codes. Through the utilization of drop-on-demand printheads, ink jet printers launch individual droplets electromagnetically, magnetostrictively, or by means of piezoelectric elements. The propelled ink droplets are then selectively deflected between the moving product and an ink return channel. Figure 3-55 illustrates the operation of an ink jet printer.

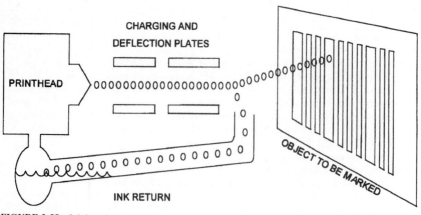

FIGURE 3-55   Ink jet printer operation.

Advantages	Disadvantages
Bar codes can be directly marked onto product packaging	Dot placement accuracy is affected by product motion if mechanization is utilized
Widely used on high-speed production lines for coding products and cartons with bar codes, human-readable data, and lot codes	Variable bar density results from splatter and spurious dots in spaces
	Ink spread due to capillary action restricts the material that can be marked
	Bar codes printed on corrugated cartons have low contrast because of the dark background
	Bar codes are printed 5 to 10 times larger than normal to avoid dot placement accuracy problems
	Specially configured scanners are required to read larger bar codes

FIGURE 3-56   Advantages and disadvantages of ink jet printing.

Although an ink jet system can mark "on the fly" at very high line speeds, there are limitations on print quality that restrict its usefulness for bar coding. These limitations include the following:

- Dot placement accuracy is affected by product motion
- Variable bar density results from splatter and spurious dots in spaces
- Ink spreading by capillary action restricts the material that can be marked
- Bar codes printed on corrugated cartons have low contrast due to the dark background

To overcome the drawbacks, bar codes printed by ink jet are intentionally made 5 to 10 times larger so that dot placement accuracy becomes less critical and are scanned with specially configured scanners that allow for increased read reliability. The advantages and disadvantages of ink jet printing are contained in Fig. 3-56.

Laser Etching.   As depicted in Fig. 3-57, laser etching, another method of direct marking, is accomplished by ablating a surface coating onto an item to be etched with a high-powered laser. For example, a black anodized object can be directly bar coded by utilizing a laser beam to vaporize the black anodization and expose the reflective aluminum (which represents the bar code spaces) underneath. The same process can also be used to produce a bar code by removing black ink printed over white ink on a carton. Even though laser etching produces a high-resolution and high-durability bar code symbol, it has thus far only been implemented on a limited basis. The advantages and disadvantages associated with laser etching are included in Fig. 3-58.

Label Production.   Compared to direct marking, on-site label printing permits much tighter control over bar code quality by separating the printing process from the prod-

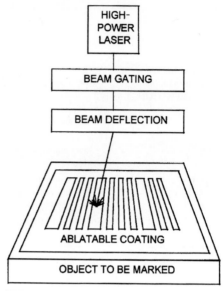

FIGURE 3-57   Laser etching.

uct. An advantage of label printing is that it typically results in smaller and denser bar codes with better scannability. Additional factors, however, that must be considered with label printing include the cost of the label, the added expense associated with its application either manually or by machine to the product, and the durability of the label and its adhesive.

Dot Matrix Printing.   Dot matrix printing is one of the oldest on-site printing techniques. As shown in Fig. 3-59, solenoid-driven print hammers impact an ink-coated nylon ribbon and thereby transfer liquid ink onto a label. While the print hammers and paper are moved relative to one another, the bar code image is created dot-by-dot within an established matrix. Because most dot matrix printers require pin-fed stock and are designed with the print carriage situated far below the media exit, they are most frequently used for printing batches of large labels containing low-density bar codes.

Advantages	Disadvantages
High-density bar codes are produced	Specular bar codes result from process
Durable bar code labels are created	Noncontact scanning is preferable due to rough surface resulting from etching process
Bar codes can be directly marked onto product packaging	

FIGURE 3-58   Advantages and disadvantages of laser etching.

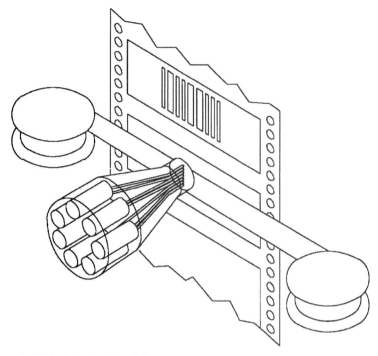

FIGURE 3-59   Dot matrix printing.

Benefits commonly cited for the use of a dot matrix system are that it operates at very low cost per label, can make multiple passes over the ribbon, and can print wide-web, multipart forms containing bar codes. The greatest limitation of dot matrix printing, however, is that the dot size (generally ranging from 0.008 to 0.020 in) limits the narrow element (bar) size and hence the bar code density. Other drawbacks of dot matrix printing entail the need to monitor the ribbon condition continuously, which is degraded by reuse, to ensure bar code contrast and the requirement to utilize a higher-cost carbon-loaded ribbon for labels to be scanned with an infrared light source. A more detailed list of the advantages and disadvantages of dot matrix printing is available in Fig. 3-60.

Laser Printing.   Many of the same characteristics that allow laser printers to create high-quality text and graphics, as well as be today's major document printing technology, also permit them to produce good bar codes. As illustrated in Fig. 3-61, the printing of a bar code by a laser printer is initiated by the cleaning or recharging station removing any excess toner and putting a uniform charge on a photosensitive drum or belt. Next, the processor-controlled laser beam scans across the drum as it rotates and discharges the "white" printing areas. When the remaining charged areas of the drum pass the developer station, they attract oppositely charged toner particles. The toner particles are then transferred to the paper by contact and electrical attraction at the image transfer station. Finally, the toner is fused to the paper by heat and pressure.

Although laser printers are excellent for producing bar code–labeled documents and forms on plain paper, they are less suited to labeling. Commonly available sheet-

Advantages	Disadvantages
Inexpensive method of label creation	Low to medium density (x = 15 to 20 mil) bar codes are created
Multiple passes across the ribbon extend the ribbon life	Continuous monitoring of the ribbon condition is required to ensure continued bar code contrast
Ability to print wide-web, multipart forms containing bar codes	Poorly suited to one-at-a-time label printing
	Carbon-loaded (OCR-grade) ribbons are needed to produce labels to be read with an infrared light source

FIGURE 3-60   Advantages and disadvantages of dot matrix printing.

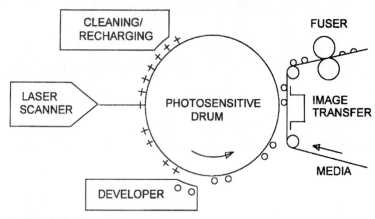

FIGURE 3-61   Laser printing.

fed label stocks, for instance, have been proven to be inadequate or wasteful when used in warehouse applications for the following reasons.

- A label adhesive that can endure the heat and pressure of the fuser is mandatory to eliminate any extrusion of the glue onto the print mechanism. Unfortunately, many of the adhesives that are formulated to remain stable at high pressure and heat do not exhibit acceptable adhesion.
- It is virtually impossible to produce a single small label. A minimum of a half page of media is typically required for a laser printer to maintain control of a sheet; unless a label is at least that size or multiple labels are needed at once, the rest is wasted.
- Ordinary toner becomes a cost issue with laser printing of bar codes. Because the typical bar code label is 15 to 30 percent "black" while normal word processing is 5 percent "black," toner usage rises dramatically during bar code printing.

Even though there are specialized laser printers that are equipped with straight-through paper paths, continuous roll feed, and flashlamp fusion and, therefore, are

Advantages	Disadvantages
Capability to produce high-quality text and graphics	Fuser head and pressure can extrude label adhesive and curl label edges
Labels with excellent infrared scannability are produced	Sheet-fed label media is wasteful unless a label is a minimum of half a page or multiple labels are needed
High-density bar codes are created	Toner usage is a cost issue with bar code printing
Excellent for document printing	Some of the high-durability label stock is incompatible with certain toners or not available in sheet form

FIGURE 3-62   Advantages and disadvantages of laser printing.

intended for label printing, the typical office laser printer is best suited for occasional use on the relatively narrow range of label stocks designed for them. The advantages and disadvantages of laser printing are delineated in Fig. 3-62.

Direct Thermal Printing.   Direct thermal printing, which had originally been developed as a technology for low-cost copiers and facsimile machines, has since been transformed into a highly successful technology for bar coding. Figure 3-63 shows one printing element in a thermal printhead. Typically, a thermal printhead includes a long linear array of tiny resistive heating elements (normally 100 to 300 per inch) arranged perpendicular to the paper flow. Each activated thermal printing element in the printhead locally heats an area on the chemically coated paper directly underneath it. This induces a chemical reaction that causes a dot to form in that area. As the media passes beneath the active edge of the printhead, the image is formed through the correct placement of dot rows.

The technology of thermal printing is controlled both by the print mechanism and the thermal paper used. Advances in thermal paper chemistry have paralleled advances in photographic film and have resulted in "faster," more sensitive thermal papers with better image resolution and contrast and the capability to be printed at higher speeds.

Since the images on these thermal papers are dye rather than carbon, special infrared scannable coatings have been developed to allow bar codes created with direct thermal printing to be read with an infrared light source. Additionally, because thermal paper remains chemically active after printing, nearly all thermal papers used in labels, tickets, and tags are then coated on the top surface for greater image stability

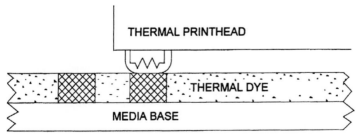

FIGURE 3-63   Direct thermal printing.

Advantages	Disadvantages
High-density bar codes are produced	Heat, ultraviolet light, chemical exposure, and/or abrasion limit the useful life of labels
High-quality print is generated	Special material is required for bar codes to be read in infrared light
Clean and quiet printer operation	Environmentally resistant label top coatings reduce the print speed
Inexpensive to operate and maintain	
Compact printer design and footprint	
Environmental resistance can be gained through the use of label top coatings	
Some thermal labels are recyclable	

FIGURE 3-64   Advantages and disadvantages of direct thermal printing.

and life. This top coating blocks ultraviolet light that can cause fading of the image or discoloration of the background and forms a barrier to water, chemicals, dirt, oil, and grease, which can similarly degrade the image. It is important to note, however, that print speed is often sacrificed in favor of heavier top coats, which act as environmental barriers and provide thermal resistance.

Due to the fact that no ribbons and toner are used and that label loading is a very simple procedure, direct thermal printers are considered to be simpler to operate compared to most other bar code print technologies. The advantages and disadvantages associated with direct thermal printing are included in Fig. 3-64.

Thermal Transfer Printing.   Thermal transfer printers use the same basic technology as direct thermal printers, with the exception of the elimination of the chemically coated thermal paper in favor of a nonsensitized face stock and a special inked ribbon. As shown in Fig. 3-65, the thermal transfer printing process involves the placement of a strong polyester ribbon film of 0.0002 in (4.5 μm) thickness, which is typically coated with a dry thermal transfer ink (a formulation of wax, resin, and carbon black), between the printhead and the label. The thermal printhead is used to melt the ink onto the label surface, where it then cools and anchors to the media surface. Finally, the polyester carrier is peeled away, leaving behind a stable, passive image.

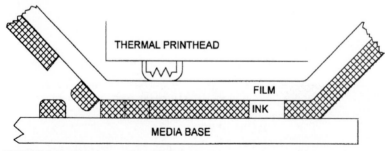

FIGURE 3-65   Thermal transfer printing.

Advantages	Disadvantages
High-density bar codes are produced	Single-pass ribbon can be wasteful if used to print a small amount of data on long labels
High-quality print is generated	Recycling of polyester ribbon is impractical
Wide selection of face stocks are available	
Ultradurable, environmentally resistant, non-fading bar code labels are created	
Clean and quiet printer operation	
Inexpensive to operate and maintain	
Compact printer design and footprint	
Labels with excellent infrared scannability are produced	

FIGURE 3-66   Advantages and disadvantages of thermal transfer printing.

A benefit of thermal transfer printing is that the polyester film between the thermal printhead and the media isolates the printhead from the ink chemistry and allows very complex and active ink formulations to be used in the creation of ultradurable labels. For this reason, thermal transfer printing has become a technology of choice for the production of labels that are exposed to hostile extremes of heat, cold, abrasion, radiation, and chemicals. Additional advantages as well as disadvantages of thermal transfer printing are delineated in Fig. 3-66.

Both direct thermal and thermal transfer printing methods create bars with extremely sharp edges and excellent uniform fill. This is why thermal printing is considered to produce some of the best-quality bar codes. The difference between the two thermal technologies is in the simplicity and environmental economy of direct thermal printers and the durability and archival stability of images produced with thermal transfer printers.

Print and Apply Systems.   Print and apply systems incorporate a dot matrix or thermal transfer printer to create on-demand, variable data labels and a vacuum or mechanical mechanism to affix the self-adhesive labels. Typically mounted on packaging equipment or a conveyor line, print and apply systems can be activated electronically or manually. Because properly designed print and apply systems provide uniform and reliable label placement, they are especially advantageous for automated systems employing bar code scanners.

ELECTRONIC DATA INTERCHANGE

The need for companies to standardize the transmission of electronic documents among manufacturers, distributors, sellers, and customers and to strive to achieve the goal of never reentering data after its initial entry into a computer system has propelled many industry segments to pursue the concept of electronic data interchange. By definition, electronic data interchange (EDI) is the intercompany, computer-to-computer communication of information in a standard data format. In theory, the EDI recipient is capable of executing standard business transactions with the transmitted data.

Guidelines including standard communications conventions, standard formats, and permissible abbreviations have been established to allow for electronic communications. Typically, specific EDI industry standards in the United States are developed from the ANSI ASC X12 standard. Internationally, the counterpart standard to ASC X12 is the United Nations EDI for Administration, Commerce and Trade or UN/EDIFACT. Because UN/EDIFACT is already accepted as the worldwide standard, the ASC X12 standard is to be discontinued in favor of UN/EDIFACT after 1997.

EDI technology is based on the following four basic concepts:

- *Industry convention.* This is the group of EDI transaction documents defined by the ASC X12 standard to be utilized by a particular industry. Examples of industries that have already established EDI standards are as follows:

  - Retail industry with the Voluntary Inter-Industry Communication Standard (VICS)
  - Electronics industry with the Electronics Industry Data Exchange (EIDX)
  - Chemical industry with the Chemical Industry Data Exchange (CIDX)

- *Transaction set.* This is a single EDI transaction document developed through the ASC X12 standard. Examples of commonly used transaction sets are depicted in Fig. 3-67.

- *Data segment.* This is an array of related transaction information. An example of a data segment is a grouping such as an order number, "ship to" information, and shipment freight terms. Data segments are arranged in a meaningful order within each transaction set.

- *Data element.* This is a single piece of data such as the order number or order line contained within a segment. Data elements, which are represented by standard industry abbreviations, are organized in a meaningful order within each data segment.

Once the correct combination of hardware and software is implemented, EDI communications between two computer systems can be achieved through three processes, namely mapping, translation, and transmission. The first step of the EDI process involves the mapping of system data to be transmitted into an EDI standard data format (i.e., elements and associated segments). The mapping process is normally accomplished through the use of commercially available conversion software.

Translation, the second step in the EDI process, is provided by software programs that encode or decode and format data and perform protocol conversion on EDI transaction sets. In addition to converting the data, some translators are also capable of executing the mapping process.

The final process in EDI communications is transmission. A modem utilizing the binary synchronous communications protocol on either a public switched telephone network or a private leased line is the normal mode for the transmission of EDI data.

Transaction set	ASC X12 transaction set number	UN/EDIFACT message
Purchase order	850	ORDERS
Purchase order acknowledgement	855	ORDRSP
Invoice	810	INVOIC
Inventory inquiry	846	INVRPT

FIGURE 3-67 Examples of common EDI transaction sets.

EDI data communications speeds are dependent on the modem and line configuration selected by the communicating parties.

In an effort to simplify the EDI communications effort and allow more users to become involved and realize the benefits of EDI, value-added networks (VANs) have been established. By contracting a VAN to establish an adequate support structure (communications protocol, data format, and so forth) among various subscribers on the VAN, a user can communicate with other computer systems on the VAN through store and forward services. Advantages of VANs include their ability to provide security, audit trails, data transmission, recovery, and redundancy as well as 24-hour-a-day access. A disadvantage of VANs is the additional cost involved with the interconnection of two VANs to permit users on separate VANs to communicate.

When constructively implemented as a strategic requirement among manufacturers, distributors, sellers, and customers, EDI in combination with bar code technology and inbound (receiving) and outbound (shipping) real-time on-line data collection provides a competitive advantage as well as strategic and operational benefits. Competitive advantage is gained through an EDI user's ability to render better service at lower costs, compete in new markets, and complete business transactions through an electronic rather than manual means. Strategic benefits that can be realized with EDI include reduced processing costs, improved information flow, and enhanced vendor relationships. Similarly, operational benefits such as increased productivity, improved customer service, and reduced inventory levels are also possible through the efficient use of EDI.

Today, the utilization of EDI in the warehouse environment is fairly common for larger companies, but very limited for smaller ones. Typically, EDI communications dealing with the warehouse are received and transmitted through the host system. For example, a vendor can send an advanced shipping notification (ASN) containing receipt information for a trailer delivery at the truck or unit level to the host computer. The host computer can then download the receipt details to a WMS for use with the receiving function. (Full automation is achievable within the WMS if UCC/EAN-128

Advantages	Disadvantages
Permits faster, more accurate information exchanges among the manufacturers, distributors, sellers, and customers	Effective implementation requires the use of automated data collection
Standardizes the transmission of electronic documents	Requires cooperation between trading partners
Eliminates rekeying of data after its initial entry	Needs to provide the capability to carry EDI information with the product
Facilitates operations such as quick response and just-in-time	Few companies have fully implemented EDI systems (i.e., many print the EDI information, and manually enter the information into the system)
Availability of VAN to aid EDI communications between partners	Cost associated with EDI technology (i.e., modems, VAN fees)
Provides competitive, strategic, and operational benefits	

FIGURE 3-68 Advantages and disadvantages of EDI.

bar codes are an integral part of the ASN information.) In a similar fashion, the warehouse can also upload outbound ASNs to the host computer for EDI transmission to a customer.

Although EDI is currently considered to have a competitive advantage, it is also viewed as required technology for future business transactions. Industry requirements such as just-in-time (JIT) and quick response (QR) have reenforced the need to glean the communication efficiencies gained with EDI. Additionally, the use of two-dimensional bar codes, which address requirements for high-information density and allow EDI information to be carried with the product, is predicted to be the next technological step for EDI communications. The advantages and disadvantages of EDI are included in Fig. 3-68.

CHAPTER 4
MATERIAL HANDLING CONSIDERATIONS

INTRODUCTION

Since many warehouses and distribution centers already use material handling equipment with a manual system, the proper time to evaluate the equipment effectiveness and decide on any changes is during the warehouse automation process design. Careful analysis is required in defining a material handling solution that has the flexibility required to manage both the current and projected product flow and volumes. Selecting the right level of material handling equipment to support warehouse processes and the warehouse management system (WMS) leads to greater operating efficiency, flexibility, and return on investment than attempting to design a system or process around existing material handling equipment.

Material handling equipment ranges from the simple and manual, such as hand pallet jacks and floor storage, to the complex and sophisticated, such as automated electrified monorails and order picking robots. With the advent of new technology, such as radio frequency (RF), real-time information can be transmitted to and from the WMS even when using the simplest material handling techniques. An RF terminal mounted on a lift truck can provide real-time information as effectively as an automated storage and retrieval system (ASRS). With the variety of material handling technologies to choose from and new technology equipment to augment the data communication, how then does one formulate the right material handling solution?

Determining the right level of mechanization for a facility depends on a complete understanding of the business objectives, including the expected level of activity over the next three to five years, the rate at which the product must be handled, the product attributes, including size, weight, shape, packaging, and environmental constraints, and any labor and space constraints. Many of these factors are tied to the distribution trends discussed in Chap. 1. For example, the rate at which the product must be handled evolves from cycle time objectives. If the business objectives include smaller, more frequent shipments and value-added services, then the product's size, weight, and packaging attributes are impacted. An operation such as pallet picking with a lift truck may now require carton picking and conveyor transportation to move the now smaller product.

There may be several alternatives that will optimize the warehouse and distribution operations. It is important to base the material handling equipment evaluation and selection on the automation project's business objectives and process and requirement definitions in order to create a flexible, integrated solution and avoid islands of automation within the DC. These process definitions also describe the interfaces required to ensure proper communication between the material handling equipment and the warehouse management system.

OVERVIEW

There are a variety of alternatives to consider when selecting the right automated material handling equipment for a warehouse or distribution center. The material handling technology selected should complement the information tracking and management designed in the WMS. For example, with a real-time WMS, cross-docking the inventory from the receiving dock to the shipping dock is now possible. An analysis is now required to determine an efficient material handling approach for the physical movement of inventory between the docks. In some cases, the material handling equipment is an integral component of the WMS design. For example, the WMS may batch the same SKU across multiple customer orders to allow batch picking and then depend on an interface with a sortation conveyor to divert the various cartons to the correct shipping lanes.

This chapter focuses on the key material handling components that require evaluation during the warehouse automation project. These key components include the following.

Transportation systems: the automated movement of inventory through the distribution center, including belt, roller, accumulation, and sortation conveyors, automatic guided-vehicle systems, and automated electrified monorails

Pick systems: the various mechanized picking technologies including A-frames, carousels, pick-to-light systems, automated storage and retrieval machines, RF-directed picking, and order-picking robots

Storage mediums: designated inventory-storing areas using various rack alternatives including single-pallet racks, double- or triple-deep drive-in racks, shelving, and a narrow aisle rack

Lift-truck equipment: the various inventory-carrying devices associated with the movement of inventory through the distribution center including the hand pallet jack, standard pallet fork truck, man-aboard order-picking truck, and turret trucks

Each of the different technologies is described, and the main concepts and considerations in evaluating material handling equipment are presented. The goal is to ensure that proper consideration is given to the material handling strategies used at the warehouse or distribution center when developing a warehouse automation strategy. Changes to material handling equipment must be included in the benefit and justification analysis, regardless of whether the equipment is implemented now or three years later as required by increases in volume. Chapter 8 describes this benefit and justification analysis.

Distribution and industrial engineers are needed to evaluate the various alternatives to maximize distribution operation efficiency properly. In addition, professionals employed by local material suppliers or material handling consultants are highly trained and knowledgeable in the field and can make recommendations based on the unique needs. The selection of material handling equipment, however, should involve the project team's full participation; they know the operation the best and can better advise the material handling professional in the analysis of current and future requirements. For a more detailed analysis on facility design, other warehouse and distribution operations and layout books are included in the References section at the end of this chapter.

TRANSPORTATION SYSTEMS

Transportation systems are composed of mechanized equipment that moves inventory through the DC without any manual intervention. In addition, these systems are fixed, requiring more careful planning and installation than mobile equipment such as lift trucks. These systems, if installed properly, provide a highly efficient operation when used over long distances and high volumes. The simpler transportation systems, such as gravity or power conveyors moving unit loads from point A to point B, usually do not require integration into the WMS. The more complex systems, such as sortation conveyors and automated guided vehicles routing inventory through the DC, require an interface to the WMS to track and direct inventory movement.

Conveyors are probably the technology of choice when the product to be handled is relatively small (3 to 36 in), is lightweight (8 oz to 100 lb), and needs to be handled at high rates (5 to 200 pieces per minute). Belt and roller conveyors can transport and accumulate a wide range of packages, as well as heavy unit loads (500 lb and greater).

However, if unit loads are the primary form of the product to be handled, other technologies may be more appropriate. Automatic guided vehicles, for example, are more effective than conveyors at moving unit loads over long distances with varied routes. And automated storage and retrieval machines are better suited to moving unit loads vertically as well as horizontally with a minimum amount of space. Automated electrified monorails can be used to move both lightweight products as well as heavy unit loads and are very well suited to flexible routing, multiple stopping, and leaving the floor clear.

Conveyor

There are three basic types of powered transportation conveyors: wheel, belt, and roller. Selecting the proper type of conveyor is based on the size and stability of the package to be handled, the plane of transportation (horizontal or incline or decline), the weight of the product, its rate, and its cost.

For optimum conveyance of any product, it must be free from protruding or loose straps, strips, hardware, and corner reinforcements. Wire binding and steel straps on cases or cartons must be relatively flush with the carrying surface if a roller or wheel conveyor is to be used.

A *powered wheel conveyor* is the most economical powered transportation conveyor available. Comprised of a series of wheels supported by a frame with a powered narrow belt running the entire length of the conveyor as shown in Fig. 4-1, a powered wheel conveyor is an excellent means to transport a wide range of goods. Products well suited for this type of conveyor include flat-bottom containers, heavy-walled cartons, light-walled cartons with rigid loads, steel tote boxes, wooden boxes, and plastic trays. Cleated crates, cross-ribbed items, chimed-bottom drums or cans, thin wood crates if warped, multiwall bags, concave or convex containers, and wire-bound containers are not conveyable on powered wheel conveyors.

A powered wheel conveyor is well suited for long transportation runs. However, it does not work well if product must be slid across the width of the conveyor. Side-

FIGURE 4-1 Power wheel conveyor. (*Courtesy of Rapistan-Demag Corporation.*)

loading cartons in a full-case pick module, for example, would not be a good application of this type of conveyor.

A *belt conveyor* is the most versatile transportation conveyor. It can handle a wide variety of packages, both rigid and soft, as well as irregularly shaped goods. Belt conveyors can be used both in the horizontal plane as well as at an incline or decline to move products between elevations.

Belt conveyors generally are configured in two styles: slider bed conveyor and belt-on-roller conveyor. A slider bed belt conveyor consists of an endless fabric, rubber, plastic, leather, or metal belt operating over a suitable drive, with tail-end and bend terminals supported by and sliding across a stationary surface as shown in Fig. 4-2. A slider bed belt conveyor is generally used for shorter distances (incline or decline) or in areas where the product is manually loaded onto the conveyor. Full-case order-selection modules, for example, often use a slider bed conveyor because of the shock loading of full cases of product onto the conveyor.

The belt-on-roller belt conveyor is similar to the slider bed unit except that the belt is supported by and slides across rollers mounted in a frame (see Fig. 4-3). The rollers have less surface area in contact with the belt and rotate as the belt is pulled across the face of the rollers. With less friction between the belt and the supporting surface, longer lengths of conveyor can be achieved.

While most belt conveyors are straight units, belt turns (usually in standard arcs of 15°, 30°, 45°, 90°, and 180°) are available to complete an all-belt conveyor layout as in Fig. 4-4. A spiral variation of the belt turn also works well when elevation changes are required, but space is limited as depicted in Fig. 4-5.

A *live roller conveyor* is also widely used in mechanized warehouse and distribution centers both as a means of transportation and accumulation. A live roller conveyor is made up of a series of rollers over which objects are moved by the application of

FIGURE 4-2   Belt conveyor. (*Courtesy of Rapistan-Demag Corporation.*)

FIGURE 4-3   Belt-on-roller belt conveyor. (*Courtesy of Rapistan-Demag Corporation.*)

FIGURE 4-4   Curved belt conveyor. (*Courtesy of Rapistan-Demag Corporation.*)

FIGURE 4-5   Spiral belt conveyor. (*Courtesy of Rapistan-Demag Corporation.*)

FIGURE 4-6 Live roller conveyor. (*Courtesy of Rapistan-Demag Corporation.*)

power to all or some of the rollers. The power-transmitting medium is usually belting or a chain. (See Fig. 4-6.)

The live roller conveyor is well suited to handle a wide variety of cartons and totes. It has the advantage of allowing a product to be slid across the face of the rollers and is therefore very suitable for side loading or unloading, merging, or diverting products.

Rollers must be close enough together to have three rollers supporting the shortest length carton. For example, if rollers are on 3-in centers, the minimum length carton that can be conveyed is 9 in long. With three rollers supporting the carton at all times, there is little chance of the carton falling between rollers. Soft-bottomed, chimed, and uneven-surfaced products, however, are not usually well suited for transport on roller conveyors.

Another advantage of roller conveyors is that a roller or series of rollers can be stopped, while other rollers continue to turn and transport product. This zoned activation is called an accumulation roller conveyor as shown in Fig. 4-7. The advantage of an accumulation roller conveyor is that products can be temporarily stored between two asynchronous processes without shutting down either process. Thus, an accumulation roller conveyor can often be found between the order-selection modules and the sortation conveyor. It is also widely used between the sortation conveyor and the shipping conveyor. In this way, order selection can happen somewhat independently of sortation and sortation independently of shipping.

Sortation Conveyor

Choosing sortation conveyors is more complex than selecting transportation conveyors. Sortation is more than simply separating consolidated-picked items and matching them

FIGURE 4-7   Accumulation roller conveyor. (*Courtesy of Rapistan-Demag Corporation.*)

to orders. A mechanized sortation system uses subsystems that consolidate, identify, and continuously deliver cartons at the required rate through the sortation system in proper orientation and gap sequence. The subsystems providing these functions are merge, induct, sort, and take away. Figure 4-8 diagrams a sortation system. The proper selection of these subsystems is the key to an efficient and economical sortation system.

Merge Subsystem.   Typically, cartons enter the merge subsystem on a series of inbound conveyors (usually powered roller accumulation conveyors) from the order-selection modules or receiving areas. This subsystem consolidates the cartons for a controlled release at the induct area. Carton size, throughput rates, and the number of incoming lines each critically affect the choice of equipment in this area.

The merge provides two important functions. The first is to provide a constant, metered flow of the product to the sorter induct area. In order to maintain system throughput, the sorter induction unit can neither be starved nor overfed. The merge, therefore, must react immediately to the demands of the induct units.

The merge also needs to function in such a way that the upstream conveyors can run without impairment. The merge needs to react to possible imbalances or "line full" conditions so as to not stop operations in the upstream order-selection modules. These two merge objectives are controlled by what is commonly referred to as "merge logic." Photoelectric eye sensors placed along the inbound conveyor lines, usually at the 50 and 75 percent full mark, are used to determine the status of each line. The merge subsystem controller, usually a programmable logic controller (PLC) or other microprocessor, constantly monitors the status of each line. If all the in-feed conveyors are approximately 50 percent full, for example, the merge subsystem will cycle between the in-feed lines, allowing each to discharge for a preset amount of time before switching to the next line. This provides a constant flow of cartons to the induct subsystem and also keeps the in-feed lines at approximately the same level of fullness.

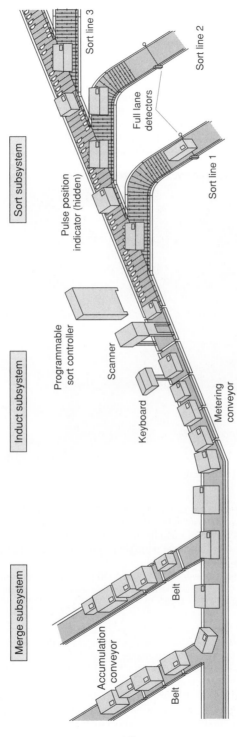

FIGURE 4-8 Sortation system.

If, however, one of the in-feed lines were to reach the 75 percent fullness condition, the merge subsystem controller would automatically prioritize that line as the next one for release. The 75 percent full line would be permitted to run until the fullness condition no longer existed. The merge subsystem would then return to its "normal" mode of operation.

The merge subsystem funnels the cartons from a wide area (the multiple in-feed conveyor lines) and discharges them onto an output conveyor. There are two configurations of merges: parallel and perpendicular designs.

In a parallel design, the in-feed conveyors are in line with the merge bed and discharge directly onto the charge end of the unit as shown in Fig. 4-9. Live roller, her-

FIGURE 4-9   Parallel merge design. (*Courtesy of Rapistan-Demag Corporation.*)

ringbone, and sliding-shoe merge conveyors are examples of parallel designed merges. The parallel merge works well when there are a limited number of input lines. This merge can realistically handle a maximum of four 30-in-wide input lines. If the number of input lines exceeds four, then multiple parallel merges will be required or a perpendicular merge can be used.

A perpendicular merge design has the in-feed conveyors feed the merge bed at an angle (90°, 45°, or 30°). The sawtooth and flexible merges are examples of perpendicular merges as depicted in Fig. 4-10.

Induct Subsystem.   The induct subsystem is the area where cartons are identified and gapped for proper sorting. The bar code label attached to the carton or tote is scanned and read by either a laser scanner or charge-coupled device (CCD). This information is sent to the conveyor computer or the WMS to determine the sort destination. With a real-time system, the WMS can transmit this information to the conveyor PC. The conveyor PC can then look up the scanned bar code and determine the sort destination. This information is sent to the sortation microprocessor, which activates the sort mechanism at the appropriate time and place along the sortation conveyor.

FIGURE 4-10   Perpendicular merge design. (*Courtesy of Rapistan-Demag Corporation.*)

The induct system also creates an appropriate gap between cartons to allow the diverting or sorting device to function properly. Figure 4-11 shows cartons in an induct system. A series of short belt conveyors, with the upstream conveyor running at a faster speed than the downstream unit, creates the gap.

The size of the gap needs to be optimized. If the gap is too large, throughput decreases; if the gap is too small, the divert mechanism misfunctions. Depending upon the system throughput requirements, induct systems can be low-, moderate-, or high-speed subsystems.

A low-rate induct subsystem (10 to 40 cartons per minute) uses two or three short belt conveyors, running at fixed but progressively faster speeds. The gap is variable in length. The rate is slow enough that manual carton identification via a keypad terminal is possible.

The moderate-rate (40 to 80 cartons per minute) induct subsystem is more complex than the low-rate subsystem. Typically, machine readers replace manual encoding, although there have been cases where manual encoders have achieved short periods of production at 80 cartons per minute. The gap between cartons is more fixed in length as a series of brake belt conveyors stop and start using hydraulic or electric clutches.

A high-rate induct subsystem (80 to 200 cartons per minute) requires the use of machine label readers and a series of electric servo motors on the brake belt conveyors. The servo-driven conveyor units momentarily accelerate and decelerate to adjust the gap between cartons. Photoelectric sensors determine the optimum gap between cartons based upon a carton's length and width. In this way, maximum throughput rates can be achieved.

Sort Subsystem.   The sortation subsystem is a conveyor with diverting mechanisms placed at the sort points. The subsystem microprocessor tracks the carton along the length of the sortation conveyor and causes the divert mechanism to activate at the appropriate moment. Divert centers are determined by the carton size, the configura-

FIGURE 4-11   Induct system. (*Courtesy of Rapistan-Demag Corporation.*)

tion of the take-away conveyors (angle to the sortation conveyor), and the speed of the sortation conveyor.

Sortation conveyors can be grouped into three classifications: positive, pop-up, and carriage types. Diverting mechanisms that push against the side of the carton, causing it to be positively displaced, are known as *positive sortation conveyors,* like the one shown in Fig. 4-12. Pop-up devices are built into the side frame of the sortation conveyor and cause the carton to divert when the device is raised above the level of the conveying surface as shown in Fig. 4-13. The carriage-type sortation conveyors place the carton onto a carriage that is transported to the proper sort point as in Fig. 4-14. The carriage is activated, and the carton is discharged at the sort point. Refer to Fig. 4-15 for additional considerations when selecting a sortation conveyor subsystem.

Take-Away Subsystem.   Sortation take-away conveyors are often an overlooked and neglected aspect of the sortation system, but they play an important role in achieving the overall system throughput rate. Take-away conveyors accept cartons diverted off of the sortation conveyor. They must do this in such a way as to not impede the throughput rate of the system. It is therefore essential that the diverted cartons, regardless of size or weight, be effectively removed from the sort conveyor so as not to come in contact with cartons still travelling on the sorter.

The second function a take-away conveyor must perform is that of slowing down the carton. The sortation conveyor is usually operating at the highest speed in the system. The product being conveyed must return to more "normal" speeds so operators in the shipping area can handle them. Therefore, the take-away conveyor must absorb a great deal of energy from the conveyed product without damaging the carton or its contents.

FIGURE 4-12 Positive divert sortation. (*Courtesy of Rapistan-Demag Corporation.*)

FIGURE 4-13 Pop-up divert sortation. (*Courtesy of Rapistan-Demag Corporation.*)

FIGURE 4-14　Carriage divert sortation. (*Courtesy of Rapistan-Demag Corporation.*)

The type of take-away conveyor used depends upon the rate, carton mix, carton weights, and fragility of the contents of the cartons. There are three types of take-away units: powered conveyor (roller and belt), gravity or nonpowered conveyor, and chutes.

Powered take-away conveyors, such as in Fig. 4-16, are typically used to facilitate the diverting action of the pop-up high-speed sorters, handling less than perfect cartons and totes. There are some vendors who supply powered take-away conveyors that are driven from the main sortation conveyor. This reduces the number of drives in the system. As noted above, however, the powered take-away conveyor needs to slow the cartons down. So, oftentimes, the take-away conveyor uses a two-speed motor or is followed by a slower running, downstream conveyor.

Gravity take-away conveyors are most commonly used with the positive sortation conveyors. This type of take-away conveyor works well especially with good-quality, uniform cartons. The gravity wheel conveyor shown in Fig. 4-17 is more effective than the gravity roller conveyor because the wheels create a lower movement inertia on the cartons than does the roller. And the gravity wheel conveyor acts as a very effective speed transition unit, allowing the diverted product to decelerate gradually.

Chute take-away conveyors are most commonly found with carriage-type sortation conveyors but can be used in place of gravity conveyors when carton orientation and product fragility are not considerations. Chutes provide a large sortation target for the diverter to hit as depicted in Fig. 4-18. The pitch of the chute facilitates the take-away action. Chutes must be installed at a steep enough angle to move the cartons away cleanly and quickly but not so steep as to cause the carton to tumble.

	Positive			Pop-up		Carriage	
	Pusher	Tube or shoe	Slat or shoe	Wheel	Roller	Tilt tray	Cross belt
Rate* (cartons per minute)	40	200	200	110	130	180	200
Minimum carton size	9×6×3 in	8×6×1 in	8×6×1 in	8×6×1 in	9×6×1 in	3×3×$\frac{1}{4}$ in	3×3×$\frac{1}{4}$ in
Maximum carton size	36×26×30 in	42×28×30 in	48×34×36 in	48×28×30 in	48×28×30 in	36×32×16 in	50×32×24 in
Minimum/maximum carton weight	2 lb/75 lb	1 lb/100 lb	1 lb/100 lb	4 lb/100 lb	4 lb/100 lb	8 oz/70 lb	8 oz/65 lb
Requires air?	Yes	Yes	No	Yes	Yes	Yes	No
Two-sided diverting?	No	Yes	Yes	No	Yes	Yes	Yes
Conveying surface	Belt	Steel tube	Aluminum slat	Belt	Belt	Wood or plastic tray	Belted carrier
Cost	Low	High	High	Moderate	Moderate	High	High

*Sustained rate; average 16-in-long carton.

FIGURE 4-15 Sortation conveyor application comparison chart.

FIGURE 4-16   Powered take-away conveyors. (*Courtesy of Rapistan-Demag Corporation.*)

FIGURE 4-17   Gravity wheel take-away conveyor. (*Courtesy of Rapistan-Demag Corporation.*)

FIGURE 4-18   Chute take-away conveyor. (*Courtesy of Rapistan-Demag Corporation.*)

Automatic Guided Vehicle System

Automated guided vehicle systems (AGVSs) can benefit operations in a number of ways. AGVSs are designed to interface with other automatic material handling systems including conveyors, automatic storage and retrieval systems, elevators, and the like. The system controller gives accountability and leaves an audit trail as to the movement of goods in the warehouse.

AGVSs are highly reliable and predictable. Damage of goods is greatly reduced as is damage to equipment including racks and the vehicles themselves. Furthermore, labor is reduced while productivity is increased. Paperwork for tracking load movement is virtually eliminated, and load movement is scheduled and regular.

AGVSs feature battery-powered driverless vehicles with programming capabilities for path selection and positioning. Vehicles are equipped to follow a flexible guide path that can be easily modified and expanded. AGVSs fall into two major categories: automatic horizontal transportation and automatically positioned stock selectors.

Automatic horizontal transportation systems use automatic guided vehicles both as towing vehicles and as unit load carriers. Towing vehicles consist of an automatic guided vehicle with trailers attached to make up a train. Often, towing guided vehicles are used to move product from receiving to put-away locations as shown in Fig. 4-19.

Unit load transport vehicles are automatic guided vehicles designed to carry individual loads onboard the vehicle. Some vehicles include automatic load or unload capability and bidirectional movement. Typical applications include transporting loads from put-away to stock-selection locations and transport to shipping locations such as Fig. 4-20. The loads can be either heavy (up to 20,000 lb) or light (up to 500 lb). And some horizontal transportation vehicles can be equipped as pallet trucks.

FIGURE 4-19   AGV—horizontal transportation. (*Courtesy of Rapistan-Demag Corporation.*)

FIGURE 4-20   AGV—unit load. (*Courtesy of Rapistan-Demag Corporation.*)

Automatically positioned stock selectors offer automatic movement in both horizontal and vertical directions to position an order selector at the proper bin or rack location.

There are two critical controls issues with the AGVSs: guidance and system control. Until recently, most AGVSs were guided by a wire embedded in the floor. This wire acts as an antenna that steers the vehicle along a guide path. But wire guidance has some limitations. Certain floor materials, steel, for example, are incompatible with wire guidance. And guide-path changes do require cutting slots in concrete floors.

In 1994, more than one-half of all AGVSs were non-wire-guided systems. Breakthroughs in technology have offered a new generation of alternate guidance systems. The three current leaders are chemical, laser, and inertial guidance systems.

The chemical system uses a transparent paint stripe, which is applied to the floor and acts as the guide path. Guide-path changes are relatively easy. Paint the new path and remove the old path. The limitation, however, is that paint wears with passing traffic and the paint has to be occasionally reapplied.

A variation of the chemical path is the free-ranging-on-grid (FROG) approach. Using a grid of contrasting colors (e.g., black and white), the vehicle is directed to follow a path on the grid. Once again, however, the grid pattern must be maintained wherever the path leads the vehicle.

Laser technology has also been incorporated into guidance systems. Using a series of strategically placed mirrors along the path, an onboard vehicle laser sends out a laser beam, light reflected from the mirrors is analyzed by the guidance system, and the vehicle's position is triangulated from this data. Again, however, mirrors must be maintained. Alignments must be accurate, mirrors clean, and vehicle-to-mirror paths must be clear and unobstructed.

Inertial guidance vehicles use gyroscopic technology to keep the vehicle on an even course. Occasional magnetic markers (or reference points) embedded in the floor give the vehicle a coordinate. The most intelligent type of these systems carries a computer-aided design (CAD)–generated file of the guide path in the onboard vehicle computer. The vehicle then calculates how to get from point A (the coordinate) to the reference point B. The inertial guided vehicle has the most flexibility for changing guide-path configurations of all guidance systems.

The other control issue is the AGVS system controller. The system controller manages the overall fleet of vehicles, dispatching them to wherever they are required. A vehicle status monitoring system tracks the movement of all vehicles in the system. The warehouse management system tells the system controller where vehicles are needed. The system controller then dispatches the most appropriate vehicle to that destination.

It is typically the practice today to simulate a proposed AGVS before system implementation. The simulation will determine the number of vehicles required to meet the most severe demands of the system. Load scheduling, i.e., how loads are identified and routed through the system, load movement, and system layout are all part of a good simulation.

Automated Electrified Monorail

The advantages of the automated electrified monorail (AEM) are its flexible path routings, transports at higher speeds than conventional pallet or heavy unit load handling conveyors, and a greater range of movement than the AGVS in both horizontal and vertical planes.

An AEM also leaves a very good audit trail. Goods placed onto the AEM carrier are usually bar coded. The code is scanned and either the WMS or an operator will route the load to the proper destination. Once the load arrives at the designated terminus, the bar code label is once again scanned. This information is transferred from the AEM computer controller to the WMS. This disciplined and identified material flow makes accountability of the movement of the goods an easy task.

AEMs use a single- or multiple-beam track either mounted overhead for underhung applications or floor supported for inverted applications. A series of independently controlled carriers are mounted to the track, which is used as a runway for such carriers. Through the use of switches, turntables, lifts, and other such path-altering devices, the carriers can be routed along the fixed track path to predetermined locations.

AEM systems use an aluminum track that is clamped into place and acts both as the runway for the carriers as well as the mounting surface for the eight (minimum) conductor bars that provide both power and communication to the carrier.

A drive trolley consisting of a motor, gear reducer, electric current collector shoe, drive wheel, guide wheels, and trolley frame is attached to the carrier. A motor brake, which must be used as a minimum emergency stopping device, is also provided for each drive trolley. There are both light-duty type 1 AEM systems that are capable of supporting up to 1100 lb on a single trolley or 2200 lb on a double trolley and heavy-duty type 1 AEM systems capable of supporting 3000 lb per single trolley or 6000 lb per double trolley.

An underhung or ceiling-suspended AEM provides a flexible routing system that is ideal for transporting loads along distances without using valuable floor space. Figure 4-21 shows a ceiling-suspended AEM. For example, an AEM system is a good means of transporting goods from receiving to put-away or from put-away to stock-picking locations. It is an especially effective material handling medium when there is a break-bulk operation, which must route products to several different locations such as put away, replenishment, or directly to shipping because of backorders.

FIGURE 4-21   Automated electrified monorail—ceiling supported. (*Courtesy of Rapistan-Demag Corporation.*)

FIGURE 4-22   Inverted monorail. (*Courtesy of Rapistan-Demag Corporation.*)

Inverted monorails are also effective in moving pallet loads from the end of aisle ASRS to the picking location(s) and returning the unpicked palleted goods back to the ASRS for storage as shown in Fig. 4-22.

PICK SYSTEMS

Most automated distribution centers today use the consolidated order-picking method where the WMS combines similar orders together to create consolidated picks, which are then sorted into their individual orders just prior to shipping. Consolidated order selection requires the information associated with a particular product to stay with that product so it can be properly identified and routed.

A product is selected either as a full case or a piece, which is picked from a full case (break pack). The piece is usually placed into some other shipping container—a tote or overpack carton. If the product picked is a full case, it is generally taken from a full pallet of like products. A bar code label is applied to the case, if it does not already have one, by the order picker. Full cases are generally placed onto a belt or roller conveyor to be transported from their place of storage and selection to the shipping dock. In addition, RF terminals attached to lift trucks can be utilized to pick full cases or pallets form storage locations throughout the warehouse.

Break-pack order selection provides a number of technology choices. This labor-intensive operation can be assisted by the use of carousels, A-frames, pick-to-light systems, RF terminals, and order-selection robots. Depending upon package characteristics (i.e., small versus large or rigid container versus polybag) and velocity (the number of picks required per order or batch) one or more of these technologies would

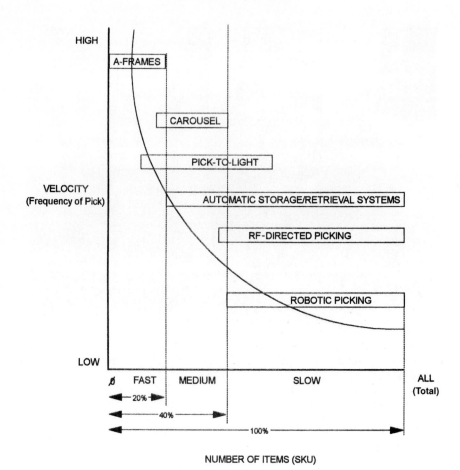

FIGURE 4-23 Picking automation guide.

be appropriate for automated order selection. Small, rigidly packaged, high-velocity items are best suited for A-frames while slow-moving, large items in polybags, on the other hand, are ideal candidates for RF-directed picking.

Figure 4-23 summarizes the technology relative to velocity movement and number of items. The following sections discuss each of these material handling technologies in depth.

A-Frame Automatic Picking Systems

A highly successful type of automatic picking system is the *A-frame*. This configuration has an A-frame with products stacked vertically in lanes (or channels) on both sides of a conveyor. This system is designed to process hundreds of orders in a short period of time, accurately, and with minimum personnel.

The A-frame system is modular and each frame has a number of channels, based on the size of the products to be picked. Frames can be bolted together, in-line or parallel, to increase capacity. The ultimate total length of an A-frame picking machine depends only on the number of channels required, calculated by the quantity of high-velocity SKUs to be stored and the corresponding replenishment strategy. An A-frame machine can be a stand-alone order picker for smaller operations, or it can be combined with other systems as part of an integrated system.

On command from the A-frame control system, individual channel dispensers eject pieces (shown in Fig. 4-24) onto the collecting belt that runs through the center of the

FIGURE 4-24   Channel dispenser ejects product. (*Courtesy of SI Handling.*)

machine. A discrete section of the belt is assigned by the system for an order; all products for that order are dispensed onto that section as it moves through the unit. At the end of the machine, the cluster of products for that order is automatically transferred into a waiting tote bin or shipping container (shown in Fig. 4-25) and then delivered by separate conveyor to other picking areas (maybe manual, mechanized, or automatic picking) or directly to shipping if the order is complete.

Some A-frame systems have dispensers that provide variable dispensing speeds. The A-frame control system contains data (packaging, contents, weight, etc.) on the product within each channel. To prevent damage to delicate products, it instructs the dispenser motor to operate at an appropriate picking speed.

Since the A-frame configuration dispenses products into the center collection belt, the channels of stacked product are exposed and accessible, making replenishment quick and easy. Replenishers simply roam the aisle looking for channels that might be low. The control system can also monitor each channel and identify a low product level to the replenisher. Reserve storage containers are usually located in flow racks (full cases) across the aisle (see Fig. 4-26).

FIGURE 4-25 Order transferred into waiting tote. (*Courtesy of SI Handling.*)

FIGURE 4-26 Dispen-SI-Matic system. (*Courtesy of SI Handling.*)

A-Frame Variations

While the standard A-frame order-picking system is designed for high volume and high speed, other variations can be provided for medium- and low-moving items. For instance, A-frame machines can be configured with double-deck (dual-tier) storage channels. Item density is increased by having two sets of channels, one mounted above the other. Less product can be stored in each channel, but there are more channels in the same linear space.

When throughput is less of a consideration, a traveling dispenser A-frame can be more cost effective. Rather than use a dispenser at each individual channel, as is the case with a standard machine, this configuration has one dispenser for each tier or bank of channels. The dispenser travels horizontally on a track beneath the channel and positions itself beneath the channel designated by the control system. It then dispenses the product onto the central collection belt (see Fig. 4-27).

The modular design allows different A-frame types to be combined to pick products of all movement categories efficiently. Standard A-frames (fast-moving items), traveling dispenser frames (medium-moving items), and multitier frames (slow-moving items) can all be integrated into a single machine.

"Dense" Automatic Picking Machines

"Dense" machines are modular, dispenser-type, automatic picking machines that are designed to have a large number of lanes in a small floor area. Rows of lanes are arranged in perpendicular or angled-back guides that allow items to feed by gravity to the bottom dispenser (one for each lane). Each row of lanes dispenses products onto a collecting belt or conveyor.

Like most automatic systems, this system is adaptable to products of various sizes and shapes, within certain limits. Replenishment is accomplished by pulling out a row of lanes and restocking each lane on that row manually. Dense machines are utilized in smaller distribution centers where space is of particular importance and where there is medium turnover of product.

Carousels

Horizontal and vertical carousels are a rotating set of shelves bringing the product to the operator, eliminating walk and search time, which accounts for as much as two-thirds of an employee's time. The vertical carousel, where the inventory moves vertically as with a ferris wheel, is a stand-alone unit providing high storage with a very small footprint. On the other hand, horizontal carousels are operated from the nose end and can be clustered, side-by-side, greatly increasing storage volume while providing a compact workstation and allowing dramatic cost efficiencies in certain applications such as split-case order picking. These carousels can also be stacked to utilize vertical space.

Either horizontal or vertical carousels can be utilized efficiently in a warehouse or distribution center operation. The implementation of a WMS allows for real-time communication with the carousel system to perform the following picking or put-away replenishment activities:

Order picking. The ideal environment for carousels is split-case picking in a distribution application. Here the carousels handle the middle-throughput range between static shelving and highly automated machines such as A-frames. Multiple pods can combine their efforts in a pick and pass or parallel pick system. A pick fence will split case, case, and pallet lots to be picked simultaneously and consolidated downstream.

FIGURE 4-27 Dual-level Dispen-SI-Matic. (*Courtesy of SI Handling.*)

Put systems.   By assigning each shelf for a specific customer or order, the carousel can be used to batch-pick SKUs from warehouse storage. The carousel directs the operator to pick the proper quantity on the shelves where that SKU is specified on the customer order. This is frequently used by retail distribution centers to sort by store and for returns processing.

Due to the moving parts with a carousel system, safety concerns need to be addressed. Carousel manufacturers provide a large number of effective safety features to help protect the operators, inventory, and nearby equipment. Most common are infrared photoelectric eyes or light curtains to protect the operator from coming in contact with moving bins. Weight-sensing safety floor mats are passive stop switches. Emergency stop buttons, pull cords, shut-down switches, audible alarms, safety markings, and fences may also be desirable depending on the application.

Horizontal Carousels.   Horizontal carousels can be driven by a motor located at the top or bottom of the carousel unit (see Fig. 4-28). The bottom-drive horizontal carousel offers the most potential in automating the distribution warehouse and its recent surge in popularity is partly due to its ability to be integrated into total systems using personal computers, microprocessor controls, and advanced sorting and picking devices, as well as other types of storage and retrieval equipment. Top-drive units are driven on and supported from a tubular track that in turn is supported by poles or stanchions. Wire bins are hung from trolleys connected by chain. The bins are guided by a bottom track. Bottom-drive units, on the other hand, have loads that are supported from a solid or tubular track welded to a bottom frame. The upper frame is supported by stanchions with an upper guide track holding the cantilevered bins in place. Typically, bottom-drive units are capable of handling heavier weights, are more dependable and easier to maintain, require less overall height, and spread the load over a greater floor area.

There is great flexibility in the design of the horizontal carousel, which is available in many sizes, configurations, and accessories. First, the bins determine much of the character of the carousel through their height, width, depth, total quantity, and spacing (see Fig. 4-29). The bins are like static shelving, except they move to the operator. They are generally made of wire that is lightweight, inexpensive, and easily adjustable. Sheet metal is sometimes added for rigidity. The only limitation is that all bins on each carousel are the same size. Widths typically range from 14 to 36 in, heights from 5 to 12 ft, and depths from 7 to 30 in. Weight capacities range up to 2000 lb and more per bin. The number of shelves is limited only to the spacing of the cross-wires used in the bins. Shelves are adjustable without the use of special hardware.

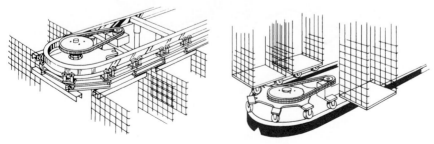

FIGURE 4-28   Top and bottom carousel drives. (*Courtesy of White Storage & Retrieval Systems, Inc.*)

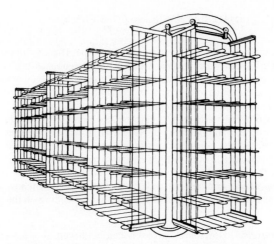

FIGURE 4-29   Carousel bins. (*Courtesy of White Storage & Retrieval Systems, Inc.*)

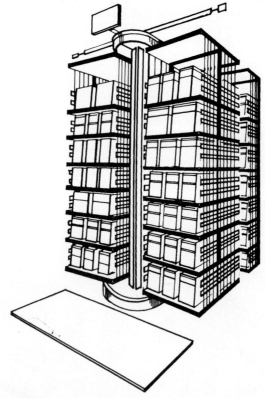

FIGURE 4-30   Dual carousel bins. (*Courtesy of White Storage & Retrieval Systems, Inc.*)

A major innovation is the dual bin, which allows the worker to pick from both the left and right sides of the bin, rather than the front (see Fig. 4-30). This doubles the volume that is available for picking at any given stop.

The second design factor is the drives, which consist of an individual motor, speed reducer, roller chain, pulleys, and sprockets. Alternating-current (AC) drives are generally used for continuous-use or very-low-use applications, such as progressive assembly. Direct-current (DC) drives provide improved stopping accuracy and smoother stops and starts. They are usually required for use with robotic interfaces and computer controls. Dual drives are often required for heavier loads or longer carousels.

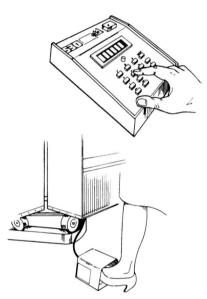

Another design factor is the range of control options available on carousels. The control options are limited only by the application. Foot-pedal controls, the most basic, allow the operator to move the carousel left or right. The operator keys in the required bin number on the keypad and a microprocessor selects the shortest route (see Fig. 4-31).

Most horizontal carousels are controlled by personal computers, either as a stand-alone or on a network integrated to the WMS (see Fig. 4-32). They have the ability to keep the item location in memory so operators simply type or scan the product number and the appropriate bin is delivered. Standard software allows the computer to keep inventory data, select storage locations for new items, batch orders for commonality, and minimize operator wait time, and performs many other tasks.

FIGURE 4-31   Keypad and pedal controls. (*Courtesy of White Storage & Retrieval Systems, Inc.*)

During the picking process, orders or requests are downloaded by the WMS to the carousel PC where the software sorts and creates batches. Sortation can be based on storage location, priority, truck route, date required, or any other criteria. The PC controls the carousel and the light display to allow a paperless environment. The light display is usually installed between carousels and displays the pick quantity at the appropriate location. While the operator picks from one carousel, another is rotating into position for the next pick (see Fig. 4-33).

Last is the carousel work center also known as the pod (see Fig. 4-34). The pod, a major factor in maximizing the carousel automation, is a picking center at the nose end of one or more carousels. Its configuration depends upon the pick rate required, the number of different items, inventory levels, packaging, and other factors. The pod typically serves two to four carousels under the command of a single PC and may integrate a lightree, sort bar, bar code scanner, manifest printer, bar code label printer, incoming and outgoing conveyors, a pick table or matrix, and, perhaps, a lift table (see Fig. 4-35). Using a carousel picking pod, pick rates can run from 200 up to 1200 line items per hour.

FIGURE 4-32 Carousel PC and sort bar. (*Courtesy of White Storage & Retrieval Systems, Inc.*)

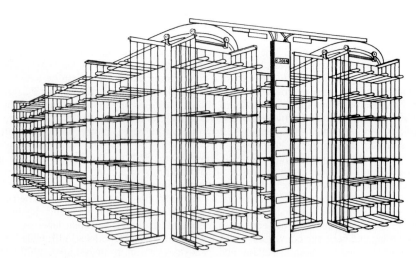

FIGURE 4-33 Dual carousel bin with light display. (*Courtesy of White Storage & Retrieval Systems, Inc.*)

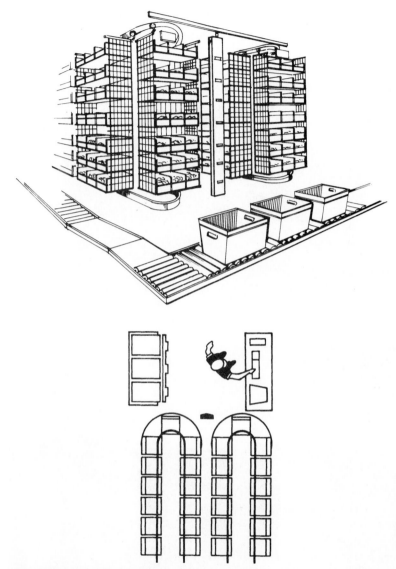

FIGURE 4-34 Carousel pod layouts. (*Courtesy of White Storage & Retrieval Systems, Inc.*)

Vertical Carousels. The use of vertical carousels (see Fig. 4-36) in distribution applications is limited because of their width. In a pod environment, there is simply too much space required to configure an efficient workstation. The operator must travel the width of the carousel to perform a pick, whereas with the horizontal carousel the operator needs only a few steps to perform the operation. There are applications, such as

FIGURE 4-35 Horizontal carousel system. (*Courtesy of White Storage & Retrieval Systems, Inc.*)

FIGURE 4-36 Vertical carousel system. (*Courtesy of White Storage & Retrieval Systems, Inc.*)

low-volume pick-to-belt and pick-to-cart types, where vertical carousels might be appropriate, allowing users to take full advantage of the vertical carousel's height and security characteristics.

Pallet Carousels. Another type of carousel is the pallet carousel. These use just a base plate at the bottom to hold pallets as shown in Fig. 4-37. This eliminates aisle space and is particularly advantageous for freezers and other environmental enclosures where cases or large items are picked from pallets.

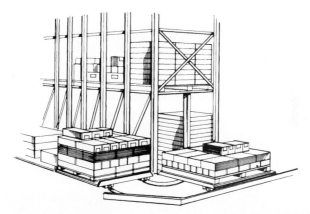

FIGURE 4-37 Pallet carousels. (*Courtesy of White Storage & Retrieval Systems, Inc.*)

Carousel Benefits. In addition to high picking rates and sequenced and batch picking, carousels provide fast replenishment and hot pick capabilities. These machines have an inherent simplicity, reliability, durability, and low routine maintenance.

Carousel operators can be easily supervised and fatigue can be sharply reduced because operator travel time is reduced and the pick rate per hour is more consistent than that with manual pick systems. Put-away operations and cycle counts can be interspersed with order picking for variety.

Today's carousel designs allow the reengineering of the warehouse process to maximize automation, giving convenience and time savings to downstream activities. Carousels integrate easily with warehouse management systems, increasing the efficiency and accuracy of order picking and inventory management.

Pick-to-Light Systems

For order selection or picking of goods requiring less than full-case quantities, pick-to-light systems not only increase productivity, but also order accuracy. Conventional split-case or break-pack order selection uses printed order lists to direct the pick operator. Much time is wasted on such unproductive tasks as reading, searching, and writing. Pick-to-light systems automate these unproductive routines.

Pick-to-light technology is especially useful for order-selection activity of moderate to moderately fast moving items. The technology can be used for very slow moving items, but cost is generally a prohibitive factor. The full-function electronic readout units are quite expensive. There are optional readout units, with less functionality at the readout unit level, which are quite a bit less expensive. Even so, putting electronic readout units on every SKU in the warehouse can be a sizable investment. Other technologies such as RF-directed order selection or robotic order selection are better suited for slower-moving products in the warehouse.

A pick-to-light system is a computer-directed, electronic order-selection technology. It uses a series of electronic pick "faces" attached to the shelf of either the static rack or flow rack where a product is stored for order selection. A pick-to-light flow rack is shown in Fig. 4-38. This electronic pick face (also known as a *readout unit* or

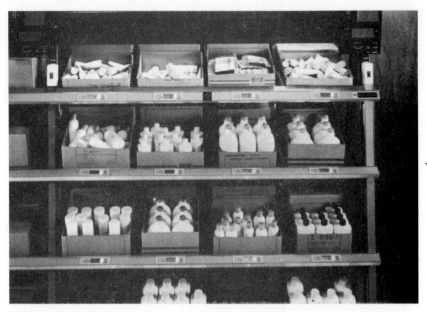

FIGURE 4-38   Pick-to-light flow rack. (*Courtesy of Rapistan-Demag Corporation.*)

slot display) is connected to either a microprocessor or microcomputer that contains the order information. The pick-to-light computer, in turn, is itself integrated to the WMS, which manages the overall activities inside the warehouse.

The pick-to-light computer manages the break-pack order-selection activity. Pick operators are usually requested to "log in" to the system. In this way, accurate management reports on picker productivity, accuracy, workload planning, zone sizing, and other management information can be obtained and used for operational improvement.

The pick-to-light computer begins order-selection activity by prompting the pick operator to choose an order, either the default order or a particular order number, or scan a bar code label representing the order number. The computer sends a signal to the readout units on which goods must be selected for the particular order. The readout unit uses a light to obtain the order selector's attention that an item is to be selected. It also displays the quantity of items to be picked. There is usually some means of communicating back to the WMS that either the pick is complete as directed, or that the pick is incomplete because that particular item is out of stock.

The pick-to-light system can also direct the replenishment activity of the break-pack area. When the system is initially brought on line, the pick-to-light computer asks for the amount of inventory loaded at each slot location. As orders are completed with the system, the computer automatically decrements the picked amount from the inventory on hand. When the amount of inventory reaches a set replenishment point, the pick-to-light computer tells the WMS to issue a replenishment order. In turn, the warehouse management computer may also issue a purchase order for more product if the quantities in the warehouse have hit a set reorder point.

There are a number of pick-to-light systems available, each with its own system features. Flexibility and cost are commonly the two major considerations buyers are concerned with when purchasing a pick-to-light system. A wide range of readout devices, flexible zoning, and good comprehensive software are features to look for when choosing a pick-to-light system.

Automatic Storage and Retrieval Systems

The automatic storage and retrieval system (ASRS) is a combination of equipment and controls that handle, store, and retrieve materials with precision, accuracy, and speed under a defined degree of automation. Systems vary from relatively simple, manually controlled order-picking machines operating in small storage structures to giant, computer-controlled storage and retrieval systems totally integrated into the distribution process.

Because the ASRS is such a combination of equipment, controls, and structures, there are important design issues that need to be addressed. Any ASRSs design starts with an analysis and definition of the loads to be handled, the number of loads to be stored, and both a material flow description and a description of the operation. Additionally, consideration must be given to certain architectural and engineering issues such as: Will the rack structure support the building or will it be free-standing? Are there environmental considerations, for example, temperature and humidity? What are the fire, seismic, and building codes that affect the entire installation? Once these issues are addressed, then the more mundane "component" selection process can begin.

More and more distribution center managers are finding ASRSs to be economically justifiable, especially when they are integrated into a totally automated facility. Better control of material and its flow helps to reduce inventory and improve labor productivity by reducing excessive and repetitive handling, as well as better utilizing the physical area of the building, thus freeing up floor space or reducing real-estate requirements. Other benefits ASRS users have received are the reduction of product loss through storage errors, damage, or theft, improved safety, security, and work environment, and saved energy consumption.

As noted above, there are both manually controlled as well as automatically controlled SR machines. Manually controlled SR machines operate on the "man-to-goods" order-picking principle as shown in Fig. 4-39. A person aboard the SR machine is taken to the appropriate stock location for retrieving a particular item. The automation eliminates seek time as well as improves the ergonomics of bending and stretching that usually accompanies a conventional manual pick from rack. Greater efficiency can further be achieved by using load-handling devices such as ball-caster tables, roller tables, loading forks, or clamping devices.

More common today are the automatic SR machines that operate on the goods-to-man principle shown in Fig. 4-40. Automatic SR machines not only store goods in a rack structure but retrieve the goods and bring them to an end-of-aisle workstation for order selection.

Commonly there are two major classes of automatic SR machines based on the weight-carrying capacity of the SR machine: unit load and miniload. Unit load SR

FIGURE 4-39 Man-to-goods ASRS. (*Courtesy of Rapistan-Demag Corporation.*)

FIGURE 4-40 Goods-to-man ASRS. (*Courtesy of Rapistan-Demag Corporation.*)

	Microload	Miniload	Unit load	Man aboard
Maximum payload	55 lb	100 to 750 lb	750 to 11,000 lb	600 to 2200 lb
Maximum mast height	20 ft	20 to 40 ft	130 to 145 ft	40 to 50 ft
Minimum aisle width	30 in	30 to 37 in	60 in	48 to 52 in
Lifting or lowering speed	60 to 100 ft/min	60 to 100 ft/min	60 to 200 ft/min	40 to 60 ft/min
Travel speed	525 ft/min	400 to 525 ft/min	200 to 525 ft/min	210 to 275 ft/min

FIGURE 4-41   ASRS summary characteristics.

machines generally handle loads greater than 750 lb and up to 5 or 6 tons. Miniload machines, on the other hand, handle loads less than 750 lb and as small as 50 lb (usually referred to as a "micro" or "micromini" load). As a general rule of thumb, the smaller the payload, the faster the cycle time of the SR machine. Figure 4-41 summarizes the characteristics of various ASRSs.

There are two levels of controls that need consideration in designing an ASRS; namely, the machine level controls and system controls. These need to be designed for the highest level of reliability and should be backed up by emergency control levels, including manual operation. The system level of controls oftentimes interfaces with the WMS. The ASRS may randomly choose a slot location for the goods, which it reports back to the WMS and updates the warehouse inventory.

The WMS also downloads the goods to be picked to the ASRS system controller. The ASRS system controller retrieves the goods to be picked and reports back to the WMS the amount picked, short, and/or backordered. The warehouse management system can in this way keep accurate inventory counts, reorder replenishment stock when levels fall to set reorder points, and create the necessary shipping manifest and billing documents to keep track of the goods movement.

Radio Frequency–Directed Picking

Radio frequency (RF) picking is a flexible option since mechanization is not required, and it can be used with any type of lift truck and storage rack. The RF terminal communicates with the WMS and presents the operator with pick directions. Fast pick rates are difficult to achieve since the RF communication with the WMS requires 1 to 2 seconds for transmission and processing time. However, RF picking is an excellent application where travel and handling time is required since the communication time can occur while the operator is traveling or actually performing the pick, thus masking the RF communication time.

RF terminals can be mounted on lift trucks, carried by pick operators, or attached to the operator to provide "hands-free" picking. Chapter 3 describes this technology further. Combined with a WMS and consolidated picking process, RF picking can provide an efficient, flexible pick operation with a relatively small investment. Figure 4-42 shows a RF terminal, mounted on a lift truck, directing a picking operator.

FIGURE 4-42   RF lift-truck picking. (*Courtesy of Sharp Electronics Corporation.*)

Robotic Order Selection

Robotic order selection is best suited for very slow moving items. In keeping with Pareto's principle, most warehouse operations see 80 percent of the SKUs account for only 20 percent of the items selected. And of that 80 percent of moderate- to slow-moving items, 50 to 65 percent can be defined as "slow-moving" product. This means that order selectors can spend a lot of time walking to find one or two items that are occasionally needed.

The robotic order selector is ideal for this latter class of product. The robot is a picking device, most often fitted with a vacuum head-end effector, mounted to a small automatic storage and retrieval (ASR) crane as depicted in Fig. 4-43. Quickly and steadily moving up and down aisles of slow-moving merchandise, the robotic order selector picks the required item(s), puts it into an onboard carousel collector basket, and delivers the goods to a waiting tote or carton at the end of the aisle.

Tied to the WMS, the robotic order selector's computer receives orders in much the same fashion as the pick-to-light computer does. But rather than illuminating readout units, the robotic order selector's computer directs the crane to move into position along the rack to the storage location of a particular item or SKU. The item is oriented to the robot by means of a simple magazine or tilted shelf, which places the item to be picked in a consistent orientation and position.

Once in position, the robotic selector end effector picks one item from the rack and places it into its onboard storage carousel. The storage carousel allows the robotic

FIGURE 4-43   Robotic order selector. (*Courtesy of Rapistan-Demag Corporation.*)

order selector to pick several orders at the same time (batch pick). Thus, if the same SKU is required to fill more than one order, the robotic order selector can pick multiple quantities of the same item, putting each item into the appropriate bin in its onboard storage carousel.

The robot reports back to its area computer whether the pick was successful, and the robotic order selector computer, in turn, reports to the WMS. Similar to the pick-to-light and carousel systems, the robotic system is capable of inventory tracking and replenishment functions.

Robotic order selection works well with a variety of items. However, each item considered a candidate for robotic order selection should be tested for its "pickability," i.e., can the robotic end effector consistently retrieve the item? Certain packaging characteristics, such as loose polybags, irregular shapes, and inconsistent sizes make an item either impossible to be picked up by the end effector or unsuitable for the magazine dispenser.

STORAGE MEDIUMS

In most projects, the storage and throughput requirements are defined based on the proposed process and projected volumes and then the goal is to optimize productivity, storage density, and capital cost. The following factors are considered when deciding the appropriate storage medium for the warehouse.

Product Characteristics. The unit load dimensions of the various products to be stored must be known. The methods used to support unit loads include pallet board, skid, slip sheet, tote, pan, or other storage containers. This unit load support method is an important ingredient in determining the storage medium and the unit load handling equipment. The product mix must be determined next. What percentage of the receipts and shipments are full pallet loads, full cases, or broken cases? These product types dictate specific types of storage racks. For example, if there is a group of fast-moving products, then dense storage systems such as double-deep rack, drive-in rack, or floor stacking are good options for reserve storage. If many products are slow movers, then a high degree of selectivity is necessary and single-deep rack and narrow aisles must be considered. Does the product have a specific shelf life? Does the product require special handling and access such as chemicals or hazardous materials? Is first in, first out (FIFO) inventory control required? These types of products require easy access to every product at any given time and affect the storage selection.

Material Handling Equipment. The type of material handling equipment used to service the storage medium affects the selection. The selected rack should not require the forfeit of square footage to accommodate the material handling equipment. Lift trucks have different turning radii that define the operating aisle widths and clearances required. Usually, a storage medium is chosen and then a compatible lift truck is selected. If there is a requirement to use existing equipment or the equipment is selected first, this information must drive the storage-medium analysis.

Throughput Requirements. The amount of unit loads required for storage and retrieval each hour or shift is calculated. As was discussed in Chap. 1, the trend is toward having real-time information and EDI available in order to increase inventory turns and allow functions such as cross-docking. It is important to take this into consideration as it reduces the amount of storage medium required. Storage itself should not be the objective but rather the movement of inventory through the distribution center. The movement of inventory to customers makes the company money, not the amount stored in the warehouse.

Layout Considerations. The storage medium is constrained by building obstacles and layout requirements, including ceiling heights, columns, air ducts, light fixtures, floor specifications, sprinkler systems, and other elements. Standards from the Occupational Safety and Health Administration (OSHA), American National Standards Institute (ASNI), or Rack Manufacturing Institute (RMI) may apply to the project. For example, areas prone to seismic activities require storage racking designed to a special set of specifications. Another factor to consider is aisle widths. Within each storage medium, the aisle widths can be varied to accommodate the level of activity. Is it important for lift trucks to be able to pass each other? What lift trucks will be operating in these aisles? How many lift trucks are required to meet the throughput rates, especially during peak periods? If the configuration is becoming complex, it might be a good candidate for simulation modeling in order to minimize the amount of rack and lift trucks.

Budget Considerations. Various alternatives are developed during the analysis process that all work well. A clear winner is not always the result. Compromises must be made in an effort to keep the system as simple as possible and still meet the project objectives. It is necessary to decide what is important in terms of the investment required and the benefits obtained. A phase-in layout plan is a practical alternative that allows the storage rack to be purchased and installed in phases as volumes dictate.

This reduces the initial capital investment and allow expenditures when additional storage rack is needed.

Below are various storage mediums, their advantages and disadvantages, and issues to consider when selecting these systems. The storage-analysis data is obtained from the proposed warehouse automation processes and future volume growth projections.

Floor Stacking

Floor stacking or block storage is storing multiple unit loads in a designated floor location that is multiple levels high, as shown in Fig. 4-44. Good candidates for floor

Advantages	Disadvantages
Avoid cost investment in racking	Cannot sequence products
Flexibility—change storage configuration	Requires investment in lift trucks, possibly with special attachments such as side clamps or barrel clamps
Good space utilization	

FIGURE 4-44   Floor stack locations. (*Courtesy of Sharp Electronics Corporation.*)

stacking are large quantities of the same SKU, where FIFO, lot control, and expiration control are not stringent and where the product itself or its packaging will support the weight of stacking. Examples of products that could lend themselves to floor stacking are major appliances, beverages, and dry groceries. Floor stacking provides flexibility in changing storage and aisle configurations easily when product types and/or volumes fluctuate.

The floor stack height is limited by ceiling height, lift-truck mast height, or the maximum stacking weight allowed by the product and its packaging. Analysis of these factors determines whether floor stacking or a racking alternative is required. Each floor stack location requires a unique location number or reference ID that is used by the WMS for tracking FIFO date, lot number, and expiration date. One difficulty with floor stack locations is determining where to affix the bar code location number for scanning into the WMS by the operator. Since the life of bar coded labels on the floor is short, it is recommended that to maintain accurate inventory location records, a reference ID is also affixed to the location. The operator can then key in this ID instead of scanning the location bar code, and then the WMS software translates this ID to verify the correct location.

In order to maximize the floor stack utilization, the WMS tracks the location's capacity by cubic feet to determine available space within a location. Products stored in floor stack locations are interchangeable when it becomes time to retrieve the product for order filling or replenishment. Lot mixing is usually not allowed within a floor stack location since unique lots are difficult to distinguish and track.

Standard Selective Pallet Rack

The standard pallet rack is designed with single rows or back-to-back rows and multiple levels as shown in Fig. 4-45. The depth and width are defined by the unit load footprint, usually a pallet, and the height is adjustable to conform to varying unit load heights. There are more applications of standard selective pallet racks than any other storage system, because this system is least expensive and is the simplest to design and configure. With an aisle between each row, this arrangement provides excellent pallet load accessibility but has a low storage density. The amount of aisles and single-deep density results in decreased warehouse space utilization.

The location capacity for a standard selective pallet rack is usually defined in the WMS based on the number of storage containers that can be stored in the rack. FIFO inventory control and picking, lot number, and expiration date tracking are allowed due to the accessibility to all locations. A unique location number for each pallet position is required for the WMS to include FIFO and lot control logic in the put-away and picking algorithms. In addition, rack weight constraints and environmental considerations can be managed with the WMS. For examples, batteries should not be stored on upper levels due to the danger of potential acid leakage. An environmental flag is stored with the SKU record to ensure proper storage locations are identified for put-away operations.

Double-Deep Rack

A double-deep rack is basically a selective pallet rack that is two pallet positions deep. The main advantage of this rack is density of storage because there are fewer aisles. The primary disadvantage is that a reach lift truck is required. This rack is loaded and unloaded from the front, requiring stocking and picking activities in the same aisle (see Fig. 4-46 for a summary of advantages and disadvantages).

The double-deep rack is used in projects that require moderate picking throughput. When additional storage capacity is needed and expansion space is limited, an existing selective rack can be replaced with a double-deep rack, which increases storage capacity by decreasing the number of aisles within the warehouse. A double-deep rack decreases picking efficiency if different SKUs are stored within the same location. If the back pal-

Advantages	Disadvantages
Low costs	Lower space utilization
Pallet selectivity	Wide aisles
Flexibility of lift trucks	
Load or unload from either direction	
Simple installation	

FIGURE 4-45   Standard selective pallet rack. (*Courtesy of Verbatim Corporation.*)

let is required, the front pallet must be moved to another location or brought down to the floor and then back to the location after retrieving the back pallet. The WMS must control the utilization of these locations to minimize double handling and honeycombing caused by partially filled locations or the inability to utilize the entire location. Again, location capacity is usually defined by the number of storage containers.

Advantages	Disadvantages
Increased space utilization	Special lift truck required
Good aisle-to-rack ratio	Last in, first out picking from location
	Less selectivity
	Crossbars needed

FIGURE 4-46   Advantages and disadvantages of double-deep rack.

Drive-In or Drive-Through Rack

A drive-in rack gives a tunnel effect where the lift truck drives in or through the storage rack. Figure 4-47 pictures a warehouse drive-in rack. The lift-truck operator drives into the rack and places the unit load in the innermost position, filling all positions

Advantages	Disadvantages
Very dense storage	LIFO rotation
Good space utilization	Safety (helmet required for drivers)
Fast in and out placement	Rack damage

FIGURE 4-47   Drive-in rack. (*Courtesy of Sharp Electronics Corporation.*)

until the rack is full. Since only the outermost loads are easily accessible for picking, last in, first out (LIFO) inventory rotation results. Primarily, this type of storage mode is used in a distribution center for medium- to fast-moving items that have nonstackable products and where FIFO, lot control, and expiration date are not required. For instance, if all that is being stored is one model or one lot number of microwaves and selectivity is not needed, then this may be a candidate for a drive-in rack.

The WMS logic can utilize a FIFO window to compensate partially for the LIFO disadvantage. A FIFO window is a range of dates that allow products to be stored together and treated as one FIFO date. Once outside this date range, a different location is required to store the product. The WMS must support the allocation of inventory to orders from a location level versus allocation from specific storage containers. The operator must be allowed to retrieve the closest storage container from the location to perform the assigned tasks; otherwise unnecessary moves are performed. The grocery industry or manufacturing storage and distribution is a good example where this type rack is utilized, especially in freezer storage for which maximum space utilization is needed. In the industry, push-back racks and pallet flow racks are being used as alternatives to drive-in and drive-through racks.

Push-Back Rack

As the newest style of storage rack, there are a growing number of push-back rack applications in the industry. Push-back racks can be used as a replenishment reserve storage medium for pallet flow or carton flow racks. The lift-truck operator brings a new unit load to the location and pushes against the existing load until space is made to drop the new unit load. The push-back rack contains either pallet gravity-flow rails or a nesting carriage that rides on two rails as shown in Fig. 4-48.

Advantages	Disadvantages
Improved selectivity	LIFO rotation
Maximize space	Higher cost
Reduce rack damage	
Increased productivity over drive-in rack	
Specialized forklift not required	
Carriage style does not require good pallets	

FIGURE 4-48   Push-back rack. (*Courtesy of Unarco Material Handling.*)

Push-back applications are usually two, three, and four loads deep. Some manufacturers make a deeper style, but fork-lift manufacturers do not guarantee the pushing capacity of the fork lifts since they are not made for this purpose. Maximizing the efficiency of this storage medium requires one SKU per location where FIFO, lot, and expiration control are not required. Again, the WMS can use a FIFO window to help offset the LIFO disadvantage. In addition, the WMS must control the utilization of these locations to minimize honeycombing caused by partially filled storage containers. The location capacity is usually calculated by the number of storage containers that fill the locations. As with drive-through racks, the WMS must be able to allocate inventory at the location level to allow the operator to perform efficiently and not perform extra moves.

Pallet Flow Rack

The pallet flow rack is a good application for full-case picking as shown in Fig. 4-49. A pallet flow rack contains at least two unit loads and more depending on velocity movement. The lift-truck operator deposits the pallet at the rear of the rack, and the pallets, sitting on skate wheel conveyor, flow to the front of the aisle. Since replenishment is from the rear of the rack, there is no congestion in the order-picker aisle. Proper installation is imperative so the system operates correctly with pallets flowing at a safe rate to the front.

The WMS calculates capacity as the quantity of storage containers per slot within a pallet rack lane. Since these racks are used primarily for medium- to high-velocity items, one SKU is allowed per pallet flow rack location, which is tracked by a unique location number in the WMS. To ensure timely replenishment and avoid stock depletion, the WMS keeps track of the location inventory at the pick unit level, typically cartons. The WMS tracks all carton picks from each pallet to identify when pallets become empty and then, based on the location replenishment trigger level and replenishment quantity, triggers the pallet replenishment for the location.

Carton Flow Rack

Commonly referred to as a gravity flow rack, this medium is ideal for break-pack, split-case, or less-than-full-case order picking. Profiled correctly, this medium can increase productivity by over 30 percent. Carton flow racks better utilize space and order-picker productivity than static shelving for medium- to fast-moving products. For products that move very slowly to slowly, modular drawers or shelving may be more cost effective. The carton flow rack is replenished from the rear, which allows order picking and stocking to take place simultaneously and eliminates any congestion between them (see Fig. 4-50).

As with the pallet flow rack, each carton flow location contains a unique location number and only one SKU is stored in each location. There are various gravity rack front configurations that allow different picking options as shown in Fig. 4-51. The flow lanes contained within the rack are adjustable to allow flexibility when carton sizes change and to maximize the number of carton flow lanes within a rack. Similar to the pallet flow rack, the WMS tracks total cartons or pieces to identify when the location requires a replenishment and triggers a carton replenishment quantity for the location. Also mezzanines, as drawn in Fig. 4-52, can be installed for efficient space utilization with carton flow racks.

Advantages	Disadvantages
FIFO rotation	More costly
Better space utilization	Good pallets needed
Better throughput and productivity	Potential safety hazard—difficult to handle large weight variances

FIGURE 4-49   Pallet flow rack. (*Courtesy of Unarco Material Handling.*)

Shelving

Shelving is normally furnished in widths from 24 to 48 in in increments of 6 in and ranges in depth from 9 to 36 in. Shelving units usually range in height from 3 to 10 ft but can be purchased as high as 20 ft in one-piece high-rise shelving. Shelving is used for small- and broken-carton storage and to pick small slow- to medium-moving products. It can be open or closed and the shelf can be subdivided by use of vertical

Advantages	Disadvantages
FIFO rotation	Flow is important so track and wheels need to be dependable
Pick face density	
Order-picker productivity	High capital investment
SKU access	Flow lanes hold small inventory quantity

FIGURE 4-50   Carton flow rack. (*Courtesy of Unarco Material Handling.*)

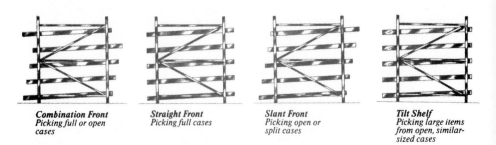

Combination Front
Picking full or open cases

Straight Front
Picking full cases

Slant Front
Picking open or split cases

Tilt Shelf
Picking large items from open, similar-sized cases

FIGURE 4-51   Gravity flow rack pick fronts.

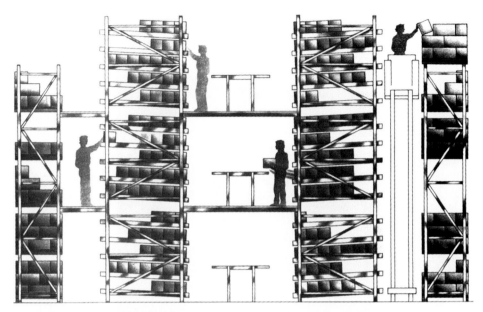

FIGURE 4-52 Mezzanine flow racks.

dividers. Bin compartments, drawers, and other accessories can be added to store products in a specific area. As the number of SKUs increases, walking time becomes significant, reducing picking productivity (see Fig. 4-53).

Shelving capacity is limited by shelving size, and usually only one SKU is stored per shelf. If any more shelves are required per SKU, then a carton flow rack could be more efficient. At a minimum, a unique location is assigned to each shelf level and sometimes to individual compartments accomplished by slots in the location ID on the shelf depending on inventory value or accuracy required. Shelving storage allows FIFO inventory control as well as lot number and expiration tracking. To increase productivity, the WMS must have the ability to group shelving locations by velocity and match the SKU velocity to the shelving location velocity or reserve storage. Typically, the middle shelves are assigned the highest velocity and the top and bottom shelves the lowest velocity. Then fast-moving SKUs are assigned to the high-velocity shelves where operator reach is minimized. The capacity of the storage space is tracked in the WMS by volume.

Cantilever Rack

Certain products require a specialized storage medium. Cantilever racks have arms that allow the storage of long awkward products such as tubular steel, pipes, and lumber (see Fig. 4-54). Furniture and boat companies are good examples where the cantilever rack is widely used.

This type of storage medium presents unique challenges to a WMS. Locations are defined by the smallest product volume or length to maximize storage capacity. As the

Advantages	Disadvantages
Versatile—many kinds of products	Walking time can become significant
Low equipment cost	Aisle interference between replenishment and picking
Many configurations and sizes	
Durable	

FIGURE 4-53   Shelving storage. (*Courtesy of McKesson Corporation.*)

product size varies and requires more than one consecutive location for storage, the WMS must have the ability to assign multiple locations to the one product based on the location and product size parameters. Since the directed put-away logic can get complex, an alternative approach is to allow the operator to determine the storage location for the inventory, referred to as *assisted put-away.*

LIFT-TRUCK EQUIPMENT

Selecting a lift truck best suited to the operation requires knowledge of a variety of lift-truck concepts and an understanding of the business processes. Today's lift trucks

FIGURE 4-54   Cantilever rack. (*Courtesy of Unarco Material Handling.*)

range from the simple hand pallet truck to a highly sophisticated turret truck, offering an array of options for operations people trying to increase productivity and decrease costs. The WMS must be designed to direct or assign work properly based on an operator's assigned truck. As trucks become more sophisticated, WMS control helps maximize utilization and thus increase the return on investment. The following factors are important lift-truck evaluation considerations:

Type of facility.   Is this an existing or new facility? If lift trucks are required for an existing building, the constraints of ceiling heights, current storage methods being used (e.g., racking), location of support columns, floor type, and other factors must be considered. A new building should be designed to accommodate the lift-truck equipment, storage medium, and material handling equipment that optimize the process.

Processes involved.   What type of activities will take place in the proposed process? Is there a dock for tractor trailer loading and unloading? Is there a staging area for materials? What is the weight of the heaviest loads? What is the ratio between a full pallet load and smaller loads (case picking or piece picking)? How will orders be filled? Will the operation handle both pallet handling and order picking? What type of storage will be used? How many levels high will inventory be stored? Determine which equipment from the current lift-truck fleet can meet the new processes and which operations will require additional equipment.

Productivity.   Analyze the amount of time required for tasks commonly done in the warehouse operations to help select the best lift truck. If a few seconds can be eliminated from a standard operation, many dollars can be saved over the course of the year. Productivity is one of the most important criteria in picking a lift truck.

Ergonomics. The fit between the operator and the machine is called ergonomics. Operators who can vary their stance and the position of their fingers and hands and who can see the load they are handling without neck or back strain are more comfortable, more productive, and less prone to cumulative trauma disorders and other injuries (see Fig. 4-55).

FIGURE 4-55 Ergonomically designed operator compartment. (*Courtesy of Raymond Corporation.*)

Space utilization. This is particularly important if the distribution facility is located in an area where real estate is especially costly. If there is limited real estate available for expansion, narrowing aisles, stacking higher, and selecting the right lift truck can "buy" the additional storage space required for growth.

Resources. The amount budgeted for a lift truck should include maintenance and operator training costs in addition to the purchase price. Also evaluate whether buying, renting, or leasing provides the best payback.

Hand Pallet Truck

The earliest lift trucks were simple, hand-operated models with two forks to move under a palletized load. The operator then pumped up the load with hand-operated hydraulics and pushed the load to the proper location. These trucks, still available today, are called hand pallet trucks and represent the most basic type of lift truck. Hand pallet trucks are generally a three-wheeled design and are used to transport lighter loads in operations with very limited space and floor storage.

Powered Pallet Truck

The next level up is the powered pallet truck or "walkie." Walkies got their name from the operator's ability to walk behind or beside the trucks as they move pallets to a storage location (see Fig. 4-56). Some walkies also have a platform for the operator to stand on-board and ride to the location. Walkies are a good solution for low-level order picking from floor storage where pallets must be transported a great distance. The walk or ride models are generally used in applications with long horizontal travel distances. Walkies generally supply 2 to 6 in of lift and are electric battery-powered.

FIGURE 4-56   Walkie truck. (*Courtesy of Raymond Corporation.*)

Counterbalanced Truck

When storage racks are required for increased space utilization and pallet accessibility, then the next level of trucks must be considered, starting with the counterbalanced truck (see Fig. 4-57). The typical counterbalanced truck requires an approximate aisle width of 12 ft, with stack height limited to four or five levels. This results in limited cubic space utilization and limited rack faces. A building with a 22-ft clear height, 180-in top beam, 92-in beam length, and 144-in aisle width is required for the typical

FIGURE 4-57   Counterbalanced truck. (*Courtesy of Raymond Corporation.*)

counterbalanced truck. This truck can stack loads weighing up to 2500 pounds, four levels high, allowing 188 in of lift height. Such a system provides storage for 1000 pallets in 10,060 ft^2 of storage space. Counterbalanced trucks are available in both stand-up and sit-down versions and are either internal combustion or electric-powered. Counterbalanced trucks are frequently used with attachments, such as carton clamps, to handle specialized loads.

Reach Truck

The next level of lift truck, the reach truck, has a scissorslike mechanism behind the forks that enables the forks to "reach" into a rack where a pallet is stored (Fig. 4-58). The reach truck operates in aisles of less than 8 ft, greatly improving the proportion of storage space to aisle space, and reducing travel distances and operation cycle times. A building with a 22-ft clear height, 180-in top beam, 92-in beam length, and 90-in aisle width is required for this lift truck. The reach truck can stack loads weighing up to 2500 lb four levels high. Such a system can provide storage for the same 1000 pallets in 7830 ft^2 of storage space.

   Taller reach trucks have 372 in of lift height and store seven levels high. Operating in a 104-in aisle, these trucks store 1000 pallets in just 4810 ft^2 of storage space.

FIGURE 4-58   Reach truck. (*Courtesy of Raymond Corporation.*)

Turret Truck

The turret truck provides even greater space utilization by working "both sides of the aisle" since the carriage can rotate the forks on the mast 180° (see Fig. 4-59). There is no need to exit the aisle and turn around, then return facing in the other direction. This is an important productivity feature. And since no right-angle stacking turns are required, the aisle can be as narrow as 5.5 ft. The building can have a 22-ft clear height, 180-in top beam, 92-in beam length, and 66-in aisle width. This truck can stack loads weighing up to 2500 lb four levels high, and with the narrower aisles 1000 pallets can now be stored in just 6840 ft². If the turret truck is a man-aboard or operator-up truck, the operator can do order picking of single cases or pieces as well as pick full pallet loads.

Deep-Reach Truck

The deep-reach truck provides access to pallets stacked "two deep," where one pallet load is stored behind another. Storage density increases since two aisles out of every

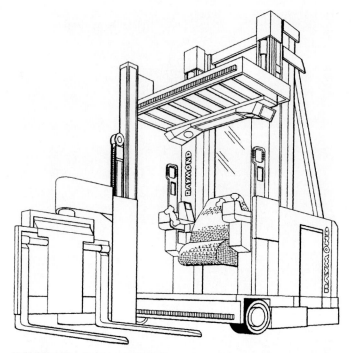

FIGURE 4-59   Turret truck illustration.

five can be eliminated. The deep-reach truck requires a building with a 27-ft clear building height, 238-in top beam, 98-in beam length (using the straddle method), and minimum 92-in aisle width. Stacking loads that weigh up to 2500 lb and are five levels high and two deep now allows storage for 1000 pallets in just 5120 ft^2 of storage space.

Taller deep-reach trucks feature 376-in of lift height and can store seven levels high. They operate in a 103-in aisle and store 1000 pallets in just 3800 ft^2 of storage space.

Hybrid Truck

Hybrid trucks, such as more sophisticated swing-reach trucks, can store from eight to thirteen levels high, using aisle widths ranging from 66 to 56 in and can store 1000 pallets in 3420 to 2020 ft^2 of storage space.

Order-Picker Truck

The order-picker truck is designed for high-level order picking (piece or case picking above the second storage level). The platform on the operator-up truck can be raised to the desired height required for order picking of cases or single items. They operate in narrow aisles and very narrow aisles where reach trucks or turret trucks are used for pallet put away (see Fig. 4-60).

FIGURE 4-60   Order-picker truck. (*Courtesy of Raymond Corporation.*)

Lift-Truck Options

Many lift trucks can be ordered with a wide range of options to provide customization for specific operations. Options include rail guidance or wire guidance, which allows automatic steering of the truck within the aisle so the operator can concentrate on other tasks. Trucks also can be equipped with on-board self-diagnostics to troubleshoot the truck automatically and alert the operator to problems. Radio frequency terminal brackets allow proper equipment mounting on the lift truck. In addition, wiring can be installed between the RF equipment and lift-truck battery to provide a continuous power source for the RF equipment. There are additional unique features designed to enhance productivity and operator comfort that can be discussed with the individual lift-truck vendors.

REFERENCES

1. Gattorna, John L., *The Gower Handbook of Logistics and Distribution Management,* 4th ed., Gower, London, 1990.

2. Mulcahy, David E., *Warehouse Distribution and Operations Handbook,* McGraw-Hill, New York, 1994.

3. Tompkins, James A., and Harmelink, Dale, *The Distribution Management Handbook,* McGraw-Hill, New York, 1994.

4. Apple, James A., *Plant Layout and Material Handling,* Krieger, Melbourne, Fla., 1991.

5. Meyers, Fred E., *Plant Layout and Material Handling,* Prentice-Hall, Englewood Cliffs, N.J., 1993.

CHAPTER 5
FUNCTIONAL CHARACTERISTICS

INTRODUCTION

In the process of designing a warehouse management system (WMS), one must consider each major functional area such as receiving, stocking, picking, and shipping warehouse inventory. The majority of functions discussed within this chapter utilize a real-time methodology to direct work within the warehouse. Batch systems and procedural methods are suggested when automation might limit productivity without significant process improvements.

Warehouse management and business requirements vary significantly across industries, across companies, and even within multiple warehouse environments. This chapter will discuss the basic functions that are common to most warehouse and distribution centers (see Fig. 5-1). Sample processes and design considerations are described in detail for use in evaluating "off-the-shelf" WMS packages or in designing a custom WMS solution.

OVERVIEW

The approach used in this chapter focuses on the design of a WMS that utilizes distributed processing to control warehouse inventory movement within the "four walls." The distributed WMS is designed to allow the end user (warehouse or DC manager) to control and monitor all inventory movement with their facility. The distributed WMS interfaces with the host system and all other corporate functions including purchasing, customer order entry, and billing.

In order to meet varied business and customer requirements, the proposed warehouse processes must encompass an array of functional areas. The following are the major topics and functional areas that will be considered:

    I. General philosophy of a real-time distributed WMS
   II. Why real-time automation?
  III. Basic requirements of a WMS
      A. Inventory creation
      B. Inventory tracking
      C. Inventory shipment
  IV. Basic master files and interfaces that are required to support a distributed WMS
      A. Receiving interfaces

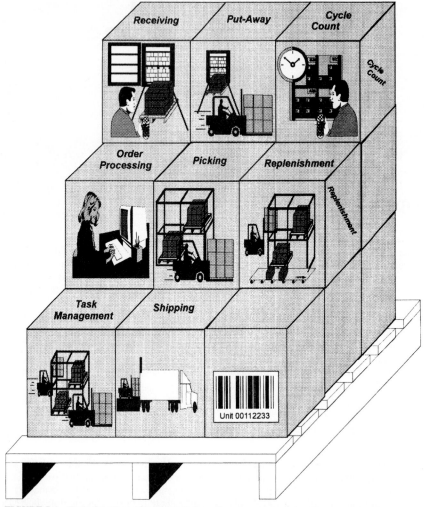

FIGURE 5-1   Basic functions of an automated warehouse management system.

> **B.** Common function interfaces
> **C.** Shipping interfaces
> **D.** Common master files
> **V.** Receiving and quality control
>   **A.** Receiving/scheduling log
>   **B.** Receiving unload
>   **C.** Receiving methods
>   **1.** Purchase order
>   **2.** ASN
>   **3.** Tentative
>   **4.** Manufacturing

    D. Receiving load
    E. Receiving adjustments
    F. Quality control
    G. Returns processing
VI. Put-away (stocking) and location management
    A. Inventory attributes
    B. Location attributes
    C. Put-away search hierarchy
    D. Put-away methods
        1. Directed
        2. Assisted
    E. Supplemental functions
VII. Cycle counting
    A. Queue generation
    B. Count verification
        1. Initial count
        2. Supervisor count
    C. Inventory adjustment
    D. Load initial inventory
VIII. Order allocation, picking, and replenishment
    A. Order processing
        1. Customer order download
        2. Order and wave planning
        3. Wave release and allocation
    B. Picking methods
        1. Individual order by RF
        2. Consolidated orders by RF and paper
        3. Label
        4. Interface
        5. Individual order by paper
        6. Assisted
    C. Replenishment
        1. Queue maintenance
        2. Directed
IX. Task management
    A. Multitasking
    B. Labor management
X. Customer compliance and shipping
    A. Pallet build
    B. Pallet transfer
    C. Carton packing
    D. Pallet wrap
    E. Load trailer
    F. Close trailer
    G. Parcel manifesting
    H. Assisted shipping

GENERAL PHILOSOPHY

Today's businesses are faced with the need to process more and more information to provide better customer service. A warehouse application running on a large corporate computer can provide the necessary inventory updates directly to corporate data bases while having direct access to purchase and customer orders. The amount of warehouse automation is, however, often limited in scope due to processing demands of other applications. The decision to automate warehouse processing and provide real-time feedback to corporate inventory files while improving customer order response times can significantly increase demand on an already burdened host computer. The cost associated with expanding the host computer center to meet these increasing demands can be prohibitive.

As such, the dominating trend in today's warehousing is to automate warehouse procedures using distributed processing and to utilize advanced technologies, such as radio frequency terminals and bar coding, to provide real-time informational control of inventory. Distributed processing provides the warehouse with a computer for tracking inventory that is separate from the host computer and thus not burdened with host computer application processing.

Distributed processing and real-time control bring significant challenges to an implementation team, such as communication and interfacing with the current host systems along with detailed understanding and documentation of the current and proposed warehouse operations and procedures. Further complications are encountered when changes that are implemented as a result of this design effort are met with reluctance by warehouse personnel.

Fortunately, low-end processors such as minicomputers and personal computers have dropped in cost while providing significant improvements in throughput. By developing a warehouse control system that has distributed processing at the warehouse, the host system need only grow to support corporate applications. A schematic of a distributed WMS is illustrated in Fig. 5-2. The host system computer can vary significantly from a mainframe computer, to a minicomputer, or even a microprocessor. The advantage of a distributed WMS design is that once the warehousing application is developed, multiple warehouses can be added with minimal impact to the host system and the distributed processors can be sized to the needs of each warehouse.

Providing real-time updates to host data bases is still a requirement with a distributed system. The distributed warehouse management system must be developed with interface programs that can provide frequent updates of inventory and order information to the host applications. With this comes the need for communication controllers and lines that can support timely updates to the host data bases.

WHY REAL-TIME WAREHOUSE AUTOMATION?

Batch systems often fall victim to inaccuracy due to preprocessing of information. In receiving, a batch system will often receive an entire purchase order as ordered and assign stocking locations in advance. If the purchase order is not complete, then inventory adjustments must be made. If these inventory adjustments are not made in a timely fashion, or not at all, customer order fulfillment may not be met and inventory write-offs maybe necessary to correct shrinkage problems. Location utilization may not be optimized due to the inaccurate inventory levels, and pickers often lose time searching for misplaced merchandise. Finally, batch systems frequently reflect inaccurate location availability and inventory levels based on logical updates that are not synchronized with physical moves.

Real-time warehouse automation allows warehouse activity tracking with timely

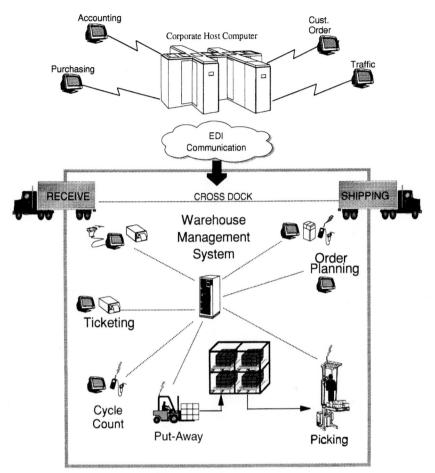

FIGURE 5-2 "Four walls" of a WMS.

updates to inventory and location files. The demand for this level of automation comes from the requirement to reduce inventory costs by reducing errors while providing better and quicker service to the customer.

The goal of real-time automation is to reduce and effectively eliminate the inaccurate and time-consuming activity of hand-keying data into computer data bases that are associated with card files that track location inventory. In addition, automation provides a consistent methodology (process) for each functional area of the warehouse. Throughput can be increased by combining work tasks while maintaining accuracy, such as multiple-pallet put-away or consolidated order-picking activities. Customer demands for bar coding and advanced shipping notification to increase their inventory accuracy while reducing shipping charges can also be met with more efficient procedures in shipping.

BASIC REQUIREMENTS OF A WMS

There are two basic requirements of a WMS. First, the WMS must create, track, and ship inventory. These requirements for inventory movement within the warehouse will be discussed in this section. Second, the WMS should automate, as appropriate, the acceptable procedures associated with the inventory creation, tracking, and shipping. This latter requirement will be explored more carefully through the discussion of each functional area. Figure 5-3 depicts the phases that inventory will go through within the warehouse.

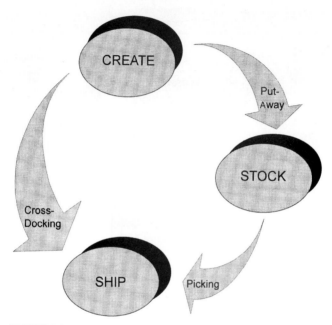

FIGURE 5-3   Basic inventory requirements of a WMS.

Inventory Creation

First, inventory must be created. It is important at this point to define clearly from whom inventory will be received and how it will be received. Typical sources for new inventory include vendors, internal manufacturing, warehouse and store transfers, and customer returns. It is common to have at least one receiving process for each source of inventory.

Vendor receipts are typically controlled through a purchase order that is submitted by a buyer on the host system and is then downloaded to the WMS. Manufacturing receipts often have an associated work order that was submitted to manufacturing through the host's manufacturing resource planning (MRP) system. Warehouse and store transfers have an associated transfer order issued on the host system by inventory control or an individual store based on order demand. Finally, customer returns have an associated return authorization (RA) submitted by a sales representative or customer service on the host system.

A WMS receiving process utilizes the appropriate receiving authorization and tracks the quantity received against it. Inventory must be received in such a way that it can be correctly transferred to the stock using put-away or allocated to a customer order using cross-docking. Therefore, the rules associated with stocking inventory and the methodology (work queues) for shipments leaving the warehouse must be known prior to designing the receiving function.

There are two basic methods for creating inventory in a WMS. The first method for creating and tracking inventory, which is often employed by batch systems, is to control inventory with slot or location logic. When an item is received, it is assigned a destination storage location. The physical receiving operation is the process of applying put-away (location) labels to the inventory and verifying that the counts match the receiving authorization. Since this method assigns the storage locations before the actual physical inventory movement is completed, there is potential for discrepancies if any receiving adjustments are required and not entered. Also, picking instructions may have inaccurate inventory locations if the put-away or cross-docking movement has not been completed in a timely and accurate manner.

The second method for creating and tracking inventory, which is the method employed by many real-time systems and adopted in this chapter, is to control inventory by assigning a unique bar coded license plate number to the inventory grouping or container (i.e., pallet, carton, or piece). As a receiver performs a detailed check-in of inventory, the system prompts the receiver to enter (scan) a license number (preprinted or on-demand) label that is applied directly to the container holding the inventory. The WMS records the quantity of each item that is assigned to the license plate number, records its location, and updates the receiving authorization. This container can then be made available for put-away or allocated to a customer order. This method of inventory tracking allows, but does not require, the advanced determination of stocking locations. However, it does *not* make the inventory update to a location in advance. Inventory availability in locations is coordinated with the physical movement, thus eliminating misdirection of pickers.

Inventory Tracking

The next basic requirement of a WMS is the ability to locate inventory in a manner that allows efficient utilization of space and provides the necessary access for order fulfillment. As discussed earlier, this is where a batch system and a real-time system begin to deviate. Either a batch or real-time system can be designed to support the desired location management philosophy (which will be discussed later in the section on put-away activity), but a real-time system supports recording actual product movement and immediate problem resolution.

A batch locator system makes location assignments during the receiving process and updates the inventory as available within the location assigned. If the product is not put-away in a timely fashion, then the inventory may not be available for picking. If the product cannot be located within the slot as directed, then it is up to the stocker to find an alternative location for the product or remove the product currently occupying the location and hold it in a discrepancy area. If location problems occur, it is up to the stocker to ensure that appropriate system updates are made to correct the problem to avoid additional stocking and picking problems in the future. Real-time systems resolve these problems by simultaneously making location assignment and inventory updates as the stocker moves inventory, thus eliminating errors associated with batch locator systems.

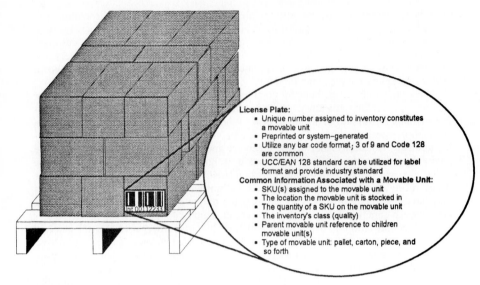

FIGURE 5-4   Movable unit and license plate concept.

Movable Units.   Batch systems track inventory by the location or slot it occupies. A real-time locator system is designed to assign a location to an inventory container (pallet, carton, and so on) when a stocker moves it to storage. A container that holds inventory is assigned a serialized (sequential) bar code license plate, known as a movable unit. The movable unit and license concept is shown in Fig. 5-4. The movable unit is used to track movement of the inventory as it is processed through the warehouse. Through the movable unit, inventory is assigned a location at all times, including when the movable unit is scanned for pickup by a stocker.

A real-time system has the added advantage of not only optimizing storage based on product attributes but takes into account stocker limitations such as the equipment the stocker is using or the building in which the stocker is working. A real-time system allows the stocker to react to storage problems by finding alternative storage locations immediately if the location is not accessible or by generating a verification request for the location for later investigation. Submitting locations for cycle count prevents the need to remove problem inventory from a location that can further compound an investigation and resolution.

Inventory Shipment

The third basic requirement of a WMS is the ability to ship inventory to fulfill customer orders. The key difference between batch and real-time systems is how and when inventory is relieved from the warehouse's ownership and how inventory problems are handled. Either a batch or real-time system can be designed to support allocating of inventory to customers. However, a real-time system gives the warehouse control of inventory and work direction from picking through truck loading.

Many batch-oriented systems are designed to allocate inventory to individual customer orders when they are made available from order entry or traffic planning. The allocated inventory is relieved from inventory immediately even though picking has not occurred and without confirming the inventory's availability. Frequently, a batch system issues a replenishment to forward picking locations during allocation, which can create a timing problem if space is not available from picking. If the inventory is not relieved at the time of allocation, replenishment to forward picking areas may not occur until after the picks are loaded onto trailers and shipping confirmation is keyed into the system. This potential delay in inventory update and the lack of picking confirmations in a batch system often result in extra checking procedures to ensure shipment accuracy.

A real-time shipping system also accepts customer orders as they are released from order entry or traffic planning. Typically, orders are made available to a workload planning or shipping supervisor, who can view the orders and plan the picking waves for future release. Planned waves can be released at the appropriate time to ensure that picks are completed in a timely fashion. Inventory is allocated when a wave of orders is released for picking.

As inventory is removed from a location, a new movable unit is created for the shipping container that transports the picked inventory to shipping. The new movable unit is updated with an "in-transit" location to provide continuous inventory tracking even when the inventory is allocated to a customer order. Any picking problems are handled by the picker by requesting reallocation of the pick by the system. This typically generates a cycle count for the problem location. Replenishment picks to forward picking locations are made available as picks are completed and space is available.

The real-time shipping process controls inventory consolidation of merchandise picked from reserve and merchandise allocated for cross-docking during the receiving process for a trailer load. Shipment checking processes can be eliminated since the picker (or receiver) is confirming the movable unit quantities as picks are executed. The real-time system then directs the loading of shipments onto trailers according to truck availability, dock assignments, and route and stop sequencing. When the shipping supervisor indicates that a truck is loaded, advanced shipping notifications (ASNs) can be generated down to the movable unit level, which are sent to the host system or directly to the customer using electronic data interchange (EDI). At the same time, the inventory within the warehouse and on the host system is updated to reflect the shipped quantities.

TYPICAL INTERFACES AND RELATED MASTER FILES

Creating a real-time distributed WMS has two fundamental data-base requirements: new and separate master files from current host data bases, and interfaces to the host data bases for periodic updates. The warehouse master files are used to allow the warehouse system to have new data fields to run the new real-time applications.

Figures 5-5 through 5-7 illustrate the transformation of current host files to the new WMS files through interface programs. The following discussion highlights common interfaces used for a distributed WMS and some master files required in a real-time WMS. More extensive discussion of master files will be noted through the functional area discussions later in this chapter.

Purchase orders are the method for requesting merchandise from vendors to be delivered to a warehouse. Purchase orders can also be used to control the internal movement of merchandise from a manufacturing plant to the warehouse and from warehouse to warehouse.

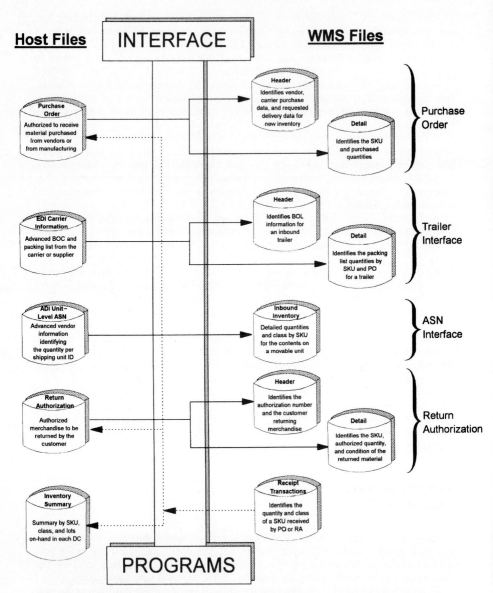

FIGURE 5-5 Receiving interfaces for a distributed WMS. The files on the left side depict host application files and information. The host files are often "flat" records with no header/detail relationship. Interface programs are written to extract the data elements required for the WMS and transmit the data into a distributed computer in which the WMS resides. The data files on the WMS, shown on the right side, are designed in a header/detail relationship. Header records hold information common to related detail records, thus reducing the number of storage requirements within the WMS and improving data integrity.

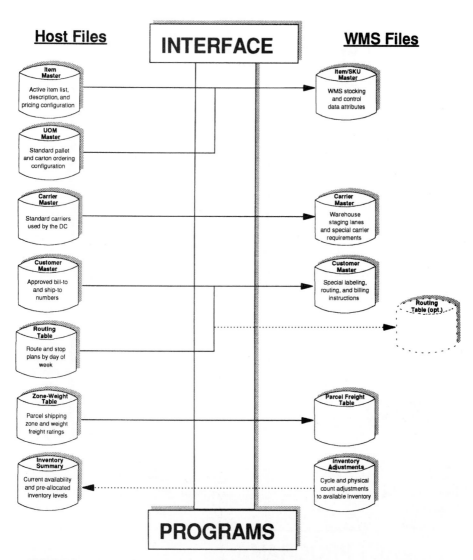

FIGURE 5-6 Control file interfaces for a distributed WMS. The files on the left side depict host application files and information. The host files are often "flat" records with no header/detail relationship. Interface programs are written to extract the data elements required for the WMS and transmit the data into a distributed computer in which the WMS resides. The data files on the WMS, shown on the right side, may be designed in a header/detail relationship. Optional data bases are indicated by a dashed file symbol and may be combined into other data bases.

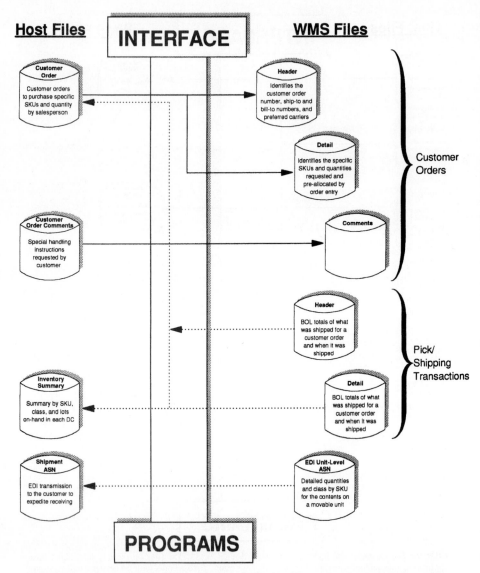

FIGURE 5-7 Shipping interfaces for a distributed WMS. The files on the left side depict host application files and information. The host files are often "flat" records with no header/detail relationship. Interface programs are written to extract the data elements required for the WMS and transmit the data into a distributed computer in which the WMS resides. The data files on the WMS, shown on the right side, may be designed in a header/detail relationship. Optional data bases are indicated by a dashed file symbol and may be combined into other data bases.

Generally, purchase-order entry and control are accomplished using host-based applications and are not considered part of a distributed warehouse system. The host computer provides corporate buyers with inventory levels from multiple warehouses as well as history files to help establish buying patterns.

Two interfaces for purchase-order control are required when designing a distributed warehouse system. First, purchase orders are downloaded to the warehouse from the host so that receipts can be verified against order quantities. The information on a purchase order will usually contain the following:

- Purchase-order number
- Purchase-order type to identify transfer orders from manufacturing work orders from vendor orders
- Buyer's name
- Vendor or supplier
- Transportation carrier
- Division for which the merchandise is being bought
- Purchase-order date
- Companies' item or stock-keeping unit (SKU) numbers, which identify the merchandise ordered
- Vendors' item or SKU numbers
- Quantity (in pieces) that is being ordered
- Expected quality of the merchandise
- Due date

The above-noted purchase-order information is downloaded from a single-record format and then split into header and detail files on the distributed system. This reduces the amount of redundant information that would be caused when multiple-line purchase orders are created.

Second, a receipt transaction file is uploaded to the host computer from the warehouse when receiving activity has been completed against a purchase order. This will identify to the host system the actual inventory received into the warehouse for a purchase order. The host system uses this information to update its inventory summary levels and to reconcile purchase-order quantities. An alternative method to a receipt file interface is to utilize an inventory adjustment interface (as discussed later).

A *trailer interface* is used to give the warehouse an advanced notification of the expected delivery schedule for inbound shipments from a carrier. The information contained in a trailer interface is created by the vendor or carrier and sent by EDI to the host computer or directly to the receiving warehouse. The information in this interface will usually contain:

- Confirmation number: a block of values given to the vendor by the warehouse or given on a demand basis
- Trailer number
- Carrier code
- Vendor number
- Two seal numbers—one for the rear door and one for the side door
- Bill of lading (BOL) number

- Number of pallets and cartons and pieces being shipped
- Expected delivery date and time
- Freight terms for the shipments

In addition, this interface may contain:

- List of purchase orders (POs)
- Progressive (PRO) number carrier tracking for each PO
- Associated PO line numbers
- Item or SKU numbers
- Shipment quantities for each line

This information is used within the warehouse system to help plan and control the receiving process. A receiving log is built using the trailer information to give the receiving supervisor estimated delivery workloads for a day and to help the supervisor schedule deliveries and assign receiving dock doors based on the content of the expected trailer.

During the detailed receiving process, carrier and vendor compliance data is captured against the expected receipt. Carrier compliance can be identified by the seal numbers and the number of damaged containers that are unloaded. Vendor compliance is captured against the quality and quantity of an item received.

In addition to the above-noted advance trailer information, unit level ASN information is used to assist in the detailed receiving process as well as created for customer shipments. A *movable unit* is defined as any shipping container that contains merchandise that will become inventory. A unique movable unit number is assigned by the vendor to each unit, which is bar coded on the shipping container. When the trailer is unloaded, the unit's bar code, or even the trailer's bar coded license number, can be scanned and the inventory automatically built into the warehouse system. This helps to speed the receiving process by not breaking down and identifying each shipping container's content.

Two ASN interfaces are required if both shipping and receiving ASN files are supported within the warehouse system. The receiving interface will download ASN information from the host computer, vendor, or manufacturing site. The shipping ASN interface will upload information to the host system or customer.

An ASN file's content is designed by the warehouse and the vendor or customer. Typical information for an ASN interface includes:

- Pallet, carton, and trailer bar code unit license plate numbers
- ASN number
- Trailer number
- Carrier code
- Division number
- PO number
- PO line number
- SKU number
- Inventory quality code
- Quantity shipped
- Manufacturing plant or vendor number
- ASN creation date

More common trends, however, are to utilize industry standard formats to allow a vendor to supply detailed shipping information to multiple customers without the need for proprietary formats for each. In addition, bar coding standards for shipping containers such as UCC/EAN-128 provide a common container unit license bar code format for all customers while identifying the container's content.

Return authorization (RA) forms, like purchase orders, identify merchandise that is expected to be incoming to the warehouse. Typically, a host application generates an RA number with an approved level of returned merchandise by a salesperson.

The RA interface is typically broken into header and detail files after it is transmitted to the distributed warehouse system. Common information includes the following:

- Division number
- RA number
- Customer number
- Store number
- Salesperson identification
- Salesperson territory
- SKU number being returned
- Approved return quantity
- Original order number, if available
- Lot and serial numbers of returned merchandise
- Return reason code to identify the cause of the return
- Condition code to identify the expected quality of the return
- A keep code to help the warehouse identify whether to scrap or keep the merchandise

An *item master file* interface is used to identify valid SKUs that are handled in the warehouse. A host application is responsible for the following information:

- Item (or SKU) numbers
- Item description
- Price
- Item's weight
- Quantity per case
- Cases per pallet
- Universal Product Code (UPC) number

An item master file is created from the host information. Purchase orders, customer orders, and return interfaces refer to this file to validate the information being downloaded. In addition to the host information that is downloaded, many control features describing how an item is stocked and picked within the warehouse system are maintained in this file. Warehouse item (inventory) attributes will be described in detail throughout this chapter and include information such as:

- Processing type (i.e., floor, conveyable, nonconveyable)
- Preferred put-away code
- Environmental restrictions

- Indication if serial number tracking is required and where it should begin (receiving or picking)
- Indication if stock rotation by FIFO date is required
- Indication if lot control (strict rotation) or lot tracking is required
- Indication if expiration date (and shelf date) controls required
- The item's ABC turn velocity
- The item's dollar-value velocity
- The cycle count velocity

Some of the item master file information directly affects the existing inventory with the warehouse and how customer orders are filled. For this reason, each item master field should be reviewed to identify if host maintenance is allowed. If the host computer is allowed to change the item master file by removing items or changing the case pack quantity, many customer orders that are in the process of being shipped, along with the warehouse's reserve inventory, may be obsolete or require repackaging before it can be used for order fulfillment.

Customer order information is transferred from the host-system order entry application to the distributed WMS. The WMS creates at least three files to manage order information. The order header identifies information such as:

- Division that will fill the order
- Order number
- Order type (regular customer, warehouse transfer, and so on)
- Shipping method (carrier)
- "Ship-to" customer number where merchandise will be shipped
- "Bill-to" customer number to identify who is paying for the merchandise
- Dollar value of all lines
- Billing method (such as regular, COD, or prepaid)
- Order cube and order weight
- Shipping date
- Cancel date

The order detail will identify information such as:

- Items (SKUs) ordered
- Order quantity
- Credit-approved allocated quantity
- Inventory class (first, second, refurbished) to ship
- Line type (such as a system, component, or promotional)

Finally, warehouse systems support a comments file that contains special shipping and handling instructions for the order.

A *customer master file* is used to store customer preferences of how orders are to be shipped and to prevent the need to download the same order information, such as "ship to" and "bill to," for every order. The customer reference file contains information relating to specific customer requirements such as:

- Account number
- Account name and address
- Indication if backorders should be created
- Preferred less than load (LTL) and parcel carriers
- Indication if ASN information should be created in shipping
- Preferred packing level (for the number of SKUs per shipping container)
- Preferred shipping container size for packed items
- Special labeling requirements (for cartons and pallets)
- Routing information
- Any specific invoicing level requirements (i.e., ship an order complete or consolidate orders by store)

This file requires an interface to download new customers and maintain host-computer information for current customers.

A *carrier master file* is maintained on the host system and downloaded to the distributed WMS. This file is used to validate newly downloaded orders to ensure they use a carrier that is currently under contract. In addition, information maintained by the WMS includes:

1. Preferred staging area or lanes
2. Maximum truck cube and weight
3. Type of carrier (LTL, parcel, or truckload)

A *routing table* is used to assist in efficiently loading a trailer in the sequence that deliveries will be made. Routing information is maintained solely at the warehouse through an interface to the host computer.

Routing tables can be as simple as a route number, a stop number, and a delivery day of the week that is maintained within the customer master file for an account and store. In addition, a routing table might be expanded to include multiple routes and stops by day of the week that an order will be shipped. Finally, complex routing logic is created when stops for a truck delivery are assigned on-the-fly based on the orders identified for release on a particular day. This chapter is based on a customer master "day of week" routing table, which is common for company-owned stores.

Zip code and zone–weight tables are files used to assist in determining freight charges for parcel carriers. This table is readily available on tape from parcel carriers. In some applications, though infrequent, these tables are downloaded from the host computer.

The zip code table identifies the shipping zone for an order using the zip code of the order. The first three digits of the zip code identify the zip code table to use, and then the last two digits identify the specific shipping zone for the order.

The zone–weight table utilizes the parcel shipping zone combined with the weight of a carton being shipped for an order. For a shipping zone, the weight of the shipping carton will fall within a range of weights supported by the parcel carrier. For each range of weights there is a shipping charge for the carton. If an item can be shipped in the same carton that it is stocked in, then predetermined weights from the item master file can be used for the parcel shipping weight. However, if an item is light, requires repacking, or fluctuates in weight, then a scale in the parcel shipping applications must be used to get exact carton weights.

An *inventory adjustment interface* is required for a distributed WMS to communicate to the host computer any changes that may occur to the warehouse inventory as a result of cycle counting or physical inventory, receiving, and shipping activities. In some instances the receiving department will have its own receipt interface (described above) if the incoming merchandise is not considered inventory until it has been put-away or the information required by the host system varies significantly from an inventory adjustment.

In many systems, manual communication using a phone or fax is used to request accounting information from host-system personnel before an adjustment can be made at the warehouse. Information that must be recorded and sent back to the host system includes:

- Division owning the inventory
- Item (or SKU) number
- Inventory class
- Reason code for the adjustment
- Document number assigned by the company's financial department
- Quantity before adjustment
- Quantity after adjustment
- General ledger account number assigned by the company's financial department
- Purchase-order number and customer order number if the interface is used for receiving and shipping adjustments, respectively

In addition to an inventory adjustment interface, an inventory summary interface is common. This interface is used to provide the host system with a "snapshot" view of the current inventory within the distributed warehouse system.

The inventory summary will identify:

- Division owning the inventory
- Item (or SKU) number
- Current on-hand (book) quantity
- Available quantity
- Allocated quantities
- Item's quality class for the inventory
- Lot number, if appropriate

This information is used to update the host-system summary levels directly and to allow the host system to make daily inventory comparisons with manual adjustments instead of conducting monthly physical inspections.

Shipping transactions are sent back to the host system using a separate interface though some applications can utilize the inventory adjustment interface discussed earlier. Generally, the information required by the host billing system and the customer is detailed enough to warrant a separate interface from inventory adjustments.

The shipping transaction interface provides the distributed WMS the ability to support and send *positive billing* information to the host system. Positive billing refers to knowing exactly what is shipped to a customer rather than *exception billing,* which bills the customer based on the portion of an order that is not sent to the customer.

Exception billing is prone to errors when communications and data transmission timing causes the host computer and the distributed WMS to be out of synchronization. Positive billing does not rely on proper communication and maintenance of multiple files; rather it simply sends actual shipment records to the host computer. However, positive billing often requires that the host system change its invoicing programs to utilize the actual shipping transactions.

Additional Master Files

Many other files are required to support a distributed WMS. Most of these files will be discussed throughout each functional area. However, two master files are common through all functions: an inventory master file and a location master file.

The *inventory master file* acts as the repository for all on-hand inventory in the warehouse that has value and can be shipped to the customer. Records in this file are created by the receiving function, made available for processing by put-away operation, validated by cycle counting, allocated by the order-processing process, transferred by the picking process, and finally removed by shipping activity. In addition, the inventory master file provides the modern warehouse with the bar coded unit license number reference (movable unit), which is fundamental to tracking inventory through the warehouse.

Common data stored within the inventory master file includes:

- Bar code unit license (movable unit) number
- Item (or SKU) number
- Location of the unit currently stocked
- Reserved location to where the unit will be moved
- On-hand inventory quantity
- Quantity allocated to a customer order
- Remaining available quantity
- Quality class of inventory
- Height and weight of the container
- Storage environment required by the SKU
- Inventory's FIFO date

Other fields will be discussed in each functional area as applicable. In addition, related files are commonly used to track control data about a piece of inventory such as serial numbers, lot numbers, and expiration dates.

The *location master file* provides the link between the WMS's logical warehouse and the warehouse's physical layout. The location master file is used by the receiving department to identify dock door and staging-lane locations that can be used to unload a trailer; by the put-away function to find available stocking locations that can stock new inventory; by the cycle counting operation to generate a counting queue of work; by the order-processing department to determine what locations are available for allocation and how to generate picking instructions to remove the inventory; by the picking function to make emptied location slots available for stocking; and by the shipping department to stage outgoing merchandise and load trailers by dock door location.

The location master file, along with the physical location, is identified by a location address that includes:

- Facility or building for the location
- Aisle
- Bay
- Level
- Slot, or partition, within each level

Other common location attributes include:

- Location reference identification (discussed in more detail in the section on put-away functions)
- Equipment type required to work at a location
- Stocking environment to identify refrigerated or caged areas
- Velocity of product (inventory) that should be stocked in the location
- Location's capacity expressed in pallets or cube
- Weight code to identify weight limitations (maximum)
- Location usage to identify it as a prime, reserve, or overflow location
- SKU number if a location is a dedicated prime (home)
- SKU's minimum, reorder, and maximum quantity for prime locations
- Identity of items by division that can be stored in a location
- Zone(s) used to group locations for put-away, picking, and cycle counting functions
- Location status to identify full and held locations

RECEIVING AND QUALITY CONTROL FUNCTIONS

The receiving and quality control processes are responsible for auditing the introduction of inventory within the distribution center and preparing the new inventory for storage or customer order fulfillment. The receiving and quality control processes are simple functions in terms of programming complexity and logic. However, these functions often involve 25 to 35 percent of the system implementation time due to the multiple sources of inventory, quality assurance requirements, and preparation of the inventory for dependent functions such as put-away and picking. (See Fig. 5-8.)

The distribution center is responsible for managing the complexity of a receiving process that is directly related to the supply stream. The simplest receiving process is often found in a manufacturing environment where the distribution center is responsible for storage of inventory coming from internal production lines. The most complex receiving process is often found in a wholesale environment where the same SKU might be supplied by many vendors and require multiple units of measure (i.e., case or inner-pack configurations).

All receiving functions within a WMS must be able to create inventory so that the output of the process (the inventory or movable unit record) has been audited for accuracy and contains the necessary information to allow the downstream functions (such as put-away and shipping) to work correctly. For this reason, the rules of inventory control and location management must be known when implementing the receiving processes. Some basic requirements for the creation of inventory might include the following:

FIGURE 5-8 Receiving functions of an automated warehouse management system.

1. *Create new inventory records known as movable units.*   A movable unit is created by assigning a quantity of a SKU to a license plate number, which is then placed on the container (pallet, carton, and so forth) that will be used to transport the merchandise through the warehouse. The SKU may be stocked in multiple units of measure (UOM) that the receiver will need to identify.

The inventory quality (first class, second class, refurbished, and so forth) must be standardized for the warehouse and assigned to the SKU during receiving. The movable unit is also created with a status that indicates if the unit has samples for quality control checking or if the unit requires validation due to unexpected receipt quantities.

The SKU might require control data capture that will need to be attached to the movable unit record. Figure 5-9 illustrates control data commonly associated with a movable unit. This information is tracked with the inventory as it is moved through the warehouse.

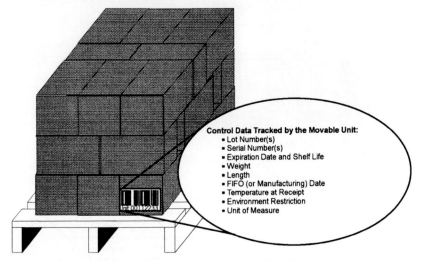

Control Data Tracked by the Movable Unit:
- Lot Number(s)
- Serial Number(s)
- Expiration Date and Shelf Life
- Weight
- Length
- FIFO (or Manufacturing) Date
- Temperature at Receipt
- Environment Restriction
- Unit of Measure

FIGURE 5-9   Inventory control data tracking using the movable unit concept.

2. *Prepare the inventory for stocking.*   When the movable unit record is created, additional attributes are typically retrieved from the item master file, or required by the locator system, and applied to the new inventory record. The receiver must identify the configuration (i.e., loose pieces or full cartons) of the inventory being received and the type of physical container onto which the SKU is being placed. The physical container might be a pallet or carton. In addition, the receiver is required to indicate the pallet's size if the locator system (warehouse racking) supports multiple pallet footprints.

Other attributes (such as for put-away, or inventory and location attributes) that are commonly applied from the item master file to the movable unit include:

- Stocking velocity
- Restricted environment
- Total weight
- Total cube
- Product processing type

These attributes are described in more detail in the section on put-away functions and location management.

3. *Reconcile the receiving authorization.*   Most receiving functions require a receiving authorization (purchase order, return material authorization, and so on) to exist before inventory can be allowed into the warehouse. The exception to this might be the receipt from internal manufacturing. In addition, as the process of receiving takes place, the quantity received is tracked against the authorization. If the authorization quantity is exceeded, the receiver is stopped (or an overage status is applied to the unit) so that a buyer can be contacted to approve or reject the excess quantity.

4. *Recognize outstanding demand.*   If the warehouse supports cross-docking, then as inventory is received and built onto a movable unit, the receiving function will typically check the demand file for the SKU. Cross-docking can be supported in a several ways. Figure 5-10 illustrates two methods of cross-docking.

One common cross-docking method is to identify a movable unit such that it is directed to a *picking zone* near the shipping lanes to prevent it being put away to normal storage. This method is used when the customer order (demand file) must be coordinated with other merchandise and shipped as a complete order pick with other goods. Another common cross-docking method is used when the inventory being received is shipped independent of other inventory. This method of cross-docking is referred to as *relaying,* where the movable unit (carton or pallet) is directly allocated to the customer order and moved directly to shipping (i.e., no picking is required).

5. *Provide quality control sampling.*   The receiving functions are used to identify if the SKU, vendor, or purchase order references a quality control sampling table. A sampling table is used to determine the amount of a total receipt that must be sampled and checked by quality control personnel. The receiving functions must identify to the receiver the sampling demand based on the total receipt and record the portion of the movable unit that is being marked to meet this demand. The movable unit is then created with a status, as described above, that will ensure that it is unavailable for picking until inspection is complete.

6. *Update host-system inventory levels.*   The receiving functions must ensure that the inventory quantities on the host system(s) are updated to reflect the new receipts. The timing of this update is dependent on the requirements for warehouse operation and may be done as each movable unit is built, after the entire authorization or trailer has been received, or as a movable unit is put into storage.

The following discussions elaborate how these functions are designed and what differentiates the various processes.

Receiving/Scheduling Log Function

Many warehouses utilize a KeyRec log book to track and schedule inbound trailers and merchandise. The receiving/scheduling log function replaces this paper log book (or KeyRec) and provides the receiving supervisor with centralized control and record keeping of all receiving operations. This function, while optional, is used to schedule trailers, assign dock doors, and assign specific receiving authorizations and quantities to a trailer that can control detailed receiving functions.

The receiving/scheduling function is considered the focal point for receiving activity and is designed to work on a CRT or workstation within a receiving office environment. The sophistication of the warehouse's receiving operation will dictate the scope of this function. The following describes some characteristics to be considered in the design of a receiving/scheduling log function:

1. *Permit flexible viewing of scheduled and in-process deliveries through the bill of lading, appointment date, carrier, and trailer number.*   The trailer interface is designed to create the initial records for inbound trailers and their contents into the receiving/scheduling log automatically. The receiving supervisor can then use this information to look up the desired information quickly.

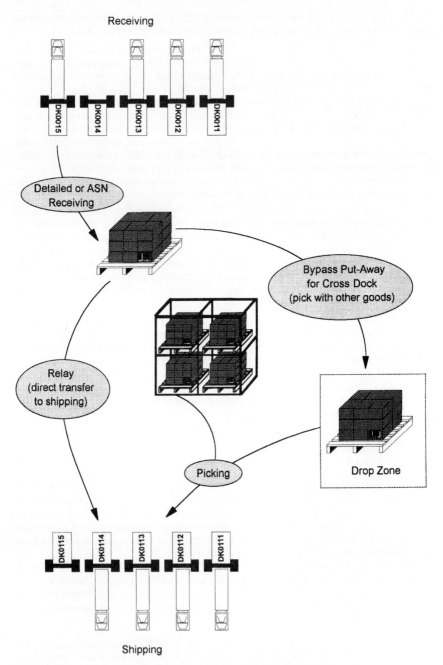

FIGURE 5-10 Two cross-docking methods: relay vs. pick with other goods.

2. *Permit manual creation of a carrier appointment by a supervisor.* A system confirmation number is generated by the WMS for a trailer when the log entry is made. If the advanced trailer information is not available electronically from the carrier, the receiving/scheduling log function gives the receiving supervisor the ability to add a new appointment into the log manually and assigns a confirmation number when the carrier calls in to schedule a delivery. The confirmation number is given to the carrier for future reference and for dock assignment when the trailer arrives and is also used for inquiries into a receipt's status.

3. *Add packing-list details after a scheduled delivery is created for an inbound delivery.* The receiving log gives the receiving supervisor the ability to add receiving authorizations to the trailer manually. The receiving authorizations assigned to a trailer represent the detailed SKUs and quantities that should be on the truck and will control what is allowed during the detailed receiving processes later. The carrier (or shipper) may be able to provide this information in advance of the trailer arriving at the warehouse via electronic data (through the trailer interface) or by fax. However, this information is often not available until after the trailer has arrived and packing lists can be pulled from the delivery.

4. *Utilize advance shipping notices.* Selecting ASN records to build a trailer record is an alternative to selecting authorizations to add to a trailer manually. An ASN number is assigned to all movable units that are contained on a trailer. The receiving log can give the receiving supervisor the ability to view outstanding ASN information and assign it to a specific delivery schedule.

5. *Assign the delivery a location after a delivery is scheduled and the trailer arrives at the warehouse.* The receiving/scheduling log gives the receiving supervisor the ability to assign a dock or yard (parking) location. If a dock door is unavailable for receiving or unloading the trailer contents is a low priority, then a yard location is assigned to a trailer that gives the receiving supervisor the ability to view the trailer's contents and stage work into the receiving docks. After a dock door location is assigned to a trailer, the trailer is unloaded and detailed receiving activity can commence.

6. *Record quantities for a bill of lading (BOL) into the receiving/scheduling log at the time the delivery is scheduled.* As the trailer is unloaded, the receiving/scheduling log is updated to reflect the BOL quantities (pallets, cartons, or pieces) actually removed from the truck along with capturing the seal number from the trailer. After unloading is complete, the receiving/scheduling log is used to confirm the signing of the BOL by reconciling the anticipated versus unloaded quantities. This level of tracking, which includes any damage comments from the unloader, can now be used to identify any carrier compliance problems.

7. *Estimate workloads for a SKU's expected receipt quantities on a trailer.* The system is designed such that an estimated receiving workload is assigned to a pallet, carton, and piece of a SKU that can be printed onto a summary report for an entire trailer or range of trailers.

8. *Include history tracking to provide compliance tracking after a trailer is completely received.* The receiving/scheduling log is used to close the delivery making the dock door available and triggering the creation of receiving transactions to update the host-system inventory. The closed log entry is removed from the active schedule and moved to a history file for future reference.

Some warehouse operations require more dock door planning and control than described in the above receiving/scheduling log, especially when high-volume cross-docking and store relay operations are in place. These systems utilize the trailer interface information to recommend and preplan the "best" dock to use for unloading based on the SKU mix, determine how long a receipt should take to receive, and queue trailers into a receiving dock. This preplanned dock schedule is then used to update the above receiving/scheduling log for tracking of the actual receipt.

The receiving/scheduling log function is also equipped to print receiving labels for the anticipated contents of a trailer. This ability is required if the design of the warehouse management system does not employ the use of RF terminals to capture receiving data. The process of printing labels (creating movable units) will need to assume a standard configuration of a SKU from the SKU master file. In addition, stocking locations can be predetermined for the anticipated receipt if there is concern with performance and the use of a RF directed put-away function. The receiving/scheduling log function is also capable of removing these predetermined movable units if discrepancies or damage exists on the trailer.

Receiving Unload Function

The receiving unload function is focused on providing the ability to expedite the unloading of incoming trailers while recording carrier compliance information such as trailer seal numbers and BOL quality and quantities into the receiving log. As the unloader removes the contents of a trailer (see Fig. 5-11) the pallets and cartons are

FIGURE 5-11 Receiving unload function and BOL quantity verification. (*Courtesy of Sharp Electronics Inc.*)

directed into a receiving lane(s) for detailed receiving at a later time. By doing this, the dock doors are made available for other trailers/receipts, and the carrier can leave without waiting for specific SKU quantities to be captured.

The use of RF terminals to capture the BOL quantities as merchandise is unloaded from a truck provides a major efficiency gain over hand tabulation that is frequently used in nonautomated warehouses. The unloader identifies transportation damage and records comments as the unloading process takes place. In designing a receiving unload process, the following characteristics should be considered:

1. *List of dock doors available for unloading.*   Available doors for work are controlled by the receiving/scheduling log function.

2. *Verification of seal numbers versus the expected seal numbers from the receiving/ scheduling log.*   If the seals are broken or damaged, this validation check will prevent the unloading of the trailer until a supervisor is contacted to resolve the problem.

3. *List of receiving authorizations to unload.*   This allows tracking of BOL quantities down to a specific receiving authorization that might be in error.

4. *Capability for an unloader to add a receiving authorization to the trailer (receiving/scheduling log) record.*   This enables adding details to the receiving/scheduling log as the unloading process is taking place and prevents the need for removing packing lists for entry in the receiving office.

5. *Ability to print unload tags to help identify merchandise belonging to a specific trailer when it is moved to a receiving lane(s).*   This will assist workers doing detailed receiving functions by having well-identified and easy-to-read markings on the cartons and pallets removed from a trailer.

6. *Temporary suspension of unloading activity.*   If the receiving supervisor determines that a dock is required for another activity (to expedite unloading of another trailer, for example), the unloader should be allowed to discontinue the current unloading and assign the trailer to a yard location. After the trailer is reassigned a dock location for continuing the unloading process, the receiving unload function bypasses the seal-number capture and continues to accumulate the quantities remaining.

As evidenced by the descriptions above, the receiving unload process is not concerned with the specific SKU being received. The process of identifying the inventory and building movable units is the function of detailed receiving by PO or by label (tentative receiving). If the trailer's contents are identified by movable unit level ASN information, then the receiving unload function can be bypassed since the process of creating inventory records will be expedited.

Receive by Purchase-Order Function

The purpose of the receive by purchase-order function is to accept inventory (by SKU) and to record the quantity and quality of the merchandise against a receiving authorization (in this case a purchase order) from suppliers who are unable to ship inventory on movable units and provide ASN information. This process is used to receive manufacturing and transfer orders if work and transfer receipts are handled through the same receiving authorization (PO) file structure.

The receive by purchase-order function is designed so that movable units are created directly from a list of open purchase orders or by a list of purchase orders that have been assigned to a trailer in the receiving log. Receiving directly from the purchase or transfer order has the advantage of simplifying the receiving process by effectively eliminating the need to know what is on the trailer before the inventory is received. The receiver uses the packing list information, or carton markings, to determine if a purchase order exists to receive merchandise against. If a PO content, however, gets split across trailers during the shipping process, difficulties could arise in knowing when a purchase order is completely received or when discrepancies (beyond overages) exist. This method still uses the receiving/scheduling log to capture the actual receipts made from a trailer.

The alternative method is to receive merchandise against the authorization(s) assigned to a trailer in the receiving/scheduling log. While this does add the complexity of having to know the trailer contents prior to receiving, it does have the advantage of resolving the receipt against the expected shipment rather than what was ordered. This method also allows splitting a PO across multiple trailers.

The use of RF terminals and bar code and scanning technologies, as shown in Fig. 5-12, when used in a real-time environment makes new inventory available as mov-

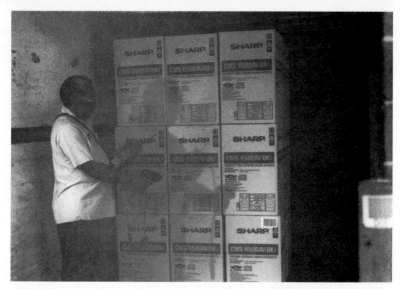

FIGURE 5-12   Receive by purchase order function using a RF terminal. (*Courtesy of Sharp Electronics Inc.*)

able units are created. It has the added advantages of accurately capturing and resolving discrepancies in the purchase-order quantities while work is being done. This function is designed so that the receiver enters "blind" counts being received and does not display the open quantities on the PO. The WMS accumulates the quantities received by SKU and notes any overage quantities with the appropriate status on the movable unit. In addition, the following characteristics should be considered in the design of the receive by PO function:

1. *Create a list of docks that have been assigned trailers for receiving functions.* This list displays all docks (or staging lanes if used with the receiving unload function) that are waiting to start receiving or that are currently in the process of receiving so that multiple receivers can unload the same trailer, if necessary. In addition, this list ensures that only the trailers that the receiving supervisor has assigned to a dock location will be received.

2. *Allow the receiver to identify the authorization (PO) to receive.* This is done by prompting the receiver to enter a PO or by allowing the receiver to select a PO

from a list of authorizations assigned to the trailer. In addition, this function incorporates receiving without a PO by allowing the receiver to request a system-generated authorization to receive inventory against. Any inventory received against a system-generated authorization is created with an overage status to ensure that a buyer or supervisor is contacted prior to the inventory being available within the warehouse.

3. *Enter the SKU to receive.*   It is important to ensure that the system can recognize multiple codes for the same SKU especially if the same SKU can be supplied from multiple vendors. The use of the UPC code, or other appropriate industry standard, is designed into the system, along with supporting the vendor's SKU reference that can be built into the purchase-order file.

4. *Allow the receiver to add a SKU to a receiving authorization while receiving is taking place.*   This function prevents the need to have a supervisor intervene in the process, and any movable units created against this line item can be assigned an overage status to ensure that the buyer is contacted prior to the inventory becoming available in the warehouse.

5. *Allow the FIFO (or manufacturing) date to be changed for a SKU.*   The WMS's current date is used as the default FIFO date for the new inventory. However, it may be necessary to use the product's actual manufacturing date (if available) especially if it is being transferred from another storage facility.

6. *Ensure inventory integrity, identify QC quantities, and capture other control data discussed earlier in the create inventory process basic requirements.*

7. *Support the ability to receive in multiple buildings.*   Depending on the warehouse's configuration, receiving may be done in a building that does not have stocking locations. If this is true, then the system should ensure that a receiving staging location is entered for the movable unit rather than directing the receiver to a stocking location directly.

8. *Allow the receiver to enter comments and reason codes if the inventory class of the new merchandise is different from what was ordered.*   This assists in resolving discrepancies with the supplier should problems occur. Also, this data is used to establish vendor compliance problems.

9. *Check the system's open demand file (outstanding customer orders and back-orders) and allow the receiver to identify an order to cross-dock or relay.*

10. *Require the receiver to close the trailer to indicate that receiving is complete.* This allows the system to identify any discrepancies associated with the trailer and create inventory transactions to update the host system's inventory records.

Receive by ASN Function

The purpose of the receive by ASN function is to utilize shipping units built by the supplier to expedite acceptance of inventory into the warehouse. A partnership must be developed with the supplier to establish electronic communications, warehouse (or industry) ASN and EDI standard specifications to be used, and quality measurements to ensure adherence to standards. This value-added service by the supplier can greatly enhance the receiving efficiency and improve inventory accuracy.

The receive by ASN function is best designed to utilize RF terminal and bar code scanning technologies (see Fig. 5-13) to allow the receiver to scan the ASN movable unit directly at the dock door. Since the containers (pallets, cartons, etc.) have been bar coded, the ASN receiving process is designed with enough efficiency and speed to

FIGURE 5-13   Receive by ASN function using a RF terminal. (*Courtesy of Sharp Electronics Inc.*)

eliminate the need of the receiving unload function. The ASN receiving function performs the same as the receive by PO function (described above) by creating inventory through the use of movable units that have associated SKUs, quantities, and quality assigned by the supplier. The result of this advanced electronic information can improve productivity in automated receiving processes (utilizing receiving unload and receive by PO functions) by more than 75 percent with even more substantial gains in nonautomated (paper) receiving processes.

As described in the receiving unload function section, a trailer is assigned a dock door from the receiving/scheduling log function, which is also used to assign an ASN to a trailer. The receiver starts the receiving process by selecting a dock door to work, and after collecting seal numbers for the trailer (if required by the receiving/scheduling log), the receiver scans a bar coded license plate to move. The system retrieves the inventory records from the ASN interface and updates the warehouse inventory file and purchase order with the quantity received. The receiver is then directed to scan a location, a staging lane, or stocking location (by utilizing put-away logic) for the container being moved. In addition, if the supplier is shipping inventory destined directly for a store (as specified by the buyer), the receive by ASN function uses the store "ship to" information as supplied on the ASN to allocate the new inventory directly. The following characteristics should be considered in the design of the receive by ASN function:

1. *Display a list of docks that have been assigned a trailer to receive.*   This list displays all docks that are waiting to start receiving or that are currently in process of receiving so that multiple receivers can work the same trailer if necessary. In addition, this list ensures that only the trailers that the receiving supervisor has assigned to a dock location will be received.

2. *Allow for the verification of seal numbers versus the expected seal numbers from the receiving or scheduling log.* If the seals are broken or damaged, this validation check will prevent the unloading of the trailer until a supervisor is contacted to resolve the problem.

3. *Prompt the receiver to scan a bar coded license plate to receive.* By allowing the receiver to indicate the ASN unit to receive in a "blind" fashion, vendor compliance validation can be achieved by ensuring the proper units were shipped.

4. *Display the expected inventory configuration to the receiver to confirm the accuracy of the ASN information.* Again, this allows for additional vendor compliance validation and allows the receiver to add any additional information to the unit that the supplier may not be required to capture on the ASN such as the container type for the unit, the size of the container (if multiple sizes of storage containers, or pallets, are supported), and the inventory class.

5. *Provide the ability to build a new bar coded container manually so that the receiver can reconcile incorrect information or indicate damaged inventory.* Any containers that are manually built by the receiver are created with a unit status that will cause additional validation by a quality control person. In addition, the receiving or scheduling data base is updated with the problem ASN unit number that was manually received to track vendor compliance.

6. *Ensure inventory integrity, identify QC quantities, and capture other control data discussed earlier in the create inventory process basic requirements.*

7. *Support the ability for receiving activity in multiple buildings.* Depending on the warehouse's configuration, receiving may be done in a building that does not have stocking locations. If this is true, then the system ensures that a receiving staging location is entered for the inventory rather than directing the receiver to a stocking location directly.

8. *Allow the receiver to enter comments and reason codes if the inventory class of the new merchandise is different from what was ordered.* This assists in resolving discrepancies with the supplier should problems occur. Also, this data is used to establish vendor compliance problems.

9. *Check the system's open demand file (outstanding customer orders and backorders) and allow the receiver to identify a customer order to cross-dock or relay.* Optionally, the ASN information is used to identify the customer (or store) order for whom the inventory is destined.

10. *Require the receiver to close the trailer to indicate that receiving is complete.* This allows the system to identify any ASN units that may be outstanding and ensure that the receiving/scheduling log is updated with a discrepancy status, along with creating inventory transactions to update the host system's inventory records.

Tentative Receiving Function

The purpose of the tentative receiving function is to accept inventory into the warehouse by the quantity expected on a receipt (or trailer). License plate (movable unit) labels are printed according to the anticipated quantities keyed in against the PO from packing-list information or by utilizing the receiving/scheduling log information. The labels are printed according to the SKU's largest container configuration (number of cartons per pallet and pieces per carton) that best fit the total quantity of a SKU being received. After labels are applied, the receiver must enter the exception labels that remain.

FIGURE 5-14   Example of a tentative receiving label with bar coded SKU and movable unit numbers. Reserved location is determined during receiving and printed on the label to expedite put-away.

The license plate labels that are printed, as shown in Fig. 5-14, show the PO being received, the SKU, and quantity. The trailer, received date, and time are also printed. The system is designed to reserve a stocking location for the inventory that is also printed on the label. This method of receiving relieves the need for the receiver to have a RF terminal and can speed put-away if the stocking location is determined in advance.

The tentative receiving function, however, is a "batch" receiving process that can result in inaccurate inventory unless appropriate procedures are enforced. Problems commonly encountered in a batch receiving process include inaccurate quantities, lost or misplaced labels, exceptions not entered, and so forth. Also, the ability to capture productivity by the receiver is lost using a batch receiving process when compared to a real-time receiving process described in the section on receive by PO function. Owing to these reasons, the receive by label function is most useful for warehouses that receive inventory in standard (pallet) configurations with only one SKU per container (pallet) and require a fast receiving process with low system utilization.

If RF terminals and bar code and scanning technologies are used to supplement the tentative receiving function, errors incurred by assuming that labels are applied correctly can be reduced by confirming their application when used in a real-time environment. When RF terminals are used, the receiver is requested to scan the label for the receiving function, and the system can display the assigned container quantity (much like ASN receiving). The receiver then confirms the displayed quantity or corrects the assigned quantity in real time. After the receiving operation is completed, the WMS prompts the receiver to scan the labels not received for confirmation and reconciliation.

If RF terminals are not used, the receiver is given access to a CRT workstation to perform the tentative receiving function. Using a CRT, the receiver scans a label to change the assigned quantity, and the receiver is required procedurally to identify inventory *not* received by scanning the labels left over from the receiving process (receive by exception). In addition, the following characteristics should be considered in the design of the tentative receiving function:

1. *Display a list of staging lanes, or docks, that have been assigned a trailer.*   This list helps the receiver to find inventory that was unloaded.

2. *Allow the receiver to identify the authorization (PO) to receive.*   This is done by prompting the receiver to enter a PO or by allowing the receiver select a PO from a list of authorizations assigned to the trailer.

3. *Enter the SKU and quantity to receive.*   Labels are generated from the quantity entered by the receiver or printed in advance from the receiving/scheduling log.

4. *Allow the receiver to override the standard system configuration of the inventory so that labels can be printed according to how the inventory was shipped.*   Procedures should be developed to ensure that multiple SKU containers are split into single SKU containers prior to printing labels.

5. *Provide the receiver the capability to reprint labels that may have been damaged or not printed correctly.* The receiver is allowed to select a label to reprint rather than just printing a new label for the quantity desired so that inaccurate inventory levels will not be incurred.

6. *If cross-docking is enforced within the warehouse, assign each printed label (and its associated inventory) to a customer order if demand exists.*

7. *Ensure inventory integrity, identify QC quantities, and capture other control data discussed earlier in the create inventory process basic requirements.* If receiving is performed without RF terminals, the WMS prompts the receiver for this information when the receiving confirmation is entered.

Receive from Manufacturing Function

The purpose of the receive from manufacturing function is to provide a means to accept inventory from an internal manufacturing source. This function is used in two basic scenarios. First, the warehouse may be supporting the storage and shipping of the "end of line" finished goods produced by manufacturing. If so, receipts are "pushed" onto the warehouse, thus reducing, or eliminating, the amount of planning the warehouse is capable of doing. Second, the warehouse may be responsible for the receipt of bulk material and letdown to a manufacturing or work in process (WIP) area and for the stocking and shipping of the finished goods. In this case, a detailed receiving function (such as the receive by PO or receive by ASN function) is used to introduce the bulk inventory into the warehouse, and the receive from manufacturing function would accept the finished goods into inventory after the WIP process is complete.

The receive from manufacturing function is designed to utilize RF equipment (see Fig. 5-15), but is equally suitable as a workstation CRT function. This function does

FIGURE 5-15 Receive from manufacturing using a RF terminal.

not require the prereceipt planning provided by the receiving/scheduling log. However, if inventory is being accepted into the warehouse's inventory at a separate manufacturing site (separate building), then it may be desirable to use the receiving/scheduling log to track the inventory onto shuttle trucks to allow scheduling into the main storage site or the warehouse.

The receive from manufacturing function is performed by manufacturing personnel as finished goods are boxed or palletized, or it is performed later at a more centralized receiving point that is common to multiple manufacturing lines. This process is initiated by the receiver by indicating the manufacturing work order and line item being received. This allows the WMS to update the total receipt quantity against the work order and to obtain the lot number and date of manufacturing (if this information is available to the WMS). If the lot number or the manufacturing date (FIFO date) is not available, the WMS prompts the receiver to input this information from paperwork or carton markings on the inventory.

After the receipt is identified, the WMS prompts the receiver to input the quantity being received and input a bar coded license plate number against which the inventory is tracked. If the warehouse is physically in the same building as the manufacturing line, then the new inventory is directed to a stocking (or staging) location. If the manufacturing site is remote to the warehouse, then the inventory is staged in receiving lanes to wait for put-away or loading onto shuttle trucks (see the receiving load function). The following characteristics should be considered in designing a receive from manufacturing function:

1. *Allow for the input of a manufacturing work order and line number against which to track the receipt.* The WMS is designed to allow overruns to be accepted but stops the acceptance of inventory in excess of the work order's allowable percent overage. If the allowable percent overage is exceeded, the receiver must contact a supervisor to validate and resolve the problem. In addition, if the WMS is utilizing the receiving/scheduling log function (described earlier) then the manufacturing order and line are added to the details of the receiving log to track the receipt of this inventory like any other vendor receipts.

2. *Retrieve the lot number (frequently this is the same as the manufacturing work order number) and the date of manufacturing from the manufacturing system (or the work order).* The manufacturing date is used to update the inventory's FIFO date to ensure proper rotation of the inventory while in storage.

3. *Provide the ability to identify cross-dock or relay customer orders.* This is a common requirement if the merchandise is manufactured based on customer demand.

4. *Ensure inventory integrity, identify QC quantities, and capture other control data discussed earlier in the create inventory basic requirements.*

5. *Allow the receiver to enter comments and reason codes if the inventory class of new merchandise is different from what was manufactured.* The proper class is applied to the manufactured inventory to ensure proper handling within the warehouse.

6. *Display the best zone (by the SKU's preferred put-away zone) where the new inventory should be stocked (located).* This allows the receiver to stage movable units according to where they will likely be located, which is useful if the inventory will be loaded onto shuttle trucks. Loading shuttles according to the inventory's preferred stocking zone can expedite put-away by reducing travel time of fork trucks.

An alternative to the preferred put-away zone is to determine the actual stocking location (reserve space) for the new container, based on the container's quantity or the total anticipated manufacturing quantity. This allows more accurate loading of shuttle trucks over the preferred put-away zone. Note, however, that if the actual location is reserved in advance, then the receiving process will be slower due to the search time associated with finding a location. If this latter method is used, the WMS stocking procedures ensure that the put-away function is done in the same sequence that inventory is received to ensure that location utilization is optimized according to the inventory's availability.

Receiving Load Function

The receiving load function is required within the WMS when a receiving function is performed in a remote building from the primary stocking warehouse (as described earlier in the receive from manufacturing function section). The primary function of the receiving load function is to track received inventory as it is loaded onto shuttle trucks that deliver the inventory to the warehouse for stocking and distribution.

Inventory that is received in a building remote to the primary warehouse is typically not made available for allocation until it is stocked in a picking location. The receiving load function, utilizing a RF terminal, is used to update the inventory's status to "in transit" as containers are placed onto shuttle trucks. If the receiving/scheduling log function is being used, this function updates the log to notify the receiving supervisor of the expected delivery. After a shuttle truck is loaded, a printed copy of the shuttle's contents is generated to meet the legal requirements of the bill of lading. The WMS can generate this listing from the inventory records created in the receiving process and updated by the receiving load function.

Since the shuttle is loaded with containers that have already been identified to the WMS, receiving at the warehouse is not necessary. The receiver at the warehouse can utilize the directed put-away function to unload the shuttle's containers directly to storage. The option exists to determine put-away locations as part of the receiving load function when the shuttle is loaded to speed the unloading process further at the warehouse. Other characteristics that should be considered when designing a receiving load function include:

1. *Allow the loader to view movable units that have been staged at the receiving building and that are ready for loading.* This assists the loader in identifying the inventory that is available for transfer.

2. *Capture the shuttle number to track the load within the receiving/scheduling log function.* This allows the receiving supervisor at the warehouse to plan dock assignments for the shuttle's arrival and to view the inventory that is in transit. If the receiving/scheduling log function is not used, the in transit inventory and shuttle can be tracked as location and status attributes on the inventory file.

3. *Allow the loader to close a shuttle, indicating it is completely loaded, and print a listing of the shuttle's contents.* This provides transportation papers (BOL) if public roads are used to get to the primary stocking warehouse.

4. *Determine stocking locations for the loaded inventory when the shuttle is closed.* This expedites the unloading at the warehouse and, by executing the put-away algorithms in batch mode, does not affect the transaction time between the loader and the system.

Receiving Adjustments Function

The receiving adjustments function provides the warehouse supervisor the ability to adjust any inventory records that may have been inadvertently or incorrectly created during the above-noted receiving processes prior to updating the host system. In addition, this function is used to review and approve the acceptance of any units that were created resulting in an overage to a PO or that may have been manually received due to an ASN problem.

A receiving supervisor knows that adjustments need to be made by viewing the receiving/scheduling log or by reviewing problem purchase orders. In addition, any problem inventory or overage containers are restricted from put-away. The following characteristics should be considered in designing a receiving adjustments function:

1. *Notify a supervisor of a discrepancy by indicating an overage condition on the PO or by preventing the removal of the receiving/scheduling log record that has a discrepant status.* Through the receiving/scheduling log, the receiving supervisor is also capable of viewing the status and location of all units received against the log entry.

2. *Allow a supervisor to use a RF or CRT terminal to update the status of the discrepant inventory record to indicate that the overage is approved or rejected.* If the container has the incorrect quantity, the overage status is removed if the unit is adjusted appropriately. If an overage is rejected, the supervisor must ensure the inventory is separated from the container and moved to shipping for return to the vendor.

3. *Allow the supervisor to update the status of a problem unit after viewing the system's inventory record.*

4. *Revise the receiving/scheduling log or PO record to reflect a "closed" status if there are no open quantities remaining to be received and all problem movable units have been resolved.* The discrepant unit is now ready for put-away. This also causes the receiving transactions to be sent to the host system for inventory updates. Note that the update to the host system is dependent upon the company's policy on whether the inventory must be in a picking (stocking) location or another location before the inventory updates are made to the host system.

Quality Control Function

An integral part of the receiving functions in a WMS is the ability to check vendor (or supplier) quality to ensure customer demands are fulfilled with first-class merchandise. It is unreasonable and costly to expect all receipts from all vendors to be checked for first-quality merchandise. Two fundamental steps can be taken by the warehouse to assist in checking the quality of receipts.

The first step in ensuring quality inventory in the warehouse is to empower the receivers to check for damage or suspect quality and change the class of inventory as part of the receiving process. This procedural check is accompanied by notification to a supervisor if the quality problem is throughout the receipt. The receiving supervisor can then make the decision if the receipt should be rejected and sent back to the supplier, or if the whole receipt should get a quality hold status until the buyer can be notified.

The second step in ensuring quality inventory in the warehouse is to utilize the WMS to direct statistical sampling (as mentioned in the receiving functions detailed earlier). The WMS is designed to utilize sampling tables that indicate the pull quantity or pull percentage over a quantity range. A sampling table is assigned to the SKU, vendor, or purchase order requiring the quality history of the supplier. As receiving is conducted, the receiver is notified of sampling requirements, and the receiving function updates the inventory container as requiring quality control checking. The sampling table also specifies if all inventory received for a SKU should be put on quality hold to prevent put-away and picking operations until the sample container is checked. (See Fig. 5-16.)

FIGURE 5-16   Quality control inspection area. (*Courtesy of Sharp Electronics Inc.*)

The quality control (QC) function provides the quality control department with the ability to inspect sample inventory pulled during the receiving process and release the inventory, hold related inventory, or remove (reject) inventory from the system for return to the vendor. This function is designed to allow the QC inspector to view containers that have sample inventory by SKU, vendor, lot number, or purchase order. The inspector has the ability to view any inventory that is related to the sample quantity, to pull additional samples, or to contact any customers that may have shipped related inventory.

If the sample quantity is inspected and is approved, then the inspector is given the ability to release the sample container for put-away into normal stock. Also, the inspector has the ability to release any related containers that may have been held during the receiving process. Finally, the inspector has the ability to pull the inventory out of storage and either adjust the inventory's class or adjust the inventory out of the warehouse if it fails to meet quality standards. Note that the standards used to inspect a SKU's quality will usually be part of the warehouse's standard operating procedures (SOPs). The following characteristics should be considered in designing a QC function:

1. *Allow the QC inspector both to view and work with samples and regular inventory by SKU, PO, lot number, and vendor.*

2. *Provide the QC inspector the ability to release sample or held containers for put-away and picking processes, hold containers related to the sample inventory, or reject QC held containers that have failed inspection.*   The inspection standards and rejection procedures (such as contacting a buyer and vendor prior to rejecting inventory) are documented in the warehouse's standard operating procedures.

3. *Reallocate picking instructions for any orders that may be ready for picking that use rejected inventory.*

4. *Provide the ability to remove sample inventory from the warehouse inventory due to destructive testing procedures.*

5. *Capture vendor compliance data as a result of inspecting the sample inventory.* This allows maintenance to be conducted to increase sampling for problem vendors and to decrease sampling of vendors who continually provide quality merchandise.

6. *Provide the inspector the ability to view shipping history so that any customers that may have been shipped poor-quality merchandise can be notified.*

Returned Goods Processing Function

The handling of goods returned from customers is often one of the least favorite topics of conversation for a warehouse manager. Customer returns can be easily spotted, even though they are frequently tucked away in a corner of a warehouse, by the characteristic pile of irregular and often crushed boxes, along with the accumulation of a layer of dust. The process of receiving returned merchandise can be time-consuming and costly due to the wide variety of merchandise (with some items not sold by the warehouse or company) and the seemingly unending variety of reasons for the return. The inability to receive and reconcile returned merchandise in a controlled manner can provide significant payback in the cost of implementing a WMS (see Fig. 5-17).

The returned goods processing function starts with the implementation of a return authorization (RA), which is issued by the sales department and authorizes a customer to return a specific SKU with the reasons associated with a return. This information is used by the WMS to validate that the customer is returning only the goods authorized after which credit transactions can be generated to the customer. The disposition of the returned merchandise is made by the warehouse and a quality class assignment is given if the goods are to be kept as inventory.

Like other receiving processes, the returned goods process has the requirement of creating inventory in the warehouse. However, returned goods processing uses the RA to replace the PO of the other receiving processes and has the ability to disposition the return by identifying if the goods are to be kept, scrapped, or rejected. Merchandise that is kept or scrapped will cause a credit transaction to be generated to the customer. Rejected merchandise is sent back to the customer with no credit given.

Inventory that is identified as "keep" within the returns process is assigned a movable unit to track the inventory. The company must establish policies regarding the classification of inventory that is received from a return. Frequently, companies require returned inventory to be assigned an inventory class other than first class even though the condition of the inventory is equal to first-class inventory. Other examples of inventory classes include a *rework* inventory class to indicate that repair is necessary, a *refurbished* inventory class for inventory that has been repaired, and so forth.

The returned goods processing function is designed to be performed in a workstation environment where containers of returned goods are staged at a workbench at which a receiver can open a carton for inspection. A workstation environment, combined with the required information to process a return, makes a CRT terminal the best equipment to use. A receiver initiates the returns process by indicating the RA. This number is typically written on a carton or has attached paperwork. The WMS displays a list of SKUs and associated quantities that have been approved by the salesperson.

After a SKU is selected for work, the WMS allows the receiver to indicate the disposition of the items. If the disposition is "keep," then the WMS prompts the receiver to enter the inventory's quality class, license plate number, and the location where the inventory is staged.

If the company wants to limit the amount of second-class inventory in storage, a "keep" quantity is maintained on the SKU master file to allow the WMS to warn the

FIGURE 5-17 Caged area for returned goods storage. (*Courtesy of Sharp Electronics Inc.*)

receiver when second-class inventory levels are exceeded. Another common way to limit the second-class inventory of a SKU is to specify a keep code directly on the RA line. Both the keep quantity and the keep code are values maintained by the host system through the standard interfaces described above. The following characteristics should be considered when designing a returned goods processing function:

1. *Allow the receiver to view the open and in-process RAs that can be received.* If a SKU has not been authorized for return, the receiver has the ability to add a "manual" RA for the customer. The manual RA, if used within the warehouse, is kept separate from authorized returns in case the sales group chooses not to credit the return.

2. *Allow the disposition of the SKU to be determined by the receiver.* Valid dispositions include keep, scrap, and reject (return to customer). The disposition screen is designed to allow the receiver to enter all three disposition quantities for the SKU at one time. This allows the receiver to inspect the entire item and then enter the total dispositions together. Any merchandise that will be kept as inventory should meet the requirements discussed earlier in the section on create inventory basic requirements.

3. *Provide the receiver the ability to credit the full contents of an RA to the customer in advance of disposition receiving.* Standard disposition descriptions should be contained within the standard operating procedures. A return can take a significant amount of time to process completely. This option gives the warehouse the ability to improve customer service by advanced crediting that can be reconciled after the return is completely received and dispositioned.

4. *Identify inventory to be ready for put-away after the movable unit is assigned.* The standard put-away algorithms, described in the directed put-away function section, are utilized to stock returned merchandise according to its class.

5. *Create a transaction to the host system identifying the credit quantities and received quantities for an RA.* Crediting to the customer is typically handled by the host system after receiving a transaction from the WMS. The WMS must also identify if the return was "manually" created at the warehouse so that the host system knows to create the RA number and to notify the sales group of the additional merchandise received.

PUT-AWAY (STOCKING) FUNCTION AND LOCATION MANAGEMENT

Put-away is responsible for managing how inventory is stocked within the warehouse storage space. The put-away function matches attributes of the SKU and of the location to ensure that stocking is optimized and to provide optimum placement for efficient picking operation. The put-away algorithm ensures that any stocker will store inventory in a consistent manner. This function typically constitutes 10 to 15 percent of the system implementation time depending on the location mapping and logic required for the warehouse (see Fig. 5-18).

Designing an effective location management and put-away algorithm must take into account the variety of product attributes, locations, and equipment types utilized in the warehouse. This section describes inventory and location attributes that should be considered in implementing a locator system, hierarchies associated with these attributes, two design alternatives on how the put-away function can be implemented, and supplemental location management functions. Figure 5-19 illustrates how inventory and location attributes map to a warehouse facility and will be referenced throughout the following discussions.

Inventory Attributes

Inventory attributes are assigned to a movable unit during the receiving process. These attributes have default values that are based on the SKU in the item master file. The default values indicate the preferred control parameters on how the SKU is stocked and processed through the warehouse.

1. An *environment code* is assigned to a SKU to restrict the locations to be used for storing inventory. For instance, some inventory may require special refrigeration storage for temperature control or a caged (fenced-in) area for high-value or controlled substances; see Fig. 5-19 (group 7). This requires that the location's address scanned by the stocker matches the product's environment.

2. A *company division code* is assigned to a SKU to ensure that inventory for a division is not mixed with inventory from another division. If the same SKU can exist in multiple divisions and the inventory must be kept separate, then the division is assigned to the inventory in receiving based on information from the receiving authorization (such as the PO).

Keeping inventory separate by division impacts space utilization in the warehouse if the SKU exists in multiple divisions. This requirement is typically a carryover requirement of manual systems in which it was difficult to find inventory and identify discrepancies. The need to keep inventory separate by division can usually be eliminated with the WMS's automated and real-time tracking system.

3. A *lot number code* is assigned to a SKU to identify if a SKU requires lot number tracking and when tracking should begin. Depending on the industry, lot number

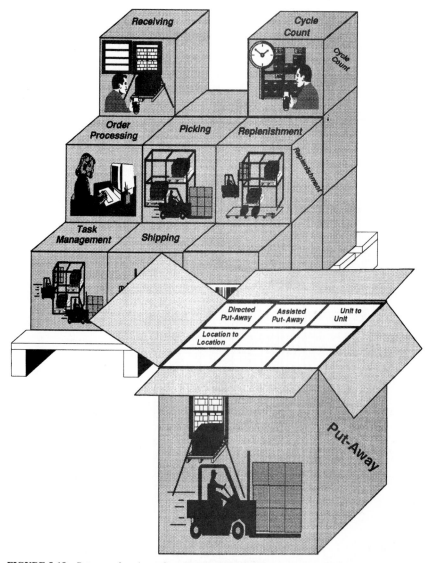

FIGURE 5-18   Put-away functions of an automated warehouse management system.

tracking and control are required from receiving through shipping. If lot number tracking occurs in the receiving operation, the put-away algorithm typically prevents mixing of lot numbers of a SKU within a location. Lot number tracking that is initiated in the receiving process will usually require special order allocation logic to ensure that strict lot rotation is accomplished to keep the number of lot numbers for a SKU to a minimum. If lot number tracking starts in the picking operation, then the put-away function will not restrict lot numbers of a SKU in a location.

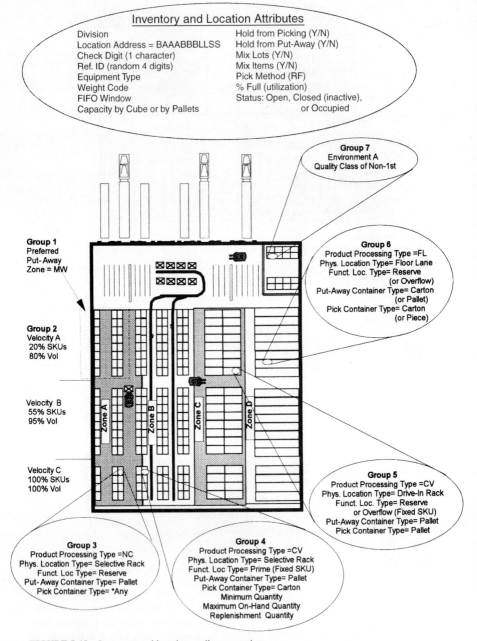

Inventory and Location Attributes

Division
Location Address = BAAABBBLLSS
Check Digit (1 character)
Ref. ID (random 4 digits)
Equipment Type
Weight Code
FIFO Window
Capacity by Cube or by Pallets

Hold from Picking (Y/N)
Hold from Put-Away (Y/N)
Mix Lots (Y/N)
Mix Items (Y/N)
Pick Method (RF)
% Full (utilization)
Status: Open, Closed (inactive),
or Occupied

Group 7
Environment A
Quality Class of Non-1st

Group 1
Preferred
Put-Away
Zone = MW

Group 6
Product Processing Type =FL
Phys. Location Type= Floor Lane
Funct. Loc. Type= Reserve
(or Overflow)
Put-Away Container Type= Carton
(or Pallet)
Pick Container Type= Carton
(or Piece)

Group 2
Velocity A
20% SKUs
80% Vol

Velocity B
55% SKUs
95% Vol

Zone A   Zone B   Zone C   Zone D

Velocity C
100% SKUs
100% Vol

Group 5
Product Processing Type =CV
Phys. Location Type= Drive-In Rack
Funct. Loc. Type= Reserve
or Overflow (Fixed SKU)
Put-Away Container Type= Pallet
Pick Container Type= Pallet

Group 3
Product Processing Type =NC
Phys. Location Type= Selective Rack
Funct. Loc Type= Reserve
Put-Away Container Type= Pallet
Pick Container Type= *Any

Group 4
Product Processing Type =CV
Phys. Location Type= Selective Rack
Funct. Loc Type= Prime (Fixed SKU)
Put-Away Container Type= Pallet
Pick Container Type= Carton
Minimum Quantity
Maximum On-Hand Quantity
Replenishment Quantity

FIGURE 5-19   Inventory and location attribute mapping.

4. An *inventory quality class code* is assigned to new inventory in receiving to allow quality class separation within the warehouse; see Fig. 5-19 (group 7). The quality class identifies inventory as first class, second class, damaged, and so forth. This code ensures that a customer order is filled with the correct quality of merchandise and allows return to vendor or "scrap" orders to be processed like any other customer order. The put-away algorithm is set up to restrict the mixing of classes within a location or to ensure that only a specific inventory class is stocked in a location.

5. An *expiration date control code* is assigned to a SKU to indicate to the receiving function that the expiration date must be captured. The WMS is designed to calculate a corresponding shelf life that is used to ensure the product is shipped soon enough to allow the customer to use merchandise in a timely manner. The WMS will ensure that expiration date mixing is not allowed within a location. This ensures that the product with the oldest shelf life is picked and shipped first, and that pickers can easily access the oldest product.

6. The *maximum inventory level* is used by the WMS to limit the amount of storage in the primary stocking building occupied by a particular SKU; see Fig. 5-19 (group 4). This product attribute is useful when off-site storage is used to hold excess inventory due to bulk buying or large manufacturing production runs. For example, if a particular SKU has a seasonal trend where demand peaks, then the warehouse will typically build up a large reserve of the item during off season when demand is low. The space within the primary stocking (picking) building is at a premium due to demand for other SKUs. The off-site storage often consists of floor storage lanes that are not easily picked. The warehouse may want to store the seasonal SKU inventory off-site except for a few pallets. Any inventory received in excess of the few pallets indicated on the SKU master file will be transferred to off-site storage, thereby minimizing the use of premium picking space.

7. *Velocity* is specified for a SKU to allow fast-moving products ("A" velocity items) to be assigned a location close to shipping dock doors; see Fig. 5-19 (group 2). A SKU's velocity is determined by the frequency with which it is picked to fill customer orders. Slow-moving products ("C" velocity items) are typically located in areas that are less accessible by pickers and are further from the shipping area.

8. *Maximum reserve locations* are assigned to a SKU to limit the number of reserve locations that are utilized by a SKU before inventory is directed to overflow locations; see Fig. 5-19 (groups 3 through 6). Reserve locations within high-velocity, easy to access warehouse zones are usually limited in number. It is common to limit the number of high-velocity reserve locations that a single item can occupy since the number of these locations is usually limited. This is accomplished by specifying the number of reserve locations that are used within each warehouse velocity zone. When the location maximum is reached in a zone, the next velocity zone is filled. When all velocity zones are filled to the location limit for a SKU, the remaining inventory is directed to larger capacity (bulk) overflow locations that are not as accessible for picking.

9. The *total receipt quantity* is used to find a location with the largest capacity available. This is used to try to keep a receipt together and to optimize the location search algorithms. Keeping a receipt together is usually important when lot number control is required for a SKU. If the largest capacity location is found, then the number of times the full search algorithm must be executed for a receipt is reduced for subsequent containers, and the overall response time for the stocker will be improved.

10. A *product processing type* is assigned to a SKU to specify the type of processing area that is best suited for the picking process (e.g., floor bulk type for heavy or large SKUs, conveyable type for standard carton SKUs with limited weight, nonconveyable type for small or irregularly shaped SKUs, and so forth). A SKU may be restricted on how it is handled within a warehouse. This attribute helps to select the optimum storage zone in the warehouse that handles a SKU's special processing requirements.

11. A *preferred put-away zone* is assigned to a SKU to allow the put-away algorithm to locate similar products together such as electronics or a sales concept; see Fig. 5-19 (group 1). This allows SKUs related by freight class, or store department, to be grouped together within a warehouse area.

12. A *FIFO window* is specified for the SKU, or for all SKUs within the warehouse, to limit the effective stock rotation by the proximity of when a receipt is made. This value expresses the maximum number of days that inventory for a SKU is considered to be the same. For instance, if a SKU is received in March and another in April, and the FIFO window is set to 60 days, then the two receipts can be mixed in one location. If another receipt for the same SKU came in June, however, then inventory mixing within one location would not be allowed so that stock rotation is ensured.

13. *Container types* are assigned to inventory during receiving to identify what the inventory resides on and its state; see Fig. 5-19 (groups 4 to 6). For instance, if full cartons of a SKU are built onto a pallet during receiving, then the license plate associated with the inventory should be *pallet* and the inventory's state on the pallet should be assigned *carton* container type. The put-away algorithm is designed to search for a location that can hold a pallet and allow either picking by pallet or picking by carton. This allows put-away and picking operations to be optimized and reduces overall handling of the product (i.e., prevents stocking of broken cases in a full pallet or case picking area).

14. A *weight code* is assigned to inventory to ensure that the recommended stocking location can physically support the put-away container. The weight is identified on the SKU and accumulated onto the movable unit as inventory is added. A weight code is then assigned according to predetermined ranges that match the type(s) of storage available within the warehouse. For instance, the style of racking may limit the weight that is supported in the top levels compared to a floor-level location. The put-away algorithm is designed to search first for a matching weight code before looking at higher-weight locations.

15. A *container size code* is assigned to inventory to allow multiple racking configurations to exist within the warehouse. Typically, if multiple pallet sizes are supported, the SKU will specify a default container size that the inventory should be built upon in the receiving operation. The put-away algorithm is designed to search only for locations that can hold the container's size.

16. A *status type* is assigned to inventory during receiving. This is used to identify inventory that requires checking through quality control. Also, the status indicates if the inventory on the container should be held from put-away or picking processes until quality checking is complete.

17. A *group code* is assigned to inventory during receiving. If different SKUs are stocked separately but sold together, then the PO or manufacturing order will indicate the group to which the SKUs are assigned. The put-away algorithm uses this code to search for existing inventory that has the same group code and attempts to find an available location with the same zone or aisle in which the existing inventory is stocked.

Location Attributes

The attributes associated with inventory discussed above also apply to locations to allow matching by the put-away algorithm. A location typically has additional attributes that affect put-away logic but are not restricted to the inventory. These additional location attributes include the following:

1. A *stocking location code* is assigned to a building. The stocking location found by the put-away algorithm must be in the same building as specified by the user's profile. This ensures that the system does not direct the stocker to a location that cannot be physically serviced by the stocker.

2. A *hold from put-away flag* is assigned to a location to prevent the location from being used for stocking. If a location has an inventory discrepancy, is full, or is not usable due to maintenance, then this flag is set to restrict its use.

3. A *capacity type* is assigned to all stocking locations. A location's capacity is specified as the cube for a location or as the number of pallets that can fit in a location. Location cube should be considered for any locations that are not limited to a pallet's footprint to control its capacity. If the location is to be used only for pallet storage, then the number of pallets should be used to control its capacity.

Location capacities that are commonly handled by cube include floor lanes, shelving, bins, and so forth. The location cube must be specified in the same unit of measure as the cube on the SKU to allow the system to match values. Location capacities that are commonly handled by specifying the number of pallets include selective (single) pallet racking, push-back racking, drive-in pallet racking, and so on. Pallet storage locations usually have an accompanying container size attribute as discussed earlier in the section on inventory attributes.

4. An *equipment type* is assigned to a stocking location if special equipment is required to address the location. The stocker's profile identifies the equipment type being used. The put-away algorithm limits the location search to only locations matching the stocker's equipment type. If a location cannot be found, the stocker can obtain a different piece of equipment and change the profile, or notify another stocker with a different piece of equipment to locate the unit.

5. A *location usage type* is assigned to a location to identify if the location is used for primary picking, reserve storage, or overflow (second reserve) storage; see Fig. 5-19 (groups 4 to 6). Locations designated for primary picking, or home locations, are also assigned a dedicated SKU. Primary picking locations are used to pick broken pallets (cases) or broken cases (pieces) and are usually specialized storage locations (such as flow racks or shelving) and may have dedicated material handling equipment (such as a conveyor for cartons).

Locations designated for reserve or overflow contain stock for higher-volume orders or for replenishment to primary pick locations. For instance, order picking for a full pallet of a SKU is performed most efficiently if the inventory is available in full pallet quantities in a reserve or overflow location rather than removing individual carton quantities from a carton picking location. Reserve locations are typically located in close proximity to prime locations and have a higher capacity, full- or multiple-pallet capacity, than prime locations. Overflow locations may be out of aisle and possibly in a remote or off-site warehouse and typically have higher capacity (floor lanes or drive-in racks) than reserve locations.

6. An *address* is assigned to a location to allow it to be found by a stocker or picker. Aisle, bay, level, and slot coordinate attributes are used in whole or in part to identify where the location physically exists in the warehouse. These attributes are typically printed in human readable and bar coded formats onto a label that is attached to the location for scanning and system confirmation when inventory is moved in the warehouse.

Coordinate attributes that constitute the location's address may have associated put-away logic. For instance, the aisle attribute is frequently used to determine preferred reserve location based on proximity to a prime location. This proximity logic is extended to searching for bays that are close to a prime location. Other logic might utilize odd and even bay assignments to determine the sequence in which locations are displayed to workers. For instance, if a conveyor is physically down the middle of an aisle, then pickers can display picks from all odd bays first followed by even bays (possibly descending) to allow efficient picking.

7. A *random reference identification* (ID) code is assigned to a location where it may be difficult to scan a bar coded address label. The reference ID is three to four randomly assigned alphanumeric characters that are assigned by the system when a location is created. The stocker or picker is directed to a location by the location's address and confirms the proper location by hand-keying the reference ID rather than scanning the location's address bar code.

The reference ID and the location's address are printed in human readable characters onto a label or sign. Locations best suited for a reference ID over a bar code include floor lanes, receiving and shipping staging lanes, drop zones, and any location that may be exposed to bar code damage.

8. A *minimum quantity code* is assigned to locations used for primary picking. This quantity specifies the minimum inventory level that should be maintained in a location for a SKU. If the quantity on hand drops below the minimum quantity for a SKU, a replenishment pick request is made to move reserve or overflow stock to the prime location. Single deep pallet racks that are used as a case pick prime location will be set with a zero minimum quantity.

The minimum quantity must be determined for every SKU that has a primary picking location, and its value should be high enough to ensure that inventory is available for average picking volumes. Multiple primary pick locations may be required for any high-velocity SKU that is typically ordered in less than full-pallet quantities.

9. A *reorder quantity code* is assigned to locations used for primary picking; see Fig. 5-19 (group 4). This quantity specifies the quantity increment that should be moved from reserve or overflow locations when the minimum quantity level is reached. This quantity is usually expressed in even carton quantities if replenishing a broken case prime location, or in even pallet quantities when replenishing a case pick prime location.

10. A *maximum quantity code* is assigned to locations used for primary picking; see Fig. 5-19 (group 4). This quantity specifies the maximum quantity of a SKU that should be maintained in a prime location. This value is set to be at least the minimum quantity plus the reorder quantity.

The maximum quantity attribute is used in two ways. First, when the minimum quantity is reached due to customer order picking, replenishments are issued in reorder quantity increments until the maximum quantity is achieved. Second, end-of-day replenishments can be generated for any prime locations that are not at maximum capacity. End-of-day replenishments allow preparation for following-day picking activities.

11. A *mixed-SKU flag* is assigned to a storage location if multiple-SKU storage is permissible. Mixing of SKUs is common for locations used to stock nonfirst inventory. Mixing of SKUs may also be permissible in large-capacity locations such as floor lanes. However, mixing of SKUs within a storage location does not imply mixing of containers that keep the products physically separated.

12. A *keep movable unit flag* is assigned to a stocking location when the system should keep individual license plates assigned to containers (inventory). Movable units represent a specific inventory quantity and configuration and are generally retained for pallet (or full container) picking locations.

Locations that are set up to lose the movable unit identity transfer inventory owner-ship to the location for inventory tracking. Locations that might qualify for noncon-tainer tracking include floor-lane (or large-capacity) locations where container sequencing may be a problem for pickers and prime picking locations where the license plate may be lost due to breaking the container down for picking.

Put-Away Search Hierarchy

As inventory and location attributes are analyzed, tradeoffs are required. For example, if it is important to keep the total receipt quantity together, then the system-generated location could be in a slow-moving zone even though the SKU has a high-velocity code. However, if it more important to keep high-velocity items close to shipping areas for accessibility, then the SKU might be stored in multiple stocking locations within the high-velocity area.

The put-away algorithm utilizes the inventory and location attributes discussed above to search for and to validate locations where inventory is stocked. The sequence and hierarchy of how these attributes are utilized are what constitute the put-away algorithm. The following discussion outlines a sample put-away attribute search hier-archy and is illustrated by Fig. 5-20.

First, the primary attributes, along with related logic, that will be considered in this example are as follows:

1. *Container type.*   This logic attempts to find a location that can hold the physical container type the inventory resides upon, along with keeping the inventory at its highest picking state. If the put-away function involves a pallet of full cartons, an attempt is made to find an available location that can hold pallets and pick pallets. If a location cannot be found, then an attempt is made to find a location that can hold pallets and pick cartons. If a location cannot be found, then an attempt is made to find a location that can hold cartons, and the container that the inventory is on is dropped.

2. *Processing type.*   This logic utilizes a table to define how processing types are related. First, an attempt is made to find an available location that matches the inventory's processing type (i.e., conveyable inventory in a location that utilizes a conveyor). If a location cannot be found, then the processing type table defines the next processing type to search for. This process continues until a location is found or all related processing types are exhausted.

3. *Velocity.*   This logic utilizes a table to define how velocities are related. First, an attempt is made to find an available location that matches the inventory's velocity code. If a location cannot be found, then the velocity table defines the next veloci-ty to search for. This process continues until a location is found or all related velocities are exhausted.

4. *Preferred put-away zone.*   This logic tries to find an available location that matches the inventory's preferred put-away zone. If a match cannot be found, then the preferred put-away zone is dropped from the search criteria.

5. *Environment.*   This logic requires that an available location be found that matches the inventory's environment. If no location can be found, the search is stopped.

6. *Inventory status.*   This indicates if the inventory is available for stocking in pick-ing locations.

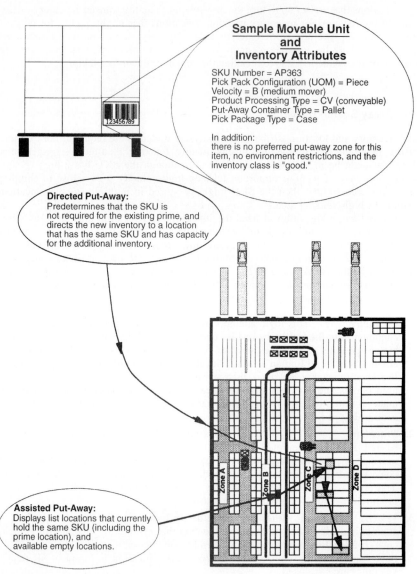

**Sample Movable Unit
and
Inventory Attributes**

SKU Number = AP363
Pick Pack Configuration (UOM) = Piece
Velocity = B (medium mover)
Product Processing Type = CV (conveyable)
Put-Away Container Type = Pallet
Pick Package Type = Case

In addition:
there is no preferred put-away zone for this
item, no environment restrictions, and the
inventory class is "good."

Directed Put-Away:
Predetermines that the SKU is
not required for the existing prime, and
directs the new inventory to a location
that has the same SKU and has capacity
for the additional inventory.

Assisted Put-Away:
Displays list locations that currently
hold the same SKU (including the
prime location), and
available empty locations.

FIGURE 5-20   Example of put-away.

The algorithm executes the following logic sequence:

1. The movable unit is checked to see if it is allocated to fill a customer order. If the inventory is allocated, then the put-away program checks the shipping lane or dock that is assigned to the order and directs the stocker to move the inventory to the shipping location. If no lane or dock is assigned to the customer order, the stocker is directed to drop the unit in a pick staging location.

2. If the movable unit has a quality control status, then put-away logic directs the movable unit to a location whose environment is designated for quarantine storage. If no location is found with a matching environment, then the stocker is directed to return the movable unit to its previous location.

If the above conditions do not apply, then the following logic hierarchy is executed:

3. The first pass is to search for an available location that matches the movable unit's environment, container type, product processing type, velocity, and preferred put-away zone. If the put-away program is unable to find a match, it drops the preferred-zone criteria and searches again.

4. If no location is found, the velocity table is used to determine the next velocity zone to search. An attempt is made to find an available location that matches the environment, container type, product processing type, and new velocity type.

5. After exhausting all available location possibilities using the velocity table, the next product processing type is obtained from the product processing type table. The original velocity is reinstated and step 4 is repeated. This process continues until a location is found or there are no more entries in the product processing type table.

6. Finally, if a location cannot be found using these tables, an error message is displayed to the operator. Whenever a suitable location is found for a unit, space for the inventory is reserved in the location file.

Put-Away Design Alternatives

The above discussion on attributes and the search hierarchy demonstrates the sophistication that can be employed by a WMS put-away function. Utilizing RF terminals to direct and record inventory moves within the warehouse adds significant benefits by reducing manual search time and recording errors.

Directed and assisted put-away are two design alternatives. Directed put-away provides optimum location utilization and inventory control by utilizing the location and inventory attributes in a predetermined search algorithm. Assisted put-away provides manual selection of stocking locations while providing real-time tracking and inventory consolidation without the overhead of a complex put-away algorithm. These functions are described in more detail below.

Directed Put-Away. Directed put-away is designed to direct a stocker to move inventory to a specific stocking location based a predefined put-away algorithm (see Fig. 5-21). The process is initiated by a stocker using a RF terminal and scanner to input the movable unit number. The WMS utilizes information on the inventory record and put-away algorithm to search for an available location and reserve location space for the movable unit. The stocker is prompted to place the movable unit in the directed location and confirm its placement by scanning the location's address label or entering the location's reference ID. The following characteristics should be considered when designing a directed put-away function:

1. *Provide a list of containers that are ready for put-away.* The stocker is prompted to scan a movable unit to move. The system searches for a location that can hold the container and display the recommended location to the stocker. The recommended location is reserved for the container to prevent other goods from using the space prior to the move being made.

FIGURE 5-21   RF directed put-away to narrow aisle selected rack.

2. *Allow the stocker to add containers to be put-away.*   This function is useful
   when the equipment and inventory configuration on a movable unit allows multiple containers to be stacked or moved together.

3. *Allow the stocker to submit a group of movable units for put-away.*   A group of
   movable units can include all containers in a receiving lane or a selectable list of
   movable units ready for put-away. This option allows the put-away algorithm to
   run in batch mode, rather than in real time, for all the movable units to minimize
   delays associated with executing the put-away algorithm for individual movable
   units.

4. *Allow the stocker to change profile.*   The profile controls such things as the
   building the stocker is in and the type of equipment being used. These attributes
   affect the locations that are found according to moves the stocker can make.

5. *Prompt the stocker to verify the inventory's description prior to searching for a
   stocking location.*   If the description is inaccurate, then the stocker can notify a
   receiving supervisor to adjust the movable unit.

6. *Direct the stocker to the location reserved for a movable unit.*   After placing the
   container in the directed location the stocker scans the location address bar code
   label or enters the location's reference ID.

7. *Allow the stocker to split a movable unit.* If the container has multiple SKUs, or the SKU on the container exists in multiple states (i.e., cartons and loose pieces), then the stocker may request to stock the inventory separately. The stocker is prompted to enter the SKU and quantity to be removed from the movable unit, and the put-away algorithm will search for a new location.

8. *Allow the stocker to identify problem locations by submitting the directed location to the cycle count queue.* If the directed location is full or has conflicting inventory to the inventory on the container, then the location can be submitted for review by a cycle counter.

9. *Allow the stocker to override the directed location or request an alternative location.* If the stocker scans a different location than the directed location, then the location entered is validated by the put-away algorithm and the override is logged. An alternative location may be requested if the directed location cannot be accessed due to other traffic. The system will direct the stocker to another location meeting the put-away criteria.

10. *Provide the stocker flexibility to indicate when a location is full, such as when the location's capacity is tracked by cube rather than number of pallets.* This causes the system to update the location as full and to reopen when the picking operations free up enough space to allow additional stocking. A system default percentage controls the amount of free space that is available prior to reopening a full location.

Assisted Put-Away. Assisted put-away is designed to help the stocker determine a location to stock new inventory. The process is initiated by scanning a movable unit number. The WMS then displays a list of locations that include the prime picking location and any location where there is matching inventory. In addition, the stocker is given the ability to view a list of empty locations or to view locations that have inventory with a matching group code (described in the section on inventory attributes).

Assisted put-away provides the ability to reconcile problem locations, to indicate when locations are full, and to split a movable unit. However, the fundamental difference between directed and assisted put-away is that assisted put-away does not use the put-away algorithm to search for and recommend a location. The assisted put-away algorithm lists locations where inventory currently exists that matches the movable unit's inventory or lists locations that are empty in the warehouse. The system checks the location scanned by the stocker to validate if the location can be used to stock the merchandise. The following characteristics should be considered when designing an assisted put-away function:

1. *Provide a list of containers that are ready for put-away.* The stocker is prompted to scan a license plate to move. The system searches for stocking locations that have the same SKU and inventory class that is on the movable unit. A list of locations is displayed to the stocker that includes any primary picking locations and the space available in the location.

2. *Allow the stocker to alternate the inventory and location look-up.* One display is an empty location list. A second, alternate, display lists locations that contain SKUs with the same group code.

3. *Allow the stocker to indicate when a location is full.*

4. *Allow the stocker to submit a location to a cycle count when an obvious inventory discrepancy exists.*

5. *Allow the stocker to split a movable unit.*

Supplemental Location Management Functions

In addition to normal put-away of new inventory, the WMS must support consolidation activity with the warehouse. The WMS allows the movement of inventory from one location to another or from one movable unit to another. These functions are called *unit-to-unit* and *location-to-location* moves. Both of these supplemental functions utilize attribute validations to ensure inventory is consolidated in an acceptable manner. These functions are designed so that new inventory that is awaiting put-away or containers with merchandise allocated to customer orders are restricted from manual consolidation and movement. This ensures that pickers and shippers can find inventory intended for a customer and that new inventory is located using the put-away algorithm.

The unit-to-unit move function is designed to allow inventory to be consolidated between containers. The stocker is prompted to scan a movable unit for the container from which to move inventory. After the inventory description is confirmed, the system prompts the stocker for a new movable unit. If the new movable unit exists, the system validates that the inventory can be combined. If the movable unit does not exist, the system will prompt the stocker for a location where the new movable unit is stocked.

The location-to-location move function is designed to allow inventory consolidation for entire locations and utilizes a RF terminal and scanner. The stocker is prompted to enter a location from which to move inventory. After an inventory description is confirmed, the system prompts the stocker for the location to which to move the inventory. The system validates that the destination location scanned by the stocker has the capacity to hold the inventory and ensures that all inventory is not improperly mixed within the location.

CYCLE COUNT

A significant problem in many warehousing operations is the requirement by the financial department to confirm inventory levels on a periodic basis in order to track the company's inventory book value. Many warehouses are required to perform physical inventory counts on a monthly basis. This frequent counting requirement is often due to the inaccuracies associated with manual processes and lack of automated tracking procedures (see Fig. 5-22).

The need to perform a physical count of the entire warehouse's inventory on a frequent basis can cause havoc with warehouse managers who are trying to fill customer demand in a timely fashion. A physical inventory requires the entire warehouse operation be shut down, with no receiving or shipping operations, ranging in duration from one day for small warehouses up to five days for large warehouses. If the physical inventory is done on a monthly basis, then loss time from order-fulfillment activity can range from 5 to 25 percent.

The purpose of the cycle count function is to reduce, and potentially eliminate, the need to stop warehousing operations to perform physical inventories. Elimination of physical inventories is usually met by resistance from the accounting or financial departments until cycle counting can prove its accuracy.

Overall, the inventory accuracy in the warehouse operation improves by implementing the positive and real-time tracking associated with an automated WMS. Cycle counting ensures inventory accuracy by periodically counting the inventory in all locations. The cycle count period is established by the accounting or financial departments and specifies how often all locations, or all SKUs, must be verified.

FIGURE 5-22   Cycle count functions of an automated warehouse management system.

The locations that require cycle counting are determined from several factors. First, locations are counted when a discrepancy is found in the put-away or picking operations. This is done by allowing the stocker or picker to submit the location to the cycle count queue, usually with a high priority, which directs a counter to verify and resolve any discrepancies that may exist. Second, when a location is picked empty, the picker is requested to verify the location as empty. This provides a quick count since there should be no inventory remaining in the location. If inventory exists in a location that the system thinks is empty, then the location is submitted as a discrepant location to the cycle count queue for verification and reconciliation by a counter. Finally, a cycle count queue generation program submits locations to the cycle count queue based on the value of the inventory (SKU) in the location or on a predetermined location count frequency.

Designing a cycle count function involves 5 to 10 percent of the overall systems design and implementation time. There are five primary functions with the cycle count module. This section will discuss the design for the following cycle count functions:

1. *Cycle count queue generation.*   This function is designed to determine the locations that should be counted on a periodic basis.

2. *RF cycle count.*   This function is designed to allow a count to work the cycle count queue and enter inventory quantities on a real-time basis.

3. *RF supervisor count.*   This function may be required if the warehouse policy requires a second count and verification for discrepant locations prior to adjusting the on-hand inventory.

4. *Inventory adjustment.*   This function is required to allow the warehouse to notify the finance department of adjustments to the on-hand inventory and book value.

5. *Load initial inventory.*   This function is required for cut-over of warehouse operations from the existing system to the automated WMS. This is an often overlooked requirement when designing a WMS.

Cycle Count Queue Generation Function

The purpose of the cycle count queue generation function is to allow the warehouse operations to generate a list of locations that require counting to satisfy the periodic count designated by the financial departments. Before designing this function, the company must agree on the methodology that will be employed to generate this queue.

The cycle count queue can be generated in two basic ways. First, the company's policy may be to count all locations on a periodic basis, such as quarterly. This requirement is met by tracking when a location is counted, and generating a cycle count queue based on the oldest count dates first. The number of locations that must be counted on a daily basis is calculated by dividing the total number of locations by the number of working days in the cycle count period.

Second, the company's policy may be to generate locations based on the value and/or turns of a SKU. This is accomplished by assigning a SKU a cycle count velocity. The table shown in Fig. 5-23 outlines how a SKU's cycle count velocity is assigned.

SKU	Turn (put-away) velocity	Value velocity	Cycle count velocity
X	A	B	A
Y	B	A	A
Z	B	B	B

FIGURE 5-23   Cycle count velocity table.

The turn velocity is the same as the put-away or picking velocity that is used to segment locations and stocking within the warehouse (as discussed in the section on put-away). The value velocity is assigned to a SKU based on the value of the SKU. The ranges for A, B, and C velocities are determined for the product handled in the warehouse. The cycle count velocity is the higher-velocity value between the turn velocity and the value velocity. The system is set up to ensure that high-value or high-turn SKUs are counted more frequently within the cycle count period than low-value and slow-moving SKUs. The cycle count queue generated from this method represents all locations that contain the SKU being counted. Thus, the number of locations to count on a daily basis will not be constant.

The cycle count queue generation program is written to be submitted in batch mode. This allows the queue to be generated on a nightly basis so that counting can be accomplished the following day. In addition, a second queue generation program version is created that submits all stocking locations to the cycle count queue to allow for physical inventory counts to be accomplished. Finally, a cycle count queue maintenance function is designed to allow the warehouse manager to submit a SKU or a selected range of locations to the cycle count queue.

RF Cycle Count Function

The RF cycle count function is designed to allow a counter to work the queue of locations that have either been submitted by the cycle count queue generation function or by stockers and pickers who have found locations that do not match the system's inventory (see Fig. 5-24). The cycle counting function is not integrated with other

FIGURE 5-24   Warehouse operator using RF terminal to verify inventory levels. (*Courtesy of Sharp Electronics Inc.*)

functional areas, such as put-away or picking, so that productivity in these other functions is not diminished. This function is designed to ensure that locations are counted based on the cycle count queue priority.

The cycle count queue priority is higher for locations submitted by stockers and pickers than locations generated for periodic counting. The cycle count priority ensures that locations being held from put-away or picking have count verification and resolution as quickly as possible. System-generated periodic counts do not restrict put-away and picking activity until the location is physically being counted.

The RF cycle count function directs the counter to a location based on the highest-priority count locations within a zone selected to count. The counter verifies that the directed location is being counted by scanning the location's address label or entering its reference ID. The counter is then prompted to identify the SKU to count and to enter the quantity that is in the location. If this "blind" count matches the system count, then the location is updated with the cycle count date and the counter is direct-ed to the next location to count. If the SKU count does not match the system count, then the WMS allows the counter to reenter the blind count. If the second count matches the system count, then the location is updated with the cycle count date and the counter is directed to the next location. If the second count does not match the sys-tem count, then the location is marked as discrepant and is held from put-away or picking activity until the inventory is recounted or adjusted by a supervisor. The fol-lowing characteristics should be considered in designing a RF cycle count function:

1. *Allow the counter to select the zone to work.*   The number of high-priority (non-periodic counts) locations within a zone is displayed to allow the counter to work problem areas first.

2. *Direct the stocker to the highest-priority location count within the selected zone.* The stocker scans the location address label or enters the location reference ID to confirm the directed location is being counted. The location being counted is held from put-away or picking activity while the count is in progress. After all high-pri-ority locations are counted, then low-priority (system-generated) locations are dis-played in sequence according to their address (aisle, bay, level, and slot sequence).

3. *Prompt the stocker to identify the SKU to count by scanning the SKU or by select-ing the SKU from the system's inventory list for the location.*   The counter has the ability to add a SKU not listed for a location.

4. *Prompt the counter to enter the quantity of the SKU in the location after the SKU to be counted has been identified.*   The system count is *not* displayed to ensure that a blind count is obtained. The count entered is the full quantity of the SKU that exists in the location. If a SKU is added to a location, the WMS automatically creates a dis-crepancy without requiring a second count prior to a supervisor's review.

5. *Prompt the counter to enter a second count if the first count does not match the system's count.*   If either the first or second count matches the system count, then the inventory level is confirmed and the counter is directed to the next location. The location counted is updated with the cycle count date to provide evidence to finance of adherence to the cycle count criteria.

6. *Hold a location from put-away and picking activity if neither the first nor second count matches the system count.*   This reduces investigation problems when a super-visor reconciles the location. The location is released if the supervisor's count match-es the system count or an inventory adjustment is made to reconcile the problem.

7. *Track all quantity counts and the person performing the count for problem investi-gation and inquiry.*

RF Supervisor Cycle Count Function

The RF cycle count function is designed to allow a supervisor to count any locations that have been identified as discrepant. A discrepant location is identified through the RF cycle count process or from locations identified by pickers as not being empty.

As in the RF cycle count function, counts entered by the supervisor are blind. Blind counting ensures that a physical count is done, and if the count matches the system count, then the location's inventory level is confirmed.

The RF supervisor cycle count function directs the supervisor to a location based on the location's sequence within a selected zone. The supervisor verifies that the directed location is being counted by scanning the location's address label or entering its reference ID. The supervisor is then prompted to identify the SKU to count and to enter the quantity that is in the location. If this blind count matches the system count, then the location is updated with the cycle count date and the supervisor is directed to the next location to count. If the SKU count does not match the system count, then the WMS allows the supervisor to reenter the blind count. If the second count matches the system count, then the location is updated with the cycle count date and the supervisor is directed to the next location. If the second count does not match the system count, then four counts (two initial and two supervisor counts) are displayed to the supervisor. The location remains discrepant, held from put-away or picking operations, until an inventory adjustment is entered for the location. The following characteristics should be considered in designing a RF supervisor cycle count function:

1. *Allow the supervisor to select the zone to work.* The number of discrepant locations is displayed next to the zone.

2. *Direct the supervisor to a discrepant location to count based on the location's address (aisle, bay, level, and slot sequence).* The supervisor confirms the location being counted by scanning the location's address label or by entering the location's reference ID.

3. *Display a list of discrepant SKUs within the location for the supervisor to count.* This ensures that only the SKUs requiring recount are verified.

4. *Prompt the supervisor to enter a blind count for the SKU selected.* If the count matches the system count, the cycle count date is updated for the location and the supervisor is directed to the next location to count. If the "blind" count does not match the system count, the supervisor is prompted to reenter the count.

5. *Display all four counts to the supervisor along with the system's count if none of the four counts match the system's inventory record.* This allows the supervisor to identify the discrepancy and note any adjustments that should be made. The location remains on hold from put-away and picking activities until the inventory is adjusted for the SKU and location.

6. *Track all quantity counts and the person performing the count for problem investigation and inquiry.*

Inventory Adjustment Functions

The purpose of the inventory adjustment function is to allow a supervisor (or authorized personnel) to adjust the on-hand book inventory value while maintaining the WMS location and inventory files. Two inventory adjustment functions are usually required in a WMS, one to support inventory control from a manager's desk using a CRT and one to allow real-time inventory updates by a cycle count supervisor using a RF terminal. The CRT version of this function allows the manager more flexibility on the accounting information captured, and the manager has access to inquiry functions that may be necessary to investigate fully any discrepancies prior to making an adjustment.

Adjustments made through these functions allow WMS inventory maintenance and create inventory adjustment transactions to update the host computer's inventory summary file. The information entered for an inventory function must reflect what is needed by a company's accounting departments to authorize an adjustment. The following is a list of common information that is required by finance departments to adjust inventory books:

- *Division.* Identifies who owns the inventory.
- *SKU.* Identifies what inventory to adjust.
- *Quantity.* Identifies how much inventory to add or remove.
- *Reason code.* A predetermined set of valid reasons for the warehouse to adjust inventory should be obtained from the finance department.
- *Account number.* Identifies who is absorbing the cost associated with the adjustment.
- *Document number.* Tracks the adjustment against the accounting books; a value assigned by the finance department.

Some additional information required for the WMS files includes:

- *Location.* Identifies where the inventory adjustment is being applied.
- *Container number.* Identifies the specific inventory record to adjust.
- *Class.* Allows the warehouse manager or supervisor to change the inventory's class.
- *Status.* Allows the warehouse manager or supervisor to hold and release specific inventory. This also allows inventory control to ensure that problem inventory can be restricted and released for general use.
- *Control data.* Allows for confirmation by lot or serial number for any adjustment if the inventory being adjusted is serialized or lot controlled.
- *Unlock location.* Allows the warehouse manager or supervisor to restrict a location from being used within the warehouse and to release locations held resulting from a cycle count discrepancy.
- *Create interface to host computer flag.* Allows the warehouse manager or supervisor to adjust the inventory without updating the host computer. This is necessary if the inventory error only exists on the WMS or the manager is using this function to prevent the inventory's movement temporarily.

The RF inventory adjustment function is designed to default many of the accounting fields to values established in WMS configuration files. Defaulting values combined with multiple screens allows the necessary information to be entered using a RF terminal. Typically, the RF inventory adjustment function defaults the account and document numbers to predetermined values. The following characteristics should be considered in designing the RF inventory adjustment function:

1. *Prompt the supervisor to scan the location's address label or enter the location's reference ID to adjust.* This ensures that the correct location is being adjusted.
2. *Prompt the supervisor to scan the bar coded container label to adjust.* It may be necessary to use other inquiry programs to look up the container number if it was lost or destroyed during handling.
3. *Prompt the supervisor to scan, or enter, the SKU to be adjusted along with the quantity that exists in the location.* The full quantity on the unit is entered, not

the quantity to adjust. The WMS will calculate the adjustment quantity that reduces manual calculation errors.

4. *Prompt the supervisor to enter a reason code associated with the adjustment if the inventory quantity or inventory class is being changed.*

5. *Allow the supervisor to change the inventory's class, change the inventory's status, and indicate whether to hold the location.* These attributes allow the supervisor to control inventory usage while walking the warehouse floor. If only the inventory status or location status is changed, then no inventory adjustment is sent to the host computer. All other adjustments will cause an adjustment to the host (unless the CRT inventory adjustment is used).

Any inventory adjustments, RF or CRT, will check customer order allocations to ensure that picking records will not be affected. If the inventory quantity adjusted is for inventory allocated to a customer order, then the inventory adjustment function ensures that pick records are reallocated or adjusted. Furthermore, if the SKU adjusted is under lot control or serial number tracking, the supervisor is prompted to confirm the control data being added or removed from the movable unit. This may only be necessary if multiple lots or serial numbers exist.

Load Initial Inventory Function

The purpose of the load initial inventory function is to provide a mechanism to turn the warehouse operations over to the new WMS. This function performs the same data capture as the RF inventory adjustment program when inventory is being added. In addition, this function defaults the reason code to a cut-over (start-up) value and submits a cycle count for every location to which inventory has been added.

This function is designed for RF terminals to allow multiple warehouse workers, not just a supervisor, to have authority to add inventory to the warehouse locations in real time rather than using secondary data-entry clerks. As inventory is added to a location, the warehouser also applies bar coded license plates to the containers to create movable units. The SKU master file is checked to ensure that the inventory is valid and if control data needs to be captured. The following characteristics should be considered when designing a load initial inventory function:

1. *Prompt the warehouser to scan the location's address label or enter the location's reference ID to ensure the proper location is being loaded.* Only one warehouser is allowed to add inventory to a location at a time.

2. *Submit the location to the cycle count queue to ensure the quantities added into a location are correct.*

3. *Prompt the warehouser to enter a movable unit number for the inventory if the location requires movable unit tracking.*

4. *Prompt the warehouser to scan or enter the SKU to be added.* The warehouser then enters the quantity on the container, the inventory class, and the inventory's FIFO date (defaulted to the system date). The WMS ensures that all location management rules outlined for put-away are adhered to.

5. *Prompt the warehouser to confirm the inventory record prior to the WMS adding the information to the system files.*

6. *Prompt the warehouser with the appropriate control data entry screen if the SKU requires lot control, expiration control, or serial number tracking.*

ORDER PROCESSING, PICKING, AND REPLENISHMENT

The goal of a warehouse, or distribution center, is to ensure accurate customer order fulfillment and shipping. Order processing, order picking, and replenishment are the primary functions required by a WMS to ensure accurate order fulfillment while managing inventory and resource utilization.

The basis for fulfilling customer orders efficiently is established by the receiving and put-away functions. The receiving process ensures accurate inventory levels while organizing the inventory onto containers for handling efficiency. Movable units were created to ensure inventory traceability and real-time inventory control. The put-away and location management processes assisted in positioning the product to allow efficient inventory picking and letdown to shipping. Inventory attributes such as the product processing type, preferred put-away zone, turn velocity, container types, and inventory groups help to optimize storage while preparing the inventory for picking.

The order fulfillment (picking) process can be broken down into three areas:

1. Order processing provides the ability to download and validate order data from the host's order entry system. Order processing then allows a picking or shipping supervisor the ability to coordinate picking and shipping activities by controlling the sequence in which customer orders are allocated and the method used to generate picking requests.

2. The picking function provides the process to direct inventory retrieval and letdown to shipping areas. Many picking methods may be utilized to provide the most efficient processing according to the quantities being picked and where the inventory is stocked.

3. The replenishment function provides the process for directing inventory movement from reserve and overflow locations to primary (home) locations. Replenishment ensures that primary pick inventory is available to fill orders efficiently.

Order processing and picking functions account for 20 to 30 percent of the total design and system implementation time based on the allocation algorithm complexity and picking varieties. In addition, customer satisfaction hinges on timely and accurate order fulfillment. The picking functions provide the processes to meet a customer's expectations.

Order Processing Functions

The purpose of the order processing functions are to provide order integrity checking and validation for order downloading from the host computer, order and wave (order group) planning, and wave release and inventory allocation. Order processing provides the mechanism for controlling workload to the warehouse door, along with timely shipment to customers (see Fig. 5-25).

Customer Order Download.   The customer order interface ensures that the data associated with the order is valid and calculates order totals. It also calculates workload totals, automatically routes orders, assists in truckload planning, and determines freight charges. First, the customer order interface program provides data integrity checking to ensure that enough data exists on the order to allow inventory allocation and shipment. The order interface provides the following integrity checks:

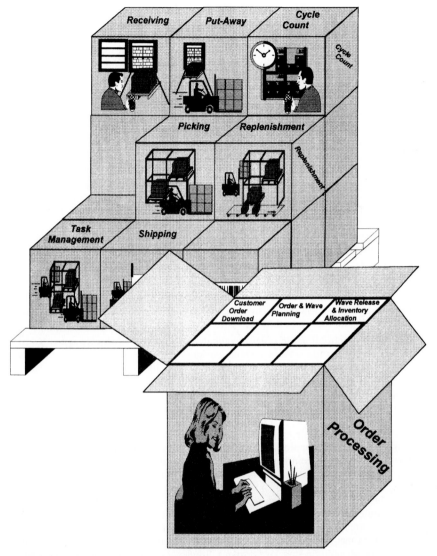

FIGURE 5-25   Order processing functions of an automated warehouse management system.

- The SKU requested is validated against the SKU master file. If the SKU is discontinued, then a substitute SKU is determined from the SKU master file. If the SKU is invalid, or a substitute SKU cannot be determined, then the order is held for review. Orders so held are managed through the order review function discussed later in this section.
- The order quantity is preallocated against the inventory summary based on the host's preallocated quantity or based on the available quantities. If the percent fill is

below warehouse shipping standards (established within the WMS), then the order is held for adjustment or until the fill percentage is satisfactory. If the percent fill meets the warehouse shipping standards, then workload estimates are calculated by the estimated number of pallets, cartons, and pieces that will meet the preallocated quantity. The workload is accumulated for all lines on the order and is presented to the order processing or shipping supervisor during wave planning.

Host preallocation is common in order entry systems to identify potential back-order conditions to the customer as early as possible. Also, host preallocation is common when lot control policies provide the customer with the ability to request a particular lot number. In these cases, the host ensures order fulfillment percentages can be met prior to releasing the order to the warehouse for shipment.

If host preallocation is not done, then the WMS preallocates inventory to an order based on the inventory summary available quantities. This preallocation reserves inventory to ensure that it is available for an order, but does not reserve inventory at the location level.

- The order is checked to see if the ship-by date has expired. If the ship-by date has expired, then the order is not preallocated and held for review.

- The order is checked to see if the cancel date has expired. If the cancel date has expired, then the order is held for review but is not preallocated.

- The customer number assigned to the order is validated against the WMS customer master file. If the customer is invalid, and proper ship-to and bill-to information does not exist on the order, then the order is held for review.

- The carrier specified on the order is validated against the WMS carrier master file. The carrier master file specifies which carriers have been contracted for use by the warehouse. If the carrier is not specified or is invalid, the customer master file is checked for a preferred parcel, less-than-truckload (LTL), or truckload (TL) carrier based on the order's size. If a valid carrier cannot be determined, the order will be held for review.

Second, the customer order interface is used to assign preplanned routes and stops to assist in truckload planning automatically. Automatic routing and truckload planning is used when the company does not use a separate traffic department to plan truckloads prior to downloading to the warehouse. The method of order routing that a warehouse utilizes is dependent on the customers that are being serviced and the carriers being used. Orders that are assigned a preplanned route and stop (from either the traffic module or the order interface) require sequenced picking, or sequenced ship lane staging, to allow trucks to be loaded in reverse route-stop sequence.

Warehouses that service company-owned stores through internal carriers can deliver by single-stop or multiple-stop loads (truck). Single-stop loads are generally used when the warehouse services only a few stores and deliveries are made when a truck is filled. Multiple-stop loads are used when servicing many stores and deliveries to a store are smaller, and deliveries are made on a weekly or a daily basis. Preplanned routes and stops can be assigned by utilizing a customer master-file-based routing table that specifies the route and stop of a shipment based on the shipment day of week (using the day the order is downloaded or using the expected shipping date specified on the order).

Warehouses utilizing contract carriers to service a variable customer base (external customers) will not use preplanned routes and stops. Two common routing methods utilized are to allow the contract carrier to route orders or use an order pooling based on the order's zip code.

1. Warehouses that utilize a common LTL carrier to determine order routing only require the order interface to calculate the order's cube and weight. This calculation is done for all customer orders and is used by the order interface to convert the LTL carrier to a TL carrier or by the shipping department to assist in determining LTL truckloads.

2. Order pooling is used in place of routes and stops to group orders by zip code when a carrier's remote terminal or another warehouse is used to service a distribution region. Grouping orders by pool can lower freight charges by alleviating the need to double handle orders through a local and remote carrier terminal and by allowing larger shipments (many orders) to distant distribution regions.

Finally, the order interface function provides the ability to calculate freight charges associated with an order. Freight calculations are based on the carrier (determined above), the order's weight and volume based on the preallocated quantities, and the shipping zone (by zip code) to which the order is being delivered.

Freight calculations are made during the order download process to expedite the shipping process. These freight calculations may require recalculation if the actual inventory allocations or picked quantities are less than the preallocated quantities, or if the carrier is changed during the shipping process.

Order and Wave Planning.   The purpose of the order and wave planning function is to provide an order processing or shipping supervisor with the ability to work with problem orders held during the downloading process. Also, he or she can plan the picking activity by working with open orders. The ability to plan the order picking provides the supervisor with the information required to manage resources to achieve efficient shipping.

First, the order planning supervisor has the ability to work with, or review, orders that were put on hold during the order download function. The order interface program will place an order on hold and assign a reason code associated with the problem found. The order comment file is used to give an expanded problem explanation based on predetermined reasons. This order review process allows expedited handling of problem orders by warehouse personnel rather than relying on error stamping problems back to the host system. Problem handling by host application may significantly impact current application designs and cause additional processing delays due to batch-oriented systems.

The order planning supervisor uses the order review function to correct the problem associated with the order through manual data entry. If an order's problem is associated with invalid or missing data, the supervisor can contact the order entry department for the necessary information. After the supervisor corrects the data problem, the WMS verifies the corrected information and makes the order available within the open order pool for order/wave planning.

If an order's problem is due to a lack of inventory, then the supervisor can view available inventory and decide whether to mark the missing SKUs for cross-docking (or relay), move preallocated inventory from one order to another based on priority or shipping date, or leave the order on hold until new inventory is received. Order entry or traffic must be contacted for any order quantity-related changes if the host (order entry or traffic) is responsible for preallocation. If the WMS is responsible for preallocation, then the order review function must have the ability to recalculate preallocation for all orders or give the supervisor the ability to adjust preallocated quantities across orders.

Next, the order and wave planning function provides the order processing or shipping supervisor the ability to plan the picking activity based on the open order pool. A supervisor might use several order characteristics to determine how to group orders for picking. Order characteristics include:

- *Order number.* This allows the supervisor to work with and release an individual order for picking. This is used to expedite an order through the picking process.
- *Carrier.* This allows the supervisor to view and group all orders that will be shipped using the same transportation carrier. This provides optimization of shipping activity by concentrating picking activity to complete orders for a carrier (or truckload).
- *Account (or customer) number.* This allows the supervisor to view and group all open orders for an account. This allows optimization of shipping charges by allowing the maximum number of orders to be combined for shipment.
- *Order type.* This allows the supervisor to view and group all open orders by type. An order's type might include customer demand, store replenishment, return to vendor, or a transfer order (store to store or warehouse to warehouse). This allows the supervisor to focus the picking activity on the type of customer being serviced.
- *Order priority.* This allows the supervisor to view and group all open orders by priority. High-priority orders might include normal-demand-based, rush or overnight, and backorders.
- *Route.* This allows the supervisor to view and group all open orders for a route. Like grouping by carrier, this provides optimization of shipping activity.

The above-noted groupings are designed to allow the supervisor to select single or multiple characteristics to view open orders. Additional characteristics, such as SKU group or SKU style, may be required depending on the business requirements.

After selecting an order or group of orders to work with, the WMS will display the workload associated with the picking and shipping of this group. The workload displayed reflects the estimated worker-hours according to the number of pallets, cartons, and piece picks. The workloads are also summarized by stocking zone to allow the supervisor to control resource placement. The workload associated with an order is calculated during the preallocation and download process by using a SKU's velocity and stocking configurations (cartons per pallet and quantity per carton).

Along with the open order group's workload, the supervisor is shown the workload for orders that have been grouped into waves and are ready to release for picking and the remaining workload for orders that have been released for picking. This combination of open group, planned wave, and active order workloads allows the supervisor to plan and control picking and shipping activities. The supervisor uses this information to modify the selected open order grouping, and a wave number is assigned by the system at the supervisor's request. The wave number is used for control when the group of orders is released and the sequence in which picking occurs.

Wave Release and Inventory Allocation. The purpose of the wave release and inventory allocation function is to provide an order processing or shipping supervisor with the ability to:

- Work with waves ready for release
- Specify how a wave of orders will be picked
- Allocate inventory to orders based on predetermined rules and generate a picking queue of work

First, the supervisor can work with the waves that have been created and are ready for release. The supervisor is given the ability to:

1. *View the orders that have been grouped within the wave that allows the supervisor to hold a particular order from the release process.* In addition, the supervisor can reprioritize an order within a wave to allow it to be expedited in the picking and shipping processes.

2. *Add a new order to a wave to accommodate new orders arriving from the host system.* In addition, the supervisor can remove an order from the wave and return it to the open order pool and host maintenance.

3. *View orders that have been combined with other orders from the create wave process.* This allows the supervisor to override the combine order logic manually for freight savings to expedite a particular order to the customer.

4. *Specify the dock to be used for a load.* This allows the WMS to select the preferred conveyor spur or shipping lane according to the specified dock.

5. *Create invoice transactions for the host system prior to picking.* This option allows the supervisor to prebill according to the preallocated quantities that were reserved for the order.

Next, the order processing or shipping supervisor is provided with the ability to specify how the wave should be picked. During the allocation routine, the location where the inventory is stocked will specify the standard picking method to be used for the location. These standard picking methods are RF, label, and interface (refer to inventory allocation and picking sections). The order processing function, however, allows the supervisor to specify manually how the wave should be picked:

1. *Individual order.* This option ensures that an order is picked complete. This option is used when the warehouse does not have the ability to consolidate an order after picking (using pallet build or packing stations) or when orders within the wave must be expedited through the shipping process (customer pickup or rush order). However, a picker may be required to make several trips to a location to pick all orders for an item within a wave.

2. *Consolidate wave.* This option is used to optimize picking by combining orders for the same item. This reduces picker travel time by maximizing the quantity of an item picked from a single location. This option requires a consolidation area to collect items for a customer order. In addition, a single order cannot be expedited until after it has been consolidated.

3. *Picking list.* This option allows the supervisor to override the picking method specified by the location and is used in conjunction with the above-noted individual order and consolidate wave options. This printed picking list is bar coded with a pick list number to allow picking confirmation at a workstation. This option is used as a backup process to the standard picking technologies (RF, label, and interface) associated with the location.

If the supervisor specifies to pick by order with a list, then a pick list will be printed for every order on the wave. If the supervisor specifies to consolidate the wave with a list, a consolidated pick list is printed for all orders by item along with a printed list by order to allow order consolidation.

Finally, the supervisor is given the ability to indicate when to release a wave. When a wave is released, inventory is allocated and picking instructions (the picking queue) are created.

The inventory allocation process is responsible for ensuring that the stock within the warehouse is rotated to reduce carrying cost and manage warehouse space utilization and that picking is accomplished in an efficient manner according to the customer's requirements. The means to achieve each of these benefits are discussed below:

- Stock rotation within the warehouse is accomplished by using the inventory's FIFO date to direct pickers to the oldest product first. However, if the product being allocated is lot controlled, then FIFO becomes a secondary consideration to ensure that the minimum number of lots are stocked within the warehouse (lot rotation). In addition, expiration date (and shelf date) control will override the FIFO date rotation to ensure that the oldest-shelf-life product is shipped first.

- Space utilization is managed in the warehouse by allocating product from a location with the UOM required to fill the order but has the lowest space utilization within the warehouse. Picking from the lowest utilized locations ensures that locations are made available for other product storage. Space utilization allocation must first meet either the FIFO date or stock rotation criteria.

- Picking efficiency is controlled by allocating product based on the UOM configurations stocked within the warehouse. The WMS uses the preallocated quantity on the order detail record to determine the number of full pallets, cartons, and pieces required. The WMS is also capable of splitting full pallets during picking for which the number of cartons required constitutes a high percentage of a pallet. Additional characteristics that should be considered in designing a WMS picking allocation algorithm include:

     1. *The quantity to allocate is based on the preallocation quantity on an order.* If the wave is designated for consolidated picks, then the quantity of an item to be allocated is based on the preallocated quantity across all orders for the same item.

     2. *The primary pick location should be used when the picking container type (pallets, cartons, or pieces) matches the UOM required.* Primary pick locations are set up for efficient picking for a container type. If a sufficient quantity is not available within the primary pick location, then the allocation algorithm will check for a reserve or overflow location that allows picking with the required container type required. If no other location types exist for picking the container type, then a replenishment order is generated to move product to the primary pick location from reserve and overflow locations. The inventory designated for replenishment is then used for allocation.

     The primary pick location may not always be capable of holding enough products to fill order demands. The WMS utilizes a "blank prime" location concept to manage peak demand requirements. The blank prime is a location marked for primary picking of a particular container type but does not have an associated item. When demand exceeds capacity for the primary pick location, the WMS will look for a blank prime location to use for temporary picking. The blank prime location is not replenished but is selected by the allocation algorithm for picking prior to the primary picking location to ensure it is made available for other allocation problems.

     3. *Picking from a reserve or overflow location is managed based on which location is active.* If two reserve locations contain the same product with the same FIFO date (and utilization), then the allocation algorithm ensures that the location used for the last pick is allocated from first.

     4. *The allocation algorithm searches for available product in the UOM that best fits the preallocated quantity.* If product is not available in the "best" UOM, including replenishment of a larger container for breakdown, then the next lowest UOM is allocated.

5. *If product is not available to fill the order, then the allocation algorithm checks the customer master file to determine if backorders are acceptable.* If the customer is set up to allow backorders, then allocation is completed for the available quantity and backorder transactions are generated to the host computer when the order is shipped. If the customer does not allow backorders, then the order is put onto a new wave that is held for supervisor review.

6. *In addition, if product is not available, the allocation algorithm checks the customer master file to determine if substitutions are allowed.* If the customer allows substitution, and a substitution item is specified on the SKU master file, then a new order line is created for the remainder of the existing line that could not be allocated.

7. *Allocation excludes products in locations that are not regular stocking (picking) locations.* These locations may include receiving lanes, drop zones, conveyors, dock doors, and "in-transit" locations. The exclusion of inventory in these locations from allocation is dependent on when ownership of inventory is accepted by the warehouse and timing of host data base updates.

8. *The allocation algorithm allocates a specific lot number or serial number if specified by the host system.* A specific lot number may be requested by customers to meet their quality requirements, or the host system may attempt to provide a single lot number to the customer to ensure a consistent product quality. Serial number allocation is used when a specific machine is required for a customer's needs.

9. *Only inventory with a "good" status and inventory in locations that are not held from picking are allocated.* In addition, the order's detail line specifies a specific inventory class to pick. This allows support for possible return to vendor and refurbishing transfer orders.

10. *Order detail lines marked for cross-dock are allocated from inventory that is staged in cross-docking drop zones.* These drop zones are located near shipping lanes, which reduce the pick and transfer time associated with primary and reserve picks.

As inventory is allocated, picking instructions are created and placed on a picking work queue. Standard pick queue information includes:

- Customer order number
- Consolidated pick control number
- Combined order number
- Wave number
- Customer order detail line number
- Pick priority
- Pick status
- SKU number
- Picking location address
- Picking zone
- Quantity to pick
- Staging (or destination) location

The picking priority is used to expedite orders that may be waiting for customer pickup or for orders that must be shipped using a priority carrier such as parcel. The pick status is used to control the time when an order becomes available to a picker. An example of when a pick record is held is when a replenishment is required before the picking can be completed. The picking zone and equipment type are fields retrieved from the location master file when the picking order is created. These fields allow picking records to be matched to the picker's profile to determine eligible picks.

Additional information is added according to the picking method that will be used. The picking type designation for the wave and the location will determine the type of picking information that is generated. Picking queue information must support the picking method that will be used by providing the information required for each technology.

As inventory is allocated, the WMS retrieves the location picking type from the location master file. This allows multiple technologies to be supported within the WMS. The picking type on the location specifies the technology and picking queue information that are used to remove inventory from the location unless the "pick by paper" option was specified for the wave.

The paper picking option specified for the wave prior to release causes the WMS to ignore the specific location type requirements and generate a printed list for an order or consolidated order. It is assumed that all locations can be picked using a paper picking list when the wave is released. The printed list is bar coded with a picking control number and specifies the order number to pick, location, SKU, and quantity to pick. In addition, if the wave is designated for consolidating picking, then an additional bar coded picking list is printed with another picking control number, the location, SKU, and combined order quantity to pick.

When the paper picking option is not used, the specific location identifies the best technology to use. There are three basic picking methods that the WMS supports:

1. *Picking by RF.* This picking method is used for all types of material handling configurations. The WMS generates an electronic picking queue that includes the order number to pick, location, SKU, and quantity to pick. If the wave is designated for consolidated picks, then an additional picking queue entry is created for the combined SKU quantity that can be picked from a location. Specific order picking queue entries for a consolidated wave are held until the related consolidated pick queue entry is completed.

RF terminals are used to direct picking activity using a real-time picking task assignment according to a picker's profile. The real-time WMS interaction provided by RF picking allows accurate and timely updates to the locations' inventory and allows immediate problem handling if inventory inaccuracies are encountered. RF terminals and scanners, however, can be expensive and may not meet the freehand requirements of some picking operations. In addition, interfacing with conveyor sortation equipment can be difficult unless the sortation equipment is capable of using existing inventory container bar coding or unless labels are applied to picked containers in addition to using RF terminals.

2. *Picking by label.* This picking method is primarily used to pick from locations that are serviced by a conveyor or from hostile environments (such as in a freezer). The WMS prints bar coded labels that contain a new movable unit number for the picking container, which provides picking confirmation and the necessary data interface to conveyor sortation subsystems. In addition, the WMS creates electronic picking queue entries that are used to confirm pick orders and update inventory levels when the bar coded label is scanned after being picked.

Labels are used to direct picking activity in a batch mode by printing the necessary pick information on a bar coded label. Label picking provides a low-cost, high-volume solution that is commonly used in many case picking operations that must interface with conveyor sortation equipment. However, pickers use the printed picking information provided on the label and only interact with the system when a label batch is started, ended, or a problem occurs. This causes delays in updates to the inventory unless pickers are equipped with RF terminals to direct picking. In addition, the timing of when a replenishment to prime locations is performed can cause pickers not to have the required inventory.

3. *Picking by interface.* This picking method is used to support locations that are picked using a secondary software system. This type of picking would be used to support picking with batch terminals, pick-to-light, automated storage and retrieval systems (ASRSs), picking from carousels, or picking with voice. These picking methods will typically have their own picking software to control the inventory within their system. In addition to identifying picking by interface, the location also identifies the program name to be used to create picking queue entries that controls the information created on the pick queue. The picking system is made responsible for confirming picks that cause the WMS to update inventory levels.

These alternative picking methods are performed in batch mode by periodically downloading information to the picking subsystem. Interface picking has inherent problems associated with batch picking: timing associated with inventory updates and replenishment to the interface storage area.

The WMS also determines if customer labels or vendor compliance labels are required. The customer master file or order will specify the type of labels to print, and the picking queue is created with the necessary customer labeling requirements. In addition, the WMS assigns a packing station for any order that requires piece picking. Packing stations are also assigned if the SKU master file identifies that a full carton quantity is not in a shippable container.

Finally, the customer master file indicates if the order should be palletized or floor-loaded. If the customer master file specifies palletized loads, then the WMS will group picks into logical pallets by product cube information for locations picked by RF or assign pallet build slots or lanes for orders picked by label or interface. If the customer master file specifies floor loads, then picking information is created to direct picked merchandise directly to shipping lanes or conveyor spurs based on staging information from the carrier master file.

Order Picking Functions

The picking process is responsible for managing how inventory is relieved from stocking locations to fill customer orders. Picking utilizes a number of technologies to provide the most efficient means to direct picking activity and ensure order shipment accuracy. This section reviews several picking methods that utilize a variety of technologies (see Fig. 5-26).

The picking functions typically constitute 10 to 20 percent of the system implementation time depending on how many picking methods are required and their complexity. The following picking functions will be reviewed in more detail:

1. RF order picking
2. RF (and paper) consolidated picks
3. Label picking
4. Interface picking
5. Order pick list picking
6. RF assisted picking

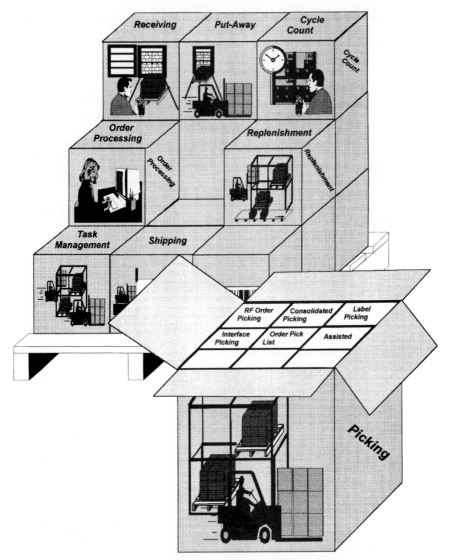

FIGURE 5-26   Picking functions of an automated warehouse management system.

RF Order Picking Function.   RF order picking is used to fill customer orders that require full pallets, full cartons, and broken cartons (pieces). In addition, RF order picking is used to control pick-pack operations where a tote is filled with piece picks and passed from one picking zone to another until the order is complete (see Fig. 5-27).

RF order picking provides real-time direction of picking activity. The picker's profile is used to identify the type of picks (full pallet, full carton, or broken carton) to display to the picker, along with the building and zone the picker is in and the picker's equipment type. Special picking instructions are displayed when a picker begins work-

FIGURE 5-27   RF directed picking of low-volume cases from selective rack.

ing on an order. The picker can identify problems by placing a location on cycle count and perform the cycle count for locations that are picked empty. The following characteristics should also be considered when designing a RF order picking function:

1. *Allow the user profile to determine the pick types that will be displayed.*   The profile should allow the selection of picking pallet, cartons, pieces, or any combination. This flexibility allows a picker to restrict picks to a particular picking type that may have high demand. In addition, the picker can select more than one picking type that is useful if the secondary picking type has low activity where a dedicated picker is not required.

2. *Allow the picker to identify the zone to pick.*   The picker can change his or her profile to indicate a specific zone in which to work. The WMS only displays picks that are within the selected warehouse zone. This allows the warehouse manager to assign responsibility to a picker for a warehouse area to ensure that picking activity is completed efficiently. Multiple pickers, or groups, can also be assigned to a zone.

3. *Allow the picker to identify the equipment type they are using.*   The WMS only displays picking locations that can be addressed by equipment type selected. Note: An equipment type table is used to specify which equipment types can be used interchangeably.

4. *Allow the picker to specify a specific order or wave to pick.*   This allows expediting of orders or waves that may not have had a priority status at the time of release.

5. *Prompt the picker to scan a movable unit to assign an order or allow the WMS to assign an order to a picker based on the order's priority and FIFO time stamp.* If the picker is working a zone within a pick-pack (piece-picking) area, then the picker can scan a movable unit for picking. If the movable unit has an order assigned and pick records exist for the order within the picker's zone, then the WMS directs the picker to the appropriate locations. If the movable unit does not have an order assigned, the WMS assigns the highest-priority, oldest order to the movable unit and prompts the picker to the locations within his or her zone for picking. If a movable unit is not entered, then the WMS will direct the picker by assigning the highest-priority, oldest order that matches the picker's profile.

6. *Display special picking instructions when an order is assigned to a picker.* Special picking instructions may indicate overpacking requirements or special labeling that is required by the customer.

7. *Direct the picker to a location to pick based on the priority and time that the pick record was created.*   After an order is selected by the WMS, locations are displayed according to their picking sequence within an aisle. A picking sequence number can be assigned to a location to control the sequence in which bays or levels are displayed. If a picking sequence number is not assigned, then locations are displayed according to their logical address (aisle, bay, level, and slot) sequence.

8. *Allow the picker to identify inventory problems by submitting the location for cycle counting.*   Submitting a location to the cycle count queue frees the picker to continue the picking activity. The WMS puts the direct location on hold with a discrepancy cycle count and attempts to reallocate any outstanding picks for the problem location.

9. *Prompt the picker to verify the unit or SKU being picked.*   The WMS directs the picker by displaying the SKU and quantity required. The picker scans either the unit to pick from or the quantity picked. If the quantity picked is less than the quantity required, then the same pick record is displayed with the remaining quantity.

10. *Prompt the picker to enter serial numbers or lot numbers if required by the SKU.* The WMS updates the picked records to track the inventory pulled for the order.

11. *Prompt the picker to confirm when a location is picked empty.*   If the location is empty, the WMS updates the location as being cycle counted. If the location is not empty, a cycle discrepancy transaction is created and the location will be locked.

12. *Prompt the picker to scan a new movable unit number for the picked inventory.* A *picked queue* record is created by the WMS that identifies the movable unit from which inventory was picked, the location picked, the new movable unit number, the order assigned to the new movable unit, and the person who picked the order.

13. *Direct the picker to a pick staging or shipping staging area to drop off the order after all picks for an order are complete that match the picker's profile.*   If the picking order is completed by the picker, then the movable unit can be dropped off in the ship staging area for loading. If the order is picked by more than one picker, then the movable unit is directed to a pick staging area for consolidation with other movable units.

14. *Generate a carrier interface if a carton is picked and will be shipped using a parcel carrier.*   The carrier master file identifies where a parcel carrier should be staged and the interface program that should be executed. This interface is used to communicate the necessary information about the customer order and movable unit to another program to handle application of freight rates. Parcel carrier subsystems use this information to print carton packing lists and a shipping manifest.

15. *Prompt the picker to enter SKU control data (lot and serial numbers) if required by the SKU and multiple inventory records exist within the location.* Capturing control data during picking is done for case and piece picking types to prevent the need to break down the unit in shipping.

16. *Allow the picker to reverse the pick sequence with an aisle to allow "snaking" through the warehouse or zone.* When the picker reaches the end of an aisle it is more efficient to work the adjacent aisle in reverse bay sequence rather than traveling back to the aisle's beginning based on the logical address.

The WMS submits replenishment requests when prime picking locations drop below the minimum quantity level. In addition, the WMS releases existing "hot" replenishments when space is available within the prime location to allow additional orders to be picked.

RF (and Paper) Consolidated Picking Function. RF consolidated picking is designed to pick multiple orders for the same SKU. This function is designed to direct picking activity by RF terminals and allows a picker to confirm consolidated picks that were completed using paper.

A picker initiates a consolidated pick by selecting (or scanning) the picking control number from a RF or CRT terminal. After a consolidated pick is selected to work, a list of picks that match the picker's profile is displayed. The picker has the option to select a pick to work or select a pick to complete. If the picker selects a pick to work, the WMS directs the picker to the location, displays the SKU and quantity to pick, and prompts the picker to enter the SKU and quantity picked. The picker has the option to place the location on cycle count if an inventory problem is encountered. A picker can select a pick to complete if he or she is working from a printed list and is confirming completed picks. The WMS assumes that the picking was performed as printed on the pick list. Finally the picker is prompted to enter the consolidated movable unit used for the inventory. In addition to the characteristics described for RF order picking, the following characteristics should also be considered when designing a RF consolidated picking function:

1. *Allow the picker to change his or her picking profile.* The picking profile allows the picker to specify a specific building or zone to work in, along with the equipment type that is being used and the picking type (full pallets, full cartons, or broken cartons).

2. *Prompt the picker to scan a bar coded pick list to initiate picking or select a consolidated pick list to work.* The consolidated pick list displayed contains picks that match the picker's profile.

3. *Prompt the picker to scan a movable unit to assign a consolidated pick.* If the picker is working from a printed list, the movable unit represents the picks associated with each pick record selected to complete. If the picker is using directed picking, then the WMS prompts the picker for the movable unit after each quantity picked.

4. *Display pick locations by the location's pick sequence or by the location's logical address sequence.* After the consolidated list is selected, a list of locations to pick is displayed by the location's picking sequence by aisle. If the locations do not have a picking sequence assigned, locations will be displayed by the logical address sequence (aisle, bay, level, and slot).

After all picks for an order that match the picker's profile are completed, the picker is directed to a pick staging location. The pick staging location is where an individual order is separated onto a new movable unit from consolidated picks. The unit build function is used to break down the consolidated picks (see the final section of this chapter on shipping functions).

Label Picking Function.   Label picking can be used to fill customer orders that call for pallets, cartons, and broken cartons (pieces). Picking labels are generated after a wave is released by the order processing function and are printed in order or SKU sequence depending on how the wave is released. If a wave is released using the pick by order option, then labels will be printed with one order per label batch and in pick location sequence. If a wave is released using the consolidated picking option, then labels are printed with one picking zone per label batch. Within a consolidated label batch, all orders requiring the same SKU within the wave are printed together according to the pick location sequence.

While label picking can be used to direct picking of pallets, cartons, and pieces, it is predominantly used to direct carton picking activities. Picking labels provide a convenient interface to material handling (carton conveyor) systems that are used to transfer cartons from picking locations to shipping docks. If the conveyor is equipped with automated scanning and sortation equipment, the efficiencies associated with consolidating picking by SKU (to reduce travel time by the picker) can be realized by allowing the material handling equipment to sort the picked cartons to the order (see Fig. 5-28).

Label picking utilizes bar code label printers and CRTs that can be located in a central area or dispersed strategically throughout the warehouse. The picking process is

FIGURE 5-28   Carton picking onto a conveyor sortation system using pick labels.

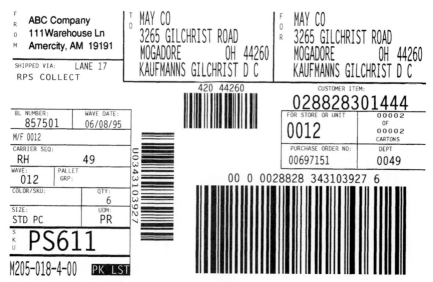

FIGURE 5-29 Pick/ship label with printed picking location, bar coded movable unit number, UCC/EAN-128 compliant zip code, and carton bar codes.

initiated by scanning a batch header label that indicates the group of labels (picks) the picker is working. The picking label indicates the SKU to pick, the quantity required, and the location where the SKU is stocked, along with customer and carrier information. In addition, a new bar coded movable unit ID is printed on the label that represents the inventory removed from the location. This new movable unit ID is associated to a specific order within the WMS and can be recognized in a conveyor or sortation system to transfer the picked movable unit to the shipping area. An example of a picking label is shown in Fig. 5-29.

Label picking requires additional label types other than picking labels. A "break" label is printed with a label batch if a replenishment to a prime location is required before subsequent picks can be completed. A picker scans the break label into the WMS to indicate that the prime location has been emptied and the replenishment can be completed. An alternative to a break label is to designate a few locations to be blank primes that allow the WMS to direct replenishments temporarily to an empty location when the prime location cannot hold enough inventory for the required picks.

Customer-specific labels are printed and applied to the picked units. These compliance labels are required by customers to expedite their receiving operations. A compliance label is printed immediately after the pick label. The picker applies the compliance label to the picked unit, which prevents additional handling in the shipping area.

When picking cannot be executed due to an inventory error, the picker scans the "unpickable" label into the WMS. The WMS submits the pick location to cycle count and allocates a different picking location, if inventory is available, and displays the new pick location to the picker. If the problem pick location is a prime, the WMS attempts to replenish a blank prime location with reserve inventory. After a label batch is completely picked, the picker scans a batch trailer label that indicates that picking is complete. The WMS updates the inventory records to reflect the inventory removed from the location and the new movable units that were created. Owing to the batch picking interface associated with label picking, serial number and lot number capture will be done in shipping. Replenishments are generated for any primes that were picked below minimum quantity. The following characteristics should be considered when designing a label picking function:

1. *Design the picking label carefully.* The information printed on the label and the formatting of the pick, customer, and carrier information, along with the bar coded information, should be carefully thought out. Industry and customer standards may be used to expedite label design.

   In addition, the information contained within the bar coded movable unit ID must be integrated with the conveyor sortation logic. If an industry standard (such as UCC-EAN 128) movable unit bar code or a customer-specific bar code is not used, the following should be considered in designing the bar coded label ID:

   a. If a "serialized" movable unit ID is used, a transaction must be sent to the sortation system for each pick that requires sortation. The transaction sent to the sortation system will identify the movable unit ID, the order number for which the pick was conducted, and the carrier or sortation lane to which the pick should be diverted. The SKU and/or the SKU's weight may be included in the sortation transaction if the conveyor system is equipped with in-line scales or the carton is bar coded with the SKU to allow picking confirmation.

   b. If an "intelligent" movable unit ID is used, a transaction must be sent to the sortation system for each order that requires sortation. An intelligent bar code can reduce the number of transactions that the sortation system must handle. When the sortation lane is capable of working an order, a transaction is sent to the sortation system indicating that the order can be diverted.

   An intelligent bar code label is encoded with the wave number of the picked unit, the order number (or order code), a carton sequence number within the order, the sortation lane to which the pick should be diverted, and a divert sequence if multiple orders are diverted to the same sortation lane. In addition, the bar code can contain the SKU or the SKU's weight for pick confirmation.

2. *Allow the picker to scan a batch header label to indicate the start of picking.* Transactions are generated for the sortation system if the location being picked indicates that picks are being made to a conveyor sortation system. Generating sortation transactions at the start of picking minimizes sortation gridlock by allowing order sortation as early as possible.

3. *Allow the picker to reprint a label batch.* This is required in case a printer jam occurs. The WMS allows the picker to reprint an entire label batch or selected labels within a batch. Reprinted labels are encoded with a new movable unit label to ensure that double picks are not performed, which could result in overshipment to the customer.

4. *Allow the picker to scan a break label when a prime location is ready for replenishment.* This indicates to the WMS that the prime location has been picked empty and replenishments can be completed to allow picking for other orders.

5. *Allow the picker to scan "unpickable" labels.* This allows the picker to identify locations that may have an inventory discrepancy. The WMS ensures that there are no outstanding replenishments that have not been completed for the pick location that may be the cause of the discrepancy. If there are no outstanding replenishments, the WMS submits the discrepant location to the cycle count queue and attempts to reallocate the order. If there is no inventory available for the order, the picker is notified that the pick label should be discarded and the order is adjusted. If alternative inventory is available, the picker is directed to the alternative pick location and a replenishment is generated if necessary.

6. *Allow the picker to scan a batch trailer label to indicate that picking is complete.* The WMS updates location inventory records with the inventory removed, creates new units for the inventory picked, and generates replenishments for prime locations picked below their minimum quantity level.

7. *Generate a carrier interface if a carton is picked and will be shipped using a parcel carrier.* The carrier master file identifies where a parcel carrier should be staged and the interface program that should be executed. This interface is used to communicate the necessary information about the customer order and movable unit to another program to handle application of freight rates. Parcel carrier subsystems use this information to print carton packing lists and a shipping manifest.

Interface Picking Function. Interface picking can be used to fill customer orders that call for pallets, cartons, and broken cartons (pieces). Interface picking is used to support filling of customer orders by using a computer system that is not integrated within the WMS. The WMS tracks the inventory levels within locations controlled by the secondary computer system (subsystem) by accepting interface transactions (see Fig. 5-30).

FIGURE 5-30 Pick-to-light used in a pick and pack carton flow rack area.

Picking transactions are created in a picking interface file after order processing allocates inventory. The location that inventory was allocated from indicates if a picking interface is required. The picking subsystem reads the interface records and directs a picker to remove the required inventory. After a pick is complete, the picking subsystem updates the WMS with the picks that were completed. The following characteristics should be considered in designing an interface picking function:

1. *The interface generated for the picking subsystem should indicate the wave number the pick is for, the order to pick, the SKU to pick, the SKU quantity required, and the location to pick.* In addition, the picking sequence and priority should be indicated to the subsystem to ensure that picks are completed according to shipping requirements. Finally, the picking interface indicates if serial number or lot number capture is required.

2. *The picking subsystem should update the WMS when a pick is complete.* The picking subsystem updates the WMS with the new movable unit that was created for the picked merchandise. In addition, the picking subsystem confirms the location, SKU, quantity, location picked, and lot numbers or serial numbers picked (if required), along with the picker's identification and the date and time the pick was completed. The WMS ensures that inventory records are updated according to the confirmed picks by the picking subsystem.

3. *The picking subsystem should update the WMS with a pick that could not be completed.* The WMS submits the location for cycle counting and attempts to reallocate the pick. If necessary, an additional picking interface transaction is generated to indicate an alternative picking location for the required inventory. The WMS must reallocate picks to avoid conflict with other orders that may have been already allocated.

Order Picking List Function. The order picking list function is designed to pick an order using a printed picking list. This function utilizes a paper picking list for direct picking activity of an entire order. Unlike other picking functions described above, this picking method does not separate pallet, carton, and broken carton picks so that picking can be performed by multiple pickers.

A picking list is generated for each order that is released. The picking list shows the SKUs requested and the quantity of each SKU allocated by location. This picking list also contains special handling instructions required by the customer. The picker initiates the picking process by scanning a bar coded picking list number using a centralized CRT. The picker uses the printed list to pick the allocated inventory. If a pick could not be completed as printed on the picking list, the picker identifies the problem to the WMS prior to completing the list. The WMS submits the problem location to the cycle count queue and attempts to reallocate the pick. If alternative inventory is available, the WMS displays the new picking location to the picker, who will manually update his or her picking list. When picking is complete, the picker scans the bar coded picking list and scans the movable unit(s) that contains the picked inventory. The following characteristics should be considered when designing an order picking list function:

1. *Prompt the picker to scan a bar coded picking list to initiate picking.* This allows the WMS to update the order's status.

2. *Allow the picker to assign specific picks (and quantities) to a movable unit number and indicate a staging location for the picked inventory.* Multiple movable unit numbers can be created for an order. The picker selects the picks and quantities to assign to a movable unit after a group of picks is completed.

3. *Allow the picker to identify inventory problems by submitting the location for cycle counting.* The WMS submits the problem to the cycle count and attempts to reallocate any outstanding picks for the indicated location. An alternative picking location is displayed to the picker, if available. The picker manually updates the picking list with new locations allocated. If necessary, the WMS generates a "hot" replenishment task to replenish reserve inventory to allow picking from prime or blank prime locations.

4. *Allow the picker to reprint a picking list that may not have been printed properly.* The reprinted picking list is printed with a new picking list number to prevent duplicate picking.

5. *Prompt the picker to enter serial numbers or lot numbers if required by the SKU.* The WMS updates the picked records to track the inventory pulled for the order.

6. *Identify picks that require replenishment before they can be completed.* This prevents the picker from searching for inventory that is not available for picking. The picker notifies a stocker to make the replenishment moves so that the pick(s) can be completed.

RF Assisted Picking Function. Assisted picking is designed to utilize picking instructions generated by the host system. The purpose of assisted picking is to help a picker find inventory, update the WMS inventory for picked merchandise, and record the customer order number as picks are made. Assisted picking does not utilize the order processing functions described above. Rather, it follows a picker using a printed picking list from preallocations made by the host system.

Assisted picking is used when the picking operation is small and simple, thus not requiring the complex RF or label picking algorithms. Assisted picking is also useful during the cut-over process to the new WMS. Assisted picking allows the warehouse to utilize efficiencies of the new receiving and locator functions and can help manage the higher risk associated with the order processing and shipping functions by phasing in the new processes.

When a customer order is entered on the host system, it is preallocated against the inventory summary that has been updated from the WMS receiving functions. The shipping supervisor can then print picking lists for selected orders from the current host application or from the customer order downloaded and printed using the assisted shipping function (described later in this chapter).

Using the printed picking list, the picker enters the customer order number, the SKU, and the quantity required onto a RF terminal. The WMS displays a list of locations that have the SKU and quantity required. After the picker confirms the location and quantity picked, the WMS prompts the picker to enter a movable unit number and location for the picked inventory. A picking history file is created to track the picked inventory by customer order, the location, and unit number from which the inventory was picked, along with the picker's identification and time stamp. After the picking is complete, the WMS updates the inventory records to reflect picked merchandise and generates replenishment if a prime location is picked below minimum quantity levels. The following characteristics should be considered when designing a RF assisted picking function:

1. *Provide the shipping supervisor with the ability to print picking lists for customer orders.* This is done by utilizing the current host-based application or by utilizing the assisted shipping function.

2. *Prompt the picker to enter the customer order number, the SKU, inventory quality, and quantity allocated to the order using a RF terminal.* The customer order number is used to create a pick tracking file that can be used by the assisted shipping function.

3. *Display a list of locations that contain the SKU and quantity required.* Prime locations for the SKU will be displayed first on the list.

4. *Prompt the picker to scan a location to pick.*   If the location retains the movable unit numbers on the inventory, the picker is prompted to enter the unit number from which to pick.

5. *Prompt the picker to enter the quantity picked and scan a new movable unit number and location for the picked inventory.*   The WMS creates a pick tracking record for the customer order and updates the location's inventory level. A replenishment is generated if a prime location is picked below minimum level.

6. *Allow the picker to indicate inventory problems by placing a location on cycle count.*   The WMS locks the location from picking and submits a cycle request.

7. *Prompt the picker to enter serial number or lot number control data if required by the SKU.*

8. *Prompt the picker to confirm when a location is picked empty.*   If the location is empty, the WMS updates the location as being cycle counted. If the location is not empty, a cycle discrepancy transaction is created and the location will be locked.

9. *Allow the picker to indicate when the picking list is complete.*   This causes the WMS to update the order's status in the assisted shipping function.

Replenishment Functions

The replenishment process is responsible for directing the movement of reserve stock inventory to prime picking locations. In addition, replenishment can be used to direct inventory movement to support reserve inventory location consolidation. The replenishment functions constitute 5 to 10 percent of the system implementation time (see Fig. 5-31).

A replenishment work queue is generated from three sources. First, order processing generates a replenishment task when the primary picking location does not have enough available inventory to satisfy the order(s) being allocated. Order demand replenishments are the highest priority to be worked since an order is waiting on the inventory movement. In addition, an order demand replenishment will cause picking tasks be "passive" until the replenishment is completed. An order demand replenishment can also be generated for the prime location of a SKU, or a blank prime location can be used for temporary picking if no space is available in the prime location and a blank prime location is available. A blank prime is a location within the primary picking zone that does not receive stock replenishment when it is picked empty since it does not have a permanent SKU assigned to it.

A second source for a replenishment task is picking. The picking functions generate replenishment tasks when a primary location is picked below the minimum quantity level for the SKU. Picking also generates a replenishment task if the picker indicates that a prime location has an inventory problem. An inventory problem is identified by the picker placing the location on cycle count. If additional inventory exists in reserve locations and is not in another prime location for the SKU, a replenishment task is generated to move the inventory to another prime location or a blank prime location (see Fig. 5-32).

The third source for a replenishment task is through replenishment queue maintenance. Replenishment queue maintenance allows manual inventory moves to be submitted by a supervisor. The replenishment queue maintenance and RF directed replenishment functions will be discussed later in further detail.

Replenishment Queue Maintenance Function.   Replenishment queue maintenance is a CRT function that allows a supervisor to work with the replenishment tasks. This

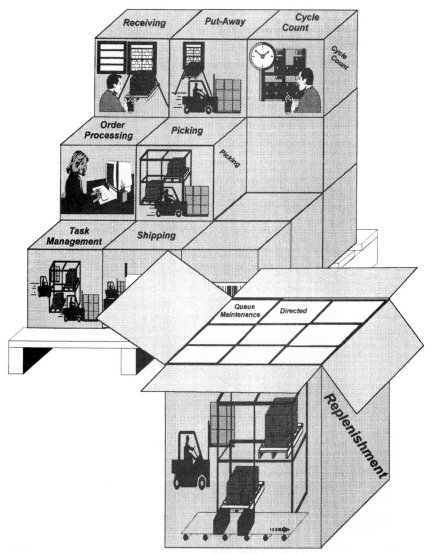

FIGURE 5-31 Replenishment functions of an automated warehouse management system.

function allows the supervisor to change the priority of existing replenishment tasks or delete them from the queue. This function also provides two methods for adding tasks to the replenishment queue.

Replenishment tasks can be manually added by selecting to refill prime locations to capacity. The WMS will attempt to find reserve inventory for all prime locations and generate replenishment tasks to fill them to their maximum capacity for a SKU. This function is useful when the prime location capacities are designed to stock enough inventory for a day-of-order demand. Replenishment tasks can also be manually added

FIGURE 5-32 RF directed replenishment of broken case prime picking area.

by entering the location to replenish (or "to" location) or entering the reserve location (or "from" location) along with the SKU, quantity, and movable unit to move. In the latter case, reserve stock consolidation can be supported since the "to" location does not need to be a primary picking slot.

Finally, the replenishment queue maintenance function provides the ability to print a replenishment list and confirm replenishment completion. This provides a paper alternative to using RF directed replenishment. The following characteristics should be considered in designing a replenishment queue maintenance function:

1. *Allow a supervisor to view all existing replenishment tasks.* The primary picking location that is being replenished is displayed, along with the SKU, "from" location, the task's status, and priority. The supervisor can select to view additional detail about a replenishment task such as the equipment type required, the task's age, "from" and "to" zones, and so forth.

2. *Allow the supervisor to delete a replenishment task that is not assigned to a stocker.* Manual moves may be entered incorrectly and may require deletion. Also, a supervisor may decide to move the primary picking location for a SKU that would make the replenishment invalid.

3. *Allow the supervisor to change a replenishment task that is not assigned to a stocker.* This allows the supervisor to reprioritize manually the sequence in

which replenishments are completed. In addition, a replenishment's status can be changed to make it passive or active.

4. *Provide the supervisor the ability to refill prime locations.*   This generates inventory moves from reserve stock to fill all prime locations to capacity if inventory exists.

5. *Provide the supervisor the ability to add a replenishment task manually.*   The supervisor must enter the "to" and "from" location, the SKU, and quantity to move. Alternatively, the SKU and quantity can be bypassed by entering the specific movable unit to move. This capability also supports inventory movement between reserve stocking locations.

6. *Allow a supervisor to select replenishment to work on and print a replenishment list.*   The status for the selected replenishments is changed to "in use" to indicate they are being worked on. The printed list is grouped by the primary picking location's zone to allow it to be divided between stockers.

7. *Provide the capability for a supervisor to confirm replenishments.* Replenishments are confirmed as complete from the stocker marking the replenishment list. The WMS updates the inventory records and releases passive picks.

RF Directed Replenishment Function.   The RF directed replenishment function directs a stocker to perform inventory moves to prime locations. The order processing and picking functions automatically generate replenishment tasks based on order demand or minimum quantity levels. Directed replenishment directs a stocker to complete moves based on the task's status, priority, and age (FIFO).

Directed replenishment is designed as a RF function to manage work based on the task's priority. High-priority tasks are created when order demand requires the inventory move. Low-priority tasks are generated for manual moves or minimum quantity replenishments. In addition, if a picking task is waiting passively for replenishment, the RF directed replenishment provides the ability to update the picking queue code to active. This intertwining of functions ensures that the directed work is accomplished after the inventory is available to complete the task. The following characteristics should be considered in designing a RF directed replenishment function:

1. *Allow the stocker to identify which replenishments to work on by their equipment type and work zone.*   The equipment type ensures the picker can work (stock) the directed locations. The zone limits which primary picking locations the stocker is responsible for filling.

2. *Prompt the stocker with the reserve ("from") location to work.*   The stocker drives to the location and scans the location's address label.

3. *Display the movable unit, SKU, and quantity required for the replenishment.*   The stocker enters the quantity and scans the movable unit to move. If a manual replenishment specified the movable unit to move, the stocker must scan the directed movable unit. Otherwise, the stocker can scan any unit that has the required inventory quantity.

4. *Allow the stocker to submit a "from" location to the cycle count queue.*   If the inventory is not available in the directed location, the WMS creates a cycle count discrepancy task and attempts to reallocate the replenishment. If the replenishment is a manually submitted move, the task's status is changed to "hold," which requires a supervisor's order to maintain the replenishment queue.

5. *Allow the stocker to add a replenishment task.*   This allows the stocker to move multiple movable units to prime locations if his or her equipment is capable.

6. *Prompt the stocker where to move the inventory.*   The stocker scans the destination location after the inventory is stocked. The WMS will check the picking queue and release passive picks that are waiting on the inventory list.

7. *Allow the stocker to submit a "to" location to the cycle count queue.*   This is required if the prime location has a different SKU or the location is full. The WMS creates a cycle count discrepancy and will attempt to find another prime or blank prime location to stock the inventory if there are picks waiting on the inventory list. Otherwise, if there are no picks waiting on the replenishment list, the task's status changes to "hold," which requires supervisor maintenance.

Task Management

Task management provides additional warehouse control spanning most procedural functions described in earlier sections. Task management utilizes the multitasking (or interleaving) function to manage the assignment of pallet tasks within the warehouse based on an operator's location. Also, task management utilizes the labor management function to display the labor standards for a task to an operator prior to starting an operation and records the operator's actual time for performance measurement. The labor management function is integrated with every procedural function used within the warehouse (see Fig. 5-33).

Multitasking Function.   The multitasking process provides the ability for a warehouse operator to perform pallet (or truck) move operations across multiple functions. RF pallet move functions (including replenishment, picking, and put-away) are intermingled by a task's priority and a stocker's location. The multitasking function reduces the possibility of "deadheading," where a warehouse picking operator is directed to a dock location with a pallet for shipping and then travels back empty for another pick assignment. Multitasking constitutes 5 to 10 percent of the system implementation time.

The multitasking function program calls for the pallet functions previously described. Pallet functions are selected based on a task's priority and the stocker's current location. The current warehouse zone for an operator, determined by the last location scanned by the operator, is used to determine which tasks are closest. The proximity of warehouse zones to one another is maintained with the WMS, which is used to determine the sequence in which zones should be searched. The table shown in Fig. 5-34 illustrates how warehouse zones are related.

The sequence in which zones are searched is based on the last location scanned by the multitasking operator. For instance, if the last location for a stocker is in zone B1, then the next zone that is searched for a task is A1. If no tasks are found in A1, then the next zone searched is C1. Finally, if no tasks are found in C1, then the next zone searched is B2.

In addition to the above-mentioned zone sequencing, multitasking assigns a task priority (or sequence) to the pallet moving functions. The table shown in Fig. 5-35 illustrates the sequence in which pallet operations are assigned.

A "hot" replenishment is any replenishment that is required before picking can be performed for a customer order. Multitasking directs a stocker to complete "hot" replenishments in his or her current zone first. If no "hot" replenishments exist within the current zone, multitasking determines the next zone to search from the zone sequence table. If no "hot" replenishments exist in any of the sequenced zones, then any "hot" replenishment that exists is selected.

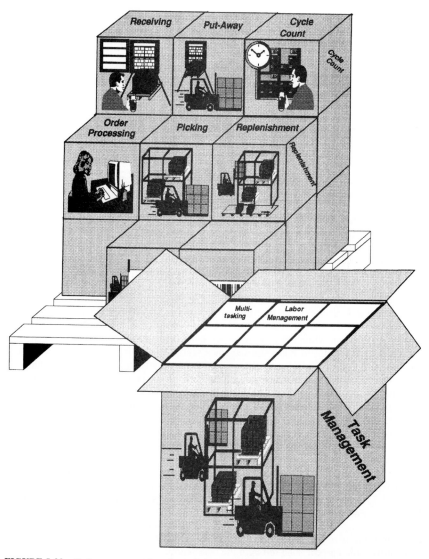

FIGURE 5-33 Task management functions of an automated warehouse management system.

Current zone	Next zone	Zone sequence
A1	A2	1
A1	B1	2
A2	A1	1
A2	B2	2
B1	A1	1
B1	B2	3
B1	C1	2
C1	B1	1

FIGURE 5-34 Multitasking zone search sequence.

Pallet task	Current zone	Next zone	Any zone
"Hot" replenishment	1	2	3
Picking	4	7	10
Normal replenishment	5	8	11
Put-away	6	9	12

FIGURE 5-35   Task search priority table for multitasking.

If there are no "hot" replenishments within the building in which the operator is located, then multitasking considers other pallet functions based on priority. First, the picking queue is checked to see if full pallet picks exist within the current zone. If there are no full pallet picks within the current zone, then multitasking checks for normal pallet replenishments within the current zone. Finally, if there are no normal replenishments within the current zone, then multitasking checks for put-away tasks within the current zone. If there are no pallet tasks within the current zone, multitasking will repeat the search for the next zone from the zone sequence table.

The flexibility provided by zone sequencing and task priority tables assists the warehouse manager in optimizing work that is performed by warehouse operators. For instance, the task priorities can be adjusted for month-end shipping activity to ensure all picks and replenishments are completed before an operator is directed to put away new inventory. However, while the zone sequence table is set up to optimize travel distance, care must be taken not to make the sequence search too extensive. For instance, there may not be significant efficiency saving by having more than four to five sequenced zones for every zone in the warehouse. If the zone sequencing function is too long, the multitasking function may spend more time searching the zone sequence than it would take to pick any task and have the operator drive to the location.

The multitasking function is initiated by a warehouse operator scanning a location address label to identify its current location. The above-noted zone sequence and task priority tables are used to select the "best" task to begin work. Multitasking then calls the appropriate function, with the selected task, which directs the operator to the next activity. After the operator scans the destination location for the assigned task, the called function passes the current zone back to the multitasking function, which searches for another task. The following characteristics should be considered in designing a multitasking function:

1. *Prompt the warehouse operator to scan a location to get a task assigned.*   This identifies the current zone to begin the task search.

2. *Utilize a zone sequence table.*   Zone sequencing identifies the preferred search sequence across zones according to their relative proximity, while location coordinates (see the discussion on labor management) are used within a zone to select between equal-priority tasks.

3. *Utilize a task priority table.*   Tasks of equal priority are assigned according to the shortest distance that an operator must travel to begin a task. If tasks have the same priority and travel distance, then an assignment is made based on FIFO.

4. *Allow the warehouse operator to change his or her multitasking profile.*   This profile identifies the equipment type being used and a zone to restrict the zone sequence searching.

5. *Call the appropriate pallet function according to the task selected.*   After the task is complete, the ending location is used to identify the current zone to search.

Labor Management Function. The labor management function provides the warehouse manager with the ability to establish and enforce labor standards for each procedural operation controlled by the WMS. Labor management is invoked with each procedural function (such as receiving, put-away, picking, etc.) by displaying the labor standard at the start of a task when it is assigned to a warehouse operator. Labor standards are established by time studies, which include personal fatigue and delay times along with regular constraints including truck speed, WMS response time, incremental operator movement, and so forth.

The foundation for the labor management function is the use of engineered labor standards to ensure consistent measurement of individual performance. Standard time measurements for each work element that comprises tasks within the warehouse are developed and grouped to create a task definition, identifying standard procedures required to complete a task.

Travel distance is a fundamental element in the calculation of a task's duration. This requires (x,y,z) coordinates for each location to be maintained within the WMS. The x and y location coordinates, as shown in Fig. 5-36, are used to calculate horizontal travel time within the warehouse, and the z coordinate is used to calculate lift time and operator travel between levels.

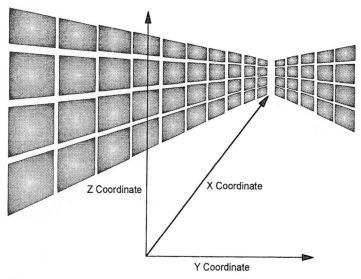

FIGURE 5-36 Schematic of (x,y,z) coordinate mapping of warehouse locations. Distance calculated using (x,y,z) for pick, (x,y,z) for put-away/letdowns, (x,y,z) for refill, (x,y) for dispatch doors, (x,y) for receiving doors, (x,y) for work offices.

Labor standards are broken down into two distinct categories: truck and nontruck standards. Truck standards include procedural tasks that require the use of a forklift and include such tasks as replenishment, put-away, pallet picking, and load trailer. Nontruck standards include procedural tasks that do not require use of a forklift such as carton and piece picking, detailed and returns receiving, and pallet wrap.

The labor standard for a task is displayed to the operator when the assignment is made. This informs the operator of his or her expected performance. At the completion of a task, the WMS records the task completion time and calculates the operator's performance. The operator's performance is displayed at the completion of each task, which informs the operator of the completed task performance along with daily and weekly cumulative totals. This feedback ensures that the operator is aware of his or her performance and gives the operator the opportunity to improve. The following characteristics should be considered in designing a labor management function:

1. *Integrate labor standards within all procedural functions controlled by the WMS.* Nonprocedural functions (such as order processing, receiving/schedule log maintenance, and so forth) do not fall within the scope of labor management due to the unplanned (no work queues) and variable activities associated with completing a task within the functions.

2. *Utilize time and motion studies to establish time standards for every work element used within the warehouse.* This ensures accurate measurements based on equipment and constraints specific to the warehouse. Predetermined industry standards provide a broad statistical base for standard operation but may not be accurate enough for the warehouse's environment and the tolerance required for labor standards.

3. *Display the standard assignment time to the operator at the beginning of a task.* This informs the operator of the expected duration and encourages the operator to exceed the standard task time.

4. *Provide immediate feedback of an operator's performance at the completion of a task.* This feedback should give the completed task's actual and standard duration, along with daily and weekly performance.

5. *Provide a reporting mechanism that identifies and quantifies productivity measurements and problem areas.* This allows the warehouse manager to review operator performance against specific tasks and help improve overall warehouse productivity.

Shipping Functions

The shipping process is responsible for ensuring that customer order quality and compliance labeling are met and ensuring the transfer of picked merchandise to the correct shipping dock, proper loading of trailers, and printing of the packing lists, manifest, and the bill of lading for a trailer. The shipping process can vary significantly in complexity depending on the quality audits that are conducted and the carrier types utilized (company-owned fleet, prerouted trailer, and common truck and parcel carriers). The shipping functions constitute 15 to 25 percent of the system implementation time (see Fig. 5-37).

The first step in designing the shipping functions is to determine what carriers will be supported and how the different carriers will be managed. Freight tables for the various carriers can be implemented within the WMS or a rating subsystem can be utilized that is resident on the WMS's platform or is PC-based. Freight tables integrated within the WMS provides a seamless integration into the shipping functions to maximize the availability of information. The updating of freight tables, however, can be cumbersome unless a process is implemented to utilize carrier-supplied tape updates. Also, high-volume processing of parcel orders can cause a significant burden on the WMS and impact its performance in other functions. High-volume parcel processing may require a separate freight rating subsystem.

A freight rating subsystem provides the ability to have freight table maintenance from the subsystem vendor. This provides the most current carrier services to the warehouse without the need to change the WMS. When used for parcel carriers, the

FIGURE 5-37   Shipping functions of an automated warehouse management system.

freight rating subsystem can be integrated with specialized material handling equipment and a separate computer processor to handle high-volume parcel shipping without impacting other WMS functions. A freight rating subsystem has the disadvantage of requiring an interface to interact with the WMS and can limit the availability of freight rating information.

This section discusses functions that can be utilized with internal or subsystem freight rating tables. Shipping functions can be broken down into the following two basic methods:

1. *Parcel shipping.*   Each carton shipped as a parcel requires a specialized carton label for ship-to and tracking information, a packing list for each carton, a manifest of each carton, and the ability to pack orders into shipping containers. In addition, high-volume parcel operation may require integration of material handling equipment such as fixed scanners, automated carton labeling, and automated packing list applicators. Parcel shipping does not require the WMS to manage trailer loading and does not require a bill of lading for a trailer.

2. *Common carrier shipping.*   This shipping requires the WMS to track the cube and weight of merchandise for a truckload, requires directed loading of trailers, shipping container (pallet) or customer order level packing lists, and a bill of lading broken down by the merchandise freight class for a trailer. In addition, a common carrier may be routed and may require stop sequence loading with a trailer manifest for each stop. Some common carrier functions, such as unit transfer and work with loads, are also used to manage parcel shipments.

The WMS provides the following functions to support the above-noted shipping methods:

- *Pallet build.*   This RF function is used to consolidate orders that are being shipped via a truck carrier onto a shipping unit (pallet). An order may require consolidation if it was picked in a batch with other orders or if it was picked by several pickers (by zone).

- *Pallet transfer.*   This RF function is used to transfer a shipping unit (pallet or carton) to its ship staging location (shipping lane or parcel drop zone). Pallet transfer can be used to move full pallet picks and carton picks, build shipping units, cross-dock, or relay units to the next appropriate shipping area depending on the order's requirements.

- *Carton packing.*   This CRT function is used to pack piece picks, SKUs that require packing (overpacking), or parcel orders into a carton for shipping. This function supports printing of a packing list and shipping label if the carton is being shipped using a parcel carrier.

- *Pallet wrap.*   This RF function is used to print customer compliance pallet and shipping labels or reprint the movable unit label. The movable unit label must be reprinted so that it can be applied on the outside of the plastic stretch wrap for ease of scanning.

- *Load trailer.*   This RF function is used to control the loading of outbound trailers with shipping containers. This function will ensure that orders are loaded in the proper sequence for a route and provides the ability to remove shipping containers and stage them back into a shipping lane.

- *Parcel manifesting.*   This CRT function supports scanning a carton to verify that it is ready for parcel shipping. A carrier-supplied tracking number is recorded and the carton weight is verified for each parcel. A parcel manifest can be printed for all orders and the cartons that have been scanned onto a manifest. A parcel manifesting subsystem will bypass this function but must provide the same services.

- *Work with trailers.*   This CRT function is used to manage outbound trailers. A group of orders, or a load, can be assigned a load dock when a trailer is available. Packing lists can be printed for orders staged for a load. After a trailer is loaded, the BOL and truck manifest can be generated and shipping transactions generated and transferred to the host system.

- *Assisted shipping.* This CRT function supports the manual picking of orders using the assisted picking function. The assisted shipping function allows the printing of picking lists, packing lists, and bills of lading utilizing open orders and output from the assisted picking function.

Pallet Build Function. The pallet build function is designed to consolidate picked, or packed, cartons for a customer order onto a pallet for truck shipping. Picked cartons are transferred to the pallet build area by a carton conveyor, dropped by an order picker, or transferred using the pallet transfer function (see Fig. 5-38).

FIGURE 5-38 Pallet build of customer orders from a high-volume carton sortation system and carousel.

This RF function is initiated by a shipper scanning a carton to build. The WMS directs the shipper to a pick staging location where the order is being consolidated. After a pallet is full or all cartons are built for an order, the shipper scans a new movable unit number for the built pallet. The built pallet is then transferred to a ship staging location using the pallet transfer function.

The pallet build process can be expedited if cartons are prelabeled with the order number, ship-to address, and pallet build slot. This information can be printed on pick labels or on labels generated from the packing function. The shipper reads the pick label applied to a carton to identify the proper pick staging location for a customer order. The individual cartons are scanned and assigned to a pallet after a layer of cartons has been built. The following characteristics should be considered in designing a pallet build function:

1. *Allow the shipper to identify a carton, or cartons, to build.* The shipper scans cartons as they arrive in the pallet build area or builds a layer of cartons onto a pal-

let for an order using information printed on a picking label. The shipper can scan several cartons if carton labels are utilized to consolidate orders.

2. *Direct the shipper to the pick staging location where an order is being consolidated.* A carton description, such as order number, SKU, and quantity, should be displayed for verification. If the carton information is incorrect, the shipper notifies a supervisor to reconcile the problem.

3. *Prompt the shipper to scan the movable unit label for the pallet being built.* If a new movable unit label is being assigned to the pallet, the shipper must scan the pick staging location to verify its location.

4. *Allow the shipper to indicate when a pallet is complete.* The WMS notifies the shipper if the last carton for an order has been built. The shipper indicates that the pallet is completed and ready for transfer to a ship staging location.

5. *Allow the shipper to inquire about an order's status.* After a carton is scanned to build, the shipper can view the order's pick status and the quantity of cartons remaining to build. This will assist the shipper by knowing when a built pallet should be completed according to the volume of cartons expected.

6. *Allow the shipper to view the cartons built on a pallet.* This allows the shipper to verify that all cartons have been scanned onto the pallet. The trailer load function will prevent an order from being loaded if all cartons have not been scanned and built onto a pallet.

7. *Prompt the shipper to enter serial numbers or lot numbers (if required by the SKU) that have not been captured in picking.* The WMS updates the picked records to track the inventory pulled for the order.

Pallet Transfer Function. The pallet transfer function is designed to direct the movement of movable units that have been picked or built for a customer order to the next order staging location. This function also supports the movement of cross-docking or relaying movable units to a ship staging location.

The proper staging location for a movable unit depends on the customer order and shipping requirements. Movable units that contain cartons for multiple orders are directed to the pallet build drop zone for order consolidation. After a pallet is built, the pallet is directed to the pallet wrap drop zone if the customer requires pallet compliance labeling (such as UCC/EAN-128) or if the carrier requires all pallets to be stretch-wrapped. The shipper is directed to a ship staging location when the pallet is ready for shipping. Finally, this function directs a shipper to the parcel staging location if the movable unit will be shipped using a parcel carrier (see Fig. 5-39).

The pallet transfer function is initiated by a shipper scanning a picked movable unit in a picking drop or pallet build location. The picked records associated with the movable unit identify if order consolidation or wrapping is required and which staging location should be used for the shipment. The shipper scans the drop or staging location to verify that a move has been completed.

Finally, this function provides the ability to recount picked movable units prior to shipping. A movable unit may require a carton count verification if a customer has experienced shipping problems or if the warehouse policy is to check all shipments. In addition, a carton count verification is used to check the quality of new pickers (established by the pickers' profiles) and to verify cross-docking or relaying movable units. The following characteristics should be considered when designing a pallet transfer function:

1. *Allow the shipper to indicate which movable unit to transfer.* The shipper has the ability to view where picked movable units are located for a customer order or

FIGURE 5-39   Pallet transfer to ship staging. (*Courtesy of Sharp Electronics Inc.*)

truckload. Movable units received and relayed for a customer order can also be moved with this function.

2. *Prompt the shipper to enter the carton count for the movable unit.*   This is done when the picker is performing the picking process to ensure customer quality or to validate cross-docking or relaying units.

3. *Prompt the shipper to move the unit to the next processing location.*   This location can be a pallet build location if multiple orders exist, a pallet wrap location if stretch wrap or pallet labels are required, or a ship staging location for a truck or parcel carrier.

4. *Allow the shipper to scan multiple movable units to transfer.*   The WMS directs the shipper to the correct staging location for each movable unit that is picked up.

5. *Allow the shipper to consolidate units for the same customer ship-to location.* This function allows the shipper to combine partial pallets or individual cartons onto a single movable unit for loading.

6. *Generate a carrier interface if a carton movable unit is being transferred and will be shipped using a parcel carrier.*   The carrier master file identifies where a parcel carrier should be staged and the interface program that should be executed. This interface is used to communicate the necessary information about the customer order and movable unit to another program to handle application of freight rates. Parcel carrier subsystems use this information to print carton packing lists and a shipping manifest.

Carton Packing Function.   The carton packing function is designed to prepare broken case and consolidated picking for shipment by truck and parcel carriers. This function directs the consolidation of an order into a shipping container for piece picking, consolidated picking, SKUs that are not stocked in shippable cartons, or for customers who limit the allowable size of a shipping container. After a shipping container is prepared, it is directed to pallet build for consolidation with other cartons onto a pallet for truck shipping or it is directed to parcel manifesting.

Merchandise that requires packing is dropped at packing stations as picking is completed or from the pallet transfer function. To assist in the packing process, the order processing function calculates the volume of all orders that have a SKU that requires packing. The order volume for carton packing is expressed in "equivalent units" based on the SKUs that require packing. The equivalent unit for a SKU is the quantity of a SKU that will fit into a standardized shipping carton. The shipper will determine the best size packing carton for an order by the total equivalent units on order or by the carton size requested by the customer.

The packing process is initiated by a shipper scanning the movable unit bar code attached to a picking tote, picking cart, or carton. This displays a list of orders with requests for the SKUs contained on the movable unit along with the estimated order's volume and the packing location where the order is being packed. From the list of orders, the shipper selects an order to pack. The WMS displays the SKUs and the quantities required for the order selected. The shipper enters the quantity of each SKU packed and scans the movable unit for the shipping container (carton) (see Fig. 5-40).

FIGURE 5-40   Carton packing for parcel shipped orders. (*Courtesy of Sharp Electronics Inc.*)

The shipper can select to view the picking status of an order along with other units that may be staged for the order that requires packing. When a shipping container is full or the order is completely packed, the shipper selects the movable unit to be closed. The WMS determines the calculated weight for the shipping container and displays it to the shipper for validation. After the weight is validated, a shipping label is printed for the carton. The following characteristics should be considered when designing a carton packing function:

1. *Allow the shipper to identify a picked movable unit pack.* The shipper scans movable units as they arrive at the packing station.

2. *Prompt the shipper to select an order to pack.* A list of orders requiring a SKU contained in the picked movable unit is displayed. The list of orders identifies the equivalent unit for the SKUs that requiring packing, along with the location where the order is being packed.

3. *Prompt the shipper with the quantity of each SKU within the picked inventory that is required for the order.* The shipper enters the quantity of each SKU packed and scans the movable unit label for the shipping container.

4. *Prompt the shipper to enter serial numbers, or lot numbers, if required by the SKU that have not been captured in picking.* The WMS updates the picked records to track the inventory pulled for the order.

5. *Allow the shipper to inquire about an order's status.* After a order is selected to be packed, the shipper can view the order's pick status, along with the remaining picked movable units that requiring packing for the order.

6. *Allow the shipper to close a packed shipping container.* The WMS displays the calculated weight for the SKUs in the shipping container. The shipper can accept the calculated weight or enter an actual weight for the carton.

7. *Generate a ship-to label, carton packing list, and carrier interface for the packed carton if it is being shipped by parcel carrier.* The carrier master file identifies where a parcel carrier should be staged and the interface program that should be executed. This interface is used to communicate the necessary information about the customer order and movable unit to another program to handle application of freight rates. Parcel carrier subsystems use this information to print carton packing lists and a shipping manifest.

8. *Generate a conveyor interface for the packed carton if it is being transferred to pallet build or parcel manifesting via conveyor.* The information allows the carton to be sorted to the correct destination.

Pallet Wrap Function. The pallet wrap function is designed to generate customer compliance pallet, ship-to, and movable unit labels for a pallet being shipped via truck carrier. A pallet is stretch-wrapped prior to shipment with plastic film to stabilize the cartons on the pallet. The movable unit label must be reprinted for the pallet since the existing label will be covered with plastic and difficult to scan. A ship-to label is printed to help the shipper and carrier to identify where the pallet is being shipped. Finally, a customer compliance label, such as the UCC/EAN-128 format, can be printed to identify the pallet's contents when coupled with ASN unit level shipping information (see Fig. 5-41).

The pallet wrap function is initiated by a shipper scanning a movable unit label that has been staged in the pallet wrap location by picking or pallet transfer. The WMS determines the labeling requirements for the pallet from information on the customer master file. Labels are directed to a bar code printer located at the pallet wrap station. The shipper is requested to verify the pallet's calculated weight and verify that labels have been applied. After the information is confirmed, the wrapped pallet is ready for transfer to a ship staging location. The following characteristics should be considered when designing a pallet wrap function:

1. *Allow the shipper to scan a movable unit label to initiate the wrapping function.* Any picked movable unit, such as relay merchandise, full pallet picks, or built pallets, qualifies for wrapping.

FIGURE 5-41   Pallet wrap of a customer shipment.

2. *Reprint the movable unit label.*   This label is applied to the outside of the wrapped pallet so that it is available for scanning.

3. *Generate a customer ship-to label to identify whom the pallet is for and where it is being shipped.*

4. *Generate a customer compliance pallet label if required by the customer.*

5. *Prompt the picker to verify the calculated weight for the pallet and confirm that labels have been applied.*   The shipper can enter the actual weight. The WMS updates the pallet shipping information to indicate that the pallet is ready for shipping.

Load Trailer Function.   The load trailer function is designed to assist a shipper in loading outbound trailers. The shipping supervisor controls which trailers can be loaded by assigning a dock to a truckload through the work with trailers function. After a dock has been assigned, this function will assist a shipper by displaying the location where the outbound freight is staged and ensure that the trailer is loaded in the proper stop sequence, if applicable. The work with trailers function is also used to print packing lists for a load that has been assigned to a dock. The shipper will procedurally apply these packing lists as the trailer is loaded (see Fig. 5-42).

The WMS provides a list of docks that have assigned loads on a RF terminal. After the shipper selects a dock for work, the shipper is prompted to scan a movable unit to load. The shipper can select to view locations where the trailer load is staged. After the shipper scans a movable unit to load, the WMS directs the shipper to scan the dock location. If the movable unit scanned is out of the loading sequence, a warning message is displayed to the shipper. Finally, the shipper has the option to unload an outbound trailer. Selecting this option allows the shipper to scan a movable unit that is at

FIGURE 5-42 RF directed loading of customer shipments. (*Courtesy of Sharp Electronics Inc.*)

the dock location and move it to a ship staging or drop zone location. The following characteristics should be considered when designing a load trailer function:

1. *Allow the shipper to select a dock door to work.* A list of docks that have assigned loads is displayed to the shipper. Multiple shippers are allowed to load at a dock, but the WMS ensures that the customer orders are in proper "reverse" stop sequence (the first stop is the last to be loaded). For LTL loads, all customer orders have the same stop sequence.

2. *Allow the shipper to view ship staging, pick staging, and drop zone locations by load sequence, where the customer orders for a trailer load are staged.* The WMS consolidates a trailer load into a ship staging lane through the pallet transfer function. However, ship staging lanes are difficult to manage due to the variable number of pallets and cube associated with picked merchandise. For this reason, the WMS allows a trailer to be staged in many ship stage locations and allows the shipper to view this information.

3. *Prompt the shipper to scan a movable unit to load.* If the movable unit is for an order that is out of the loading sequence, an error message is displayed to the shipper identifying the stop that is currently being loaded. Also, if the last movable unit for a customer order is scanned, the shipper is prompted to confirm that the packing list has been applied.

4. *Prompt the shipper to confirm that a movable unit is loaded by scanning the dock door location label.* This ensures that the loader transferred the movable unit onto the trailer.

5. *Allow the shipper to load multiple movable units.* This will expedite the loading process if the shipper can handle multiple movable units at one time.

6. *Allow the shipper to remove a movable unit from a trailer.*   This function supports the ability to adjust a customer order up to the last minute before the trailer is closed. In addition, the shipper may need to remove merchandise to reorganize a load.

7. *Allow the shipper to "split" a trailer load.*   If the outbound trailer is full, the shipper may not get all assigned customer orders onto the trailer. The shipper is warned if a customer order is partially loaded so that movable units can be removed prior to splitting the load. When a load is split, a new load number is assigned to the merchandise that was not loaded, which can then be assigned to another dock location when a trailer is available.

8. *Allow the shipper to reprint packing lists for a customer order.*   If the packing list cannot be found, or the customer order is adjusted, it may be necessary to reprint the packing list. Also, if the packing list is printed prior to completion of picking and the inventory is not available for picking, the WMS warns the shipper that the packing list needs to be reprinted.

9. *Allow the shipper to close a trailer.*   This updates the dock status and allows the bill of lading to be generated within the work with trailers function.

Build Parcel Manifest Function.   Parcel shipping allows the warehouse to expedite shipment to a customer or to handle small shipments to customers spread over a large geographic area (such as mail order). The order processing, or traffic, function assigns a truck or parcel carrier depending on the order's size and shipping costs. Also, a customer may request expedited parcel shipping at the time of order entry.

Parcel shipping differs from truck shipping due to shipping container (carton) size and weight constraints. The packing function is used to prepare shipping containers for parcel shipping. The parcel manifest function provides the ability to weigh a shipping container, apply freight charges, and generate a parcel manifest for multiple customer shipments.

The carrier master file will specify if the parcel carrier is being manifested using a parcel manifest subsystem. If a parcel manifesting subsystem is being used, the pick, pack, and pallet transfer function will create an interface record for each parcel (movable unit) being shipped to a customer. The interface to the parceling subsystem identifies the movable unit number, order number, ship-to zip code, carrier, and movable unit weight. The parceling system assigns a freight charge and tracking number to the movable unit and then updates the WMS with this data.

If a parceling subsystem is not used, then the build parcel manifest function supports assigning freight charges and tracking numbers to parcel shipments. The build parcel manifest function is initiated by scanning the shipping container that has arrived at the parcel shipping staging location. Parcel shipping containers can be directed to the parcel staging location after being processed through the packing function or directly from picking (i.e., cartons picked onto a conveyor).

Similar to the interface of a parceling subsystem, the movable unit number attached to the shipping container identifies the merchandise being shipped, where the parcel is being shipped, and the parcel carrier that has been designated for the customer order. Also, the movable unit number identifies whether the parcel requires a packing list (if coming directly from picking) and the parcel's calculated weight. After scanning a movable unit number, the shipper will be prompted to apply a shipping label or packing list if the parcel was not processed through the packing function (see Fig. 5-43).

Next, the WMS displays the calculated weight and parcel description to the shipper. The shipper confirms the carton's weight by scanning a parcel tracking number if required by the parcel carrier. The WMS then calculates the freight charge for the par-

FIGURE 5-43 Parcel weighing station. (*Courtesy of Sharp Electronics Inc.*)

cel and applies it to the active manifest for the carrier. The following characteristics should be considered when designing a build parcel manifest function:

1. *Allow the shipper to scan a movable unit to manifest.* The WMS confirms that the order is being shipped by parcel carrier and the parcel meets the carton size and weight limits set for the carrier. If an error is found, the shipper is directed to have the carton moved to packing or truck staging.
2. *Prompt the shipper if a shipping label or packing list is required for the parcel.* A shipping label is required if the carton did not have labels applied during picking or the carton was not processed through the packing function. A packing list is required if the carton was not processed through the packing function, and the movable unit is the last carton for a customer order.
3. *Allow the shipper to print a shipping label or packing list.* This supports the reprinting of these documents if they were lost or damaged.
4. *Prompt the shipper to confirm the carton's calculated weight and scan a parcel tracking number if required by the carrier.* The shipper can enter the actual carton weight if a discrepancy is found. The parcel tracking number is printed on the parcel manifest and replaces the movable unit number.
5. *Calculate the freight charge for the parcel and apply the carton to a carrier load for manifest printing using the work with trailers function.* Freight charge calculations vary between parcel carriers. Order processing determines the specific parcel shipping method that will be used for a customer order during the order download process. A parcel carrier may support ground delivery (least expensive), next-day air delivery (most expensive), or second-day air delivery for an order. The following summarizes how parcel freight charges are assigned:
    a. The parcel carrier can have a fixed charge associated with each parcel by weight range.

b. If the parcel carrier does not have a fixed freight charge, then the first three digits of the zip code are used to determine the shipping zone assigned by the carrier. The shipping zone and parcel weight are then used to determine the freight charge for the parcel.

Work with Trailers Function.   The work with trailers function is designed to assist a shipping supervisor in managing outbound customer orders and trailers. Customer orders are grouped for a carrier and route by a load number that is assigned during the order download process in order processing. All outbound loads, whether shipped by parcel or truck carriers, are managed through this function. The supervisor works with a trailer selecting a load number and can assign a load to a dock for loading, print packing lists, print the BOL and manifest for a trailer, or adjust orders assigned to a load.

The shipping supervisor views the outbound traffic from a list of load numbers that represent trailers. A load number shows the status, route, and carrier and indicates if a dock has been assigned. The supervisor can assign a dock location to release a trailer for loading. After the dock is assigned, the supervisor requests packing lists to be printed. The packing lists are given to the shipper, who will load the trailer and apply them to the last pallet loaded for each customer order.

After a dock has been loaded and the dock closed by the shipper, the trailer status indicates that it is ready for billing. Prior to billing, the supervisor can view the customer orders assigned to a load and inquire about their picking status. A customer order can be adjusted if customer service requests a last-minute change. When a trailer is complete, the supervisor can print the truck BOL and manifest. Manifests are only printed for parcel loads that have been assigned customer shipments through the parcel manifest function.

Finally, the supervisor closes a load when it is ready to leave the distribution center. The supervisor is prompted to enter the trailer seal number(s). The closing of a load number prevents any further action on assigned orders. The WMS generates shipping transactions to update the host's inventory records, generates EDI transactions for customers requesting ASN information, and moves the customer shipping information into history files. The following characteristics should be considered when designing a work with trailers function:

1. *Allow the shipping supervisor to view all trailer loads that have been created through order processing.*   The picking and shipping status, assigned dock, carrier, and route are displayed to the supervisor.

2. *Allow the supervisor to assign a ship staging lane for a load.*   This causes pickers and shippers to consolidate a load into a specified shipping lane. This is done to ensure that a load is close to a particular dock to expedite loading or to keep available lanes for other loads that may be processed first.

3. *Allow the supervisor to assign a dock to a trailer to load.*   Assigning a dock location indicates that the trailer has arrived and loading can commence.

4. *Allow the supervisor to change a trailer.*   This provides the ability to change the carrier to which orders are assigned. Also, the supervisor can change the picking priority for a load. This changes the sequence in which picks will be executed for any outstanding picking activity for orders assigned to a load.

5. *Allow the supervisor to split a trailer.*   A trailer can be split and a new load number is assigned if the carrier supplies a half-loaded trailer (most common with LTL shipments), if the supervisor needs to manage parcel manifest assignments, or if a trailer is too small for the shipment provided by the carrier.

6. *Allow the supervisor to print a packing list for a load.* This can be done after a dock has been assigned to a load.

7. *Allow the supervisor to print the BOL and/or manifest for a load.* This is done after a truck has been loaded or after parcels have been manifested. Parcel carrier loads have a manifest printed that indicates the individual cartons and their specific ship-to address and freight charge. The BOL for a truck carrier summarizes the pallets, cartons, and pieces for all merchandise, by the freight class assigned to each SKU for each order on the trailer. A truck manifest lists each movable unit for a destination, route, and stop on a trailer.

8. *Provide the supervisor with the ability to manually edit a BOL or parcel manifest and packing list.* This supports the shipment of noninventory items such as supplies, customer samples, or "no charge" items.

9. *Allow the supervisor to view individual orders assigned to a load.* This gives the supervisor the view of a customer order status and allows the supervisor to adjust an order. A customer order requires manual adjustment if it is in the process of being picked, or if picking is complete, and customer service wants to change the quantity being shipped to the customer. If a customer order is picked, the shipping supervisor must enter a new movable unit number for any merchandise removed from the shipment. If the SKU being removed has multiple serial numbers or lot numbers, the supervisor should specify the specific inventory being removed. The inventory removed from a shipment can be moved back into inventory using the quick put-away or location-to-location move function.

10. *Allow the supervisor to close a trailer load when it leaves the distribution center.* A yard location can be assigned to the trailer for staging or the trailer's location is left "blank," indicating it has left the property. When the trailer physically leaves the DC's property, the host system's inventory and customer order records are updated along with generating the ASN transactions requested for a customer. The picking transaction and customer order shipment data is moved to history files for future inquiry. Closing a trailer removes the load from this function, keeping the shipment traffic screens filled with active loads.

Assisted Shipping Function. The assisted shipping function is designed to assist a shipping supervisor in managing outbound customer orders and trailers. The assisted shipping function allows the shipping supervisor to view open customer orders and to print picking lists selectively that are used to fill orders using the assisted picking function. The combination of assisted picking and shipping functions allows manual order fulfillment while utilizing the WMS to control inventory and stocking locations while directing inbound processes.

The shipping supervisor uses the assisted shipping function to view open orders that have been downloaded from the host system. Customer orders are selected for picking based on the transportation carrier, order priority, required shipping date, and other selection criteria. After an order is selected for picking, the WMS allocates inventory based on inventory availability at the summary level. The assisted shipping function is simplified from the order processing function by not creating a picking work queue and by not allocating specific movable units. Inventory selection is handled and picks are sequenced manually by the picker. Paper picking lists are printed based on the available inventory and distributed to pickers by the shipping supervisor. Compliance labels are printed, if required by the customer, for application during the picking process.

Pickers use the assisted picking function to determine where an item is stocked and remove inventory from the warehouse according to the allocated order quantity. The shipping supervisor uses the assisted shipping function to view the picking status for an order. Packing lists and bills of lading are printed based on the picking transaction file created by the assisted picking function. A customer order is closed by indicating that it is loaded. The freight charge, trailer number, seal number, and shipment number are recorded and shipping transactions are generated to update the host system's data base. The following characteristics should be considered in designing an assisted shipping function:

1. *Allow the shipping supervisor to view all open orders that have been downloaded from the host system.* The shipping supervisor can view the open demand (order) file by requested shipping date, order priority, carrier, and so forth.

2. *Allocate inventory at the summary level for orders selected for picking.* This ensures that pickers are not directed to pick items that do not exist in inventory. In addition, the shipping supervisor is notified of backorder situations.

3. *Print picking lists for the quantities allocated to a customer order.*

4. *Print compliance labels for customer orders that require case or pallet shipping labels.*

5. *Allow the shipping supervisor to view picking status and to view where customer orders are staged.*

6. *Allow the shipping supervisor to indicate when to generate packing lists and BOLs for selected orders that have completed picking.* The packing list and BOL are printed from the actual quantities picked using the picking transaction file created by the assisted picking function.

7. *Prompt the shipping supervisor to record the PRO number, freight charge, trailer number, and seal number when a trailer is closed.* Shipping transactions are created to update the host data base after this information is captured.

8. *Create ASN transactions when a customer order, or trailer, is closed if required by the customer.* Movable unit level ASN transactions are generated using the picking transactions created by the assisted picking function.

REFERENCES

1. Mulcahy, David E., *Warehouse and Distribution Operations Handbook*, McGraw-Hill, New York, 1994.

2. Tompkins, James A., and Harmelink, Dale A., *The Distribution Management Handbook*, McGraw-Hill, New York, 1994.

CHAPTER 6
REQUIREMENTS DEFINITION AND SYSTEM EVALUATION

INTRODUCTION

In order to evaluate a third-party warehouse management system (WMS) properly, or determine if a company is better served through the internal development of a WMS, preliminary steps must be completed to define and document the system requirements. Once a concise understanding of the requirements is obtained, goals and objectives to be realized by a warehouse management system implementation, whether internal or external, must be developed. The goals, objectives, and system requirements are utilized in the preparation of a Warehouse Management System Requirements Report.

By formally documenting the warehouse management system's goals, objectives, and requirements, an evaluation of system options can be made. Information relevant to third-party WMS vendors may be solicited through a simple request for information (RFI) or a detailed request for proposal (RFP). Concurrently, while the call for vendor proposals is occurring, the warehouse manager and Information Systems (IS) department can complete resource, skill, and budget evaluations as to the feasibility of the internal development of a WMS.

After vendor proposals and an internal feasibility study are obtained, the process of evaluating potential solutions begins. Although it may seem overwhelming at the onset, by following the steps outlined in this chapter, the requirements definition, evaluation, and selection processes are greatly simplified.

OVERVIEW

The objective of this chapter is to delineate a process to define the requirements for, evaluate, and select a warehouse management system provider. By utilizing the steps outlined in this chapter, the following goals may be achieved:

1. Requirements definition for a warehouse management system
2. Creation and evaluation of RFIs and RFPs
3. Selection of an external or internal warehouse management system provider

REQUIREMENTS DEFINITION

One of the most important, yet difficult, aspects of implementing a warehouse management system is determining what the new system is to include with respect to software functionality, computing and peripheral hardware, and host and system interfaces. Everyone involved with the system, from the warehouse workers to the Information Systems people to the management team, has an opinion of what the new system must be. To coordinate these views and minimize major difficulties, a plan needs to be developed to ensure that the requirements are collected, defined, and documented in a format that is conducive to evaluation techniques.

Application Development Process

As illustrated in Fig. 6-1, the development of a warehouse management system solution is divided into three distinct phases: Design, Implementation, and Maintenance. Within the Design phase three subphases, Requirements Definition, External Design, and Internal Design, exist.

Although this chapter deals primarily with the gathering, definition, and documenting of requirements, it is important to understand the relevance of the Requirements Definition subphase in relation to the logical design progression of a WMS. The following paragraphs briefly describe the design subphases and their respective deliverables.

Requirements Definition. The objective of the Requirements Definition subphase is to identify and document the work flow, major processing steps, information, and hardware required at each step included in the warehouse management system scope. The deliverable that results from the Requirements Definition subphase is a Requirements Report, which clearly describes and documents the software functionality as well as the hardware requirements for the warehouse management system.

External Design. The objective of the External Design subphase is the determination of the manner in which the system meets the functional requirements previously defined. During the External Design subphase, design of all the interactions such as screens and reports between the user and the WMS occurs. The External Design Report that is created at the end of this subphase contains external software behavior characteristics including the flow of data and sequence of transactions, user inputs and

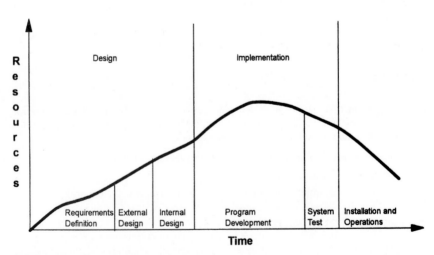

FIGURE 6-1 Application development process.[1]

outputs such as screens and reports, system hardware and software requirements, and peripheral hardware requirements.

Internal Design.   The final subphase within the Design Phase for a warehouse management system is Internal Design. The objective of the Internal Design subphase is the identification and description of program and file structures that support the WMS external components. The deliverable resulting from the Internal Design subphase is an Internal Design Report containing the internal structure and including a mapping of the system into subsystems and programs, program functional specifications, and data base specifications.

Requirements Definition Tasks

Each subphase within the application development process may be further divided into tasks and subtasks. The tasks associated with performing a successful Warehouse Management System Requirements Definition subphase are:

Task 1   Identify WMS Requirements Definition Project Leaders and Define Their Responsibilities.

Task 2   Plan and Organize the WMS Requirements Definition Project.

Task 3   Conduct a Planning Session.

Task 4   Create an Interview and Requirements Gathering Plan.

Task 5   Orient WMS Requirements Definition Project Participants.

Task 6   Determine Warehouse Requirements, Interfaces, and System Function.

Task 7   Perform a Material Handling Analysis.

Task 8   Determine Preliminary Computing and Peripheral Hardware and Software Requirements.

Task 9   Analyze the Warehouse Management System Requirements Definition Data.

Task 10   Publish the Warehouse Management System Requirements Report.

Before proceeding with detailed descriptions of each of the tasks associated with Requirements Definition, it is appropriate to mention that many project managers and/or users confronted with this task list do not envision the benefits that are gleaned from completing the Requirements Definition process. Because of this lack of vision, the following common errors occur:

1. The Requirements Definition subphase is skipped.

2. The requirements are defined without user involvement.

3. Simple functions are defined in considerable detail, while complicated functions that involve difficult decisions are barely documented.

4. Tasks are trivialized or eliminated altogether.

5. Islands of automation rather than a congruent system are created.

Other stumbling blocks occur when there is a lack of upper management support and involvement or a lack of funds. Equally detrimental is a management decision to just begin programming immediately since planning and designing are not considered progress.

Task 1

The first task, "Identify WMS Requirements Definition Project Leaders and Define Their Responsibilities," is necessary to ensure that the Requirements Definition subphase for the warehouse management system project is planned, organized, and led effectively. Three people, the project administrator, the project manager, and the technical leader are to be named at the project's onset.

Project Administrator.   This person represents upper management and the users (warehouse workers, IS, host personnel, and so forth). Often, the project administrator is referred to as the executive sponsor. Responsibilities assigned to the project administrator include:

1. Assigning warehouse, IS, and host personnel to the project team
2. Coordinating the availability of warehouse, IS, and host personnel for data gathering
3. Obtaining and providing the project manager with information, data, decisions, and approvals within the time frame agreed upon so as not to delay the project
4. Helping resolve project issues and deviations from the WMS Requirements Definition Project Plan

Project Manager.   This person is responsible for managing, planning, and controlling the overall project and for performing the following subtasks:

1. Preparing the WMS Requirements Definition Project Plan and Work Plan
2. Creating weekly status reports
3. Conducting regularly scheduled status meetings
4. Evaluating and tracking project progress against the WMS Requirements Definition Project Plan
5. Administering the change control procedure
6. Resolving deviations from the WMS Requirements Definition Project Plan

Technical Leader.   Reporting directly to the project manager, the technical leader is a person who is very knowledgeable about warehouse operations, including IS and host interactions, and who is highly respected by his or her peers within the organization. Responsibilities assigned to the technical leader include:

1. Coordinating and managing the technical activities defined in the WMS Requirements Definition Project Plan
2. Understanding the company's business needs relevant to the warehouse operations
3. Performing and/or leading pertinent research and data gathering activities
4. Obtaining user support for the WMS project

Task 2

With the proper leadership in place, Task 2, "Plan and Organize the WMS Requirements Definition Project," is now possible. Due to the fact that planning a project, especially a Requirements Definition subphase, entails detailed thought, organization, and documentation, this task is often skipped. Do not allow this to happen! Not planning and organizing the critical Requirements Definition subphase almost always results in a project team with poor morale, a lack of confidence in the end user population and upper management, and an undermining of the entire project.

The most effective way to plan a Warehouse Management System Requirements Definition subphase is to document the project's objectives in a project plan. A project plan, which is considered to be an agreement to develop a WMS that meets the company's requirements and becomes a company asset, must contain the following sections at a minimum:

1. Executive Summary
2. Project Scope
3. Approach, including phases and tasks
4. Estimated Schedule and Associated Resource Requirements
5. Project Assumptions
6. Completion Criteria

By documenting the above topics in their respective sections, the project plan validates the project and permits:

- Preliminary decisions to be determined based on relevant information
- Resources required to make the project a success to be cognizant of their responsibilities
- Project participants to understand the scope and completion criteria of the project
- The project team to be quickly educated with respect to the project and its goals

The project plan's *Executive Summary* briefly describes the project and the company's motivation for beginning such a project. It also establishes the reasons for creating a project plan and introduces the sections to follow. An example of a WMS Requirements Definition Project Plan Executive Summary is detailed below:

In today's business environment, our company has become increasingly aware of the importance of efficient warehouse operations. With greater customer demands for products and services on a more timely basis at lower costs, business trends indicate that our warehouses have become the focal point of our distribution network.

Based on customer needs, our company's recent growth, and new industry requirements, the apparent solution is a warehouse management system designed to improve inventory control, manage warehouse space, and increase employee productivity. In addition to improving internal warehouse operations, our warehouse management system must also allow access to our company's host to ensure that the latest data is available for operations at our host site and within the warehouse.

The objective of this project plan is to define the planning information required to allow for the successful undertaking of a Requirements Definition subphase for a warehouse management system. Documenting the warehouse management system's requirements, goals, and objectives is not only a crucial step, but one that once completed permits the evaluation of solution options, including third-party or internally developed warehouse management systems.

Subsequent sections of this document including the Project Scope, Approach, including phases and tasks, Estimated Schedule and Associated Resource Requirements, Project Assumptions, and Completion Criteria provide a more detailed view of the project.

The *Project Scope* section of the project plan defines the breadth of the project, delineates the warehouse functions to be included and excluded, and lists areas to be supported by the proposed system. To reenforce the practical need for a warehouse management system, this section is also used to summarize and document current problems to be solved and/or key benefits to be realized with implementation of the proposed system. An abbreviated example of a WMS Requirements Definition Project Plan Project Scope section follows:

> Success in warehousing today is measured in reduced cost and improved quality and customer service. New business is attained through the establishment of oneself as a leader in warehouse automation. In order for our company to continue evolving and managing our distribution operations, we must take steps to implement a warehouse management system. This new comprehensive and flexible system is to automate our current manual operations, while it centrally controls the diverse technologies and equipment in our warehouse.
>
> Our new warehouse management system is slated to be the required bridge between the various host-level systems such as production scheduling, purchasing, logistics planning, and order management systems, allowing us to provide the proper customer order response time critical in today's competitive market. A real-time accurate assessment of available inventory, labor, and equipment will permit our properly designed warehouse management system to issue and manage tasks and to monitor performance in each of the following functional areas:
>
> - Receiving and dock management
> - Put-away
> - Location management
> - Cycle count
> - Order processing
> - Picking
> - Replenishment
> - Packing
> - Shipping
>
> In addition to the functional areas listed, requirements are also to be defined for host and system interfaces and computing and peripheral hardware and software.
>
> Key benefits to be realized from the implementation of our warehouse management system include:
>
> 1. Improving customer service through the reduction of order cycle times and better shipping accuracy
> 2. Improving productivity by reducing product handling, optimizing travel time of storage and retrieval equipment, optimizing warehouse tasks, and eliminating errors
> 3. Increasing inventory accuracy through real-time activity confirmation
> 4. Maximizing facility utilization through location maintenance and rewarehousing

The project plan's *Approach* section contains a description of the major development steps, specifies standards and guidelines to be used during the project, and outlines the tasks associated with the project. This section, in particular, is necessary to delineate the project's structure and control mechanisms. The following paragraphs show an example of a WMS Requirements Definition Project Plan Approach section:

> The application development of our warehouse management system is to be accomplished in three distinct phases, namely, design, implementation, and maintenance. Within the Design phase, three subphases, Requirements Definition, External Design, and Internal Design, exist. The objective of this section of the WMS's Project Plan is to describe the approach to be utilized during the Requirements Definition subphase.
>
> In order to complete our warehouse management system project successfully, guidelines and procedures must be developed and implemented. Therefore, in pursuit of our goal the following formal project controls are to be employed throughout the warehouse management system's application development:

- Specified manageable tasks, namely, Task 1, Identify WMS Requirements Definition Project Leaders and Define Their Responsibilities; Task 2, Plan and Organize the WMS Requirements Definition Project; Task 3, Conduct a Planning Session; Task 4, Create an Interview and Requirements Gathering Plan; Task 5, Orient WMS Requirements Definition Project Participants; Task 6, Determine Warehouse Requirements, Interfaces, and System Function; Task 7, Perform a Material Handling Analysis; Task 8, Determine Preliminary Computing and Peripheral Hardware and Software Requirements; Task 9, Analyze the WMS Requirements Definition Data; and Task 10, Publish the Warehouse Management System Requirements Report.
- *Change Control.*   The project manager has established a change control procedure and forms that provide direction for requesting, documenting, evaluating, and approving or disapproving desired or required changes. Please see the appropriate appendix for more information.
- *Status Reporting.*   The purpose of status reports is to communicate the status of the project relative to the WMS Requirements Definition Project Plan and approved change requests. Reports are to be provided on a weekly basis by the project manager to the project administrator and are to include, at a minimum, accomplishments (planned versus actual), open problems and issues, a change control summary, and planned activities for the next reporting period.
- *Project Reviews.*   Periodic independent project reviews are to be scheduled throughout the WMS Requirements Definition subphase. These reviews are designed to evaluate project plan integrity with respect to objectives, progress, schedule, and budget.

The *Estimated Schedule and Associated Resource Requirements* section of the project plan includes the project work plan, which consists of the project schedule and estimates of personnel and other required resources. Creation of a WMS Requirements Definition Project Work Plan is significantly simplified by completing the following steps:

1. Define the tasks.
2. Divide the tasks into specific, manageable, and trackable subtasks.
3. Describe each subtask.
4. Assign each subtask a number.
5. Specify the skills required to complete each subtask.
6. Identify the type, number, and mix of personnel required to complete each subtask.
7. Estimate the effort and duration of each subtask. (Usually, effort is estimated in work-hours, and duration in work-days.)
8. Validate the estimates.
9. Create the preliminary schedule and project calendar.
10. Assign specific personnel to each subtask.

A well-thought-out work plan increases the chances of project success, and provides the basis for tracking and reporting. An example WMS Requirements Definition Project Plan Estimated Schedule and Associated Resource Requirements section is shown below:

In order to ensure the success of our Warehouse Management System Requirements Definition subphase, a work plan including an estimated schedule and associated resource requirements has been created. Figure 6-2 illustrates the preliminary schedule, while Figs. 6-3a and 6-3b show an example task description and its associated schedule from the WMS Requirements Definition Project Work Plan. Additional background information relating to the development of the work plan is available from the project manager.

ID	Task Name	Pre-Req	Duration	Effort	Timeline (resources)
1	Identify Prj Leaders and Responsibilities		2d	6h	Prj Mgr
2	Plan and Organize Req Defn Proj	1	3d	24h	Prj Mgr
3	Conduct Planning Session	2	1d	4h	Prj Mgr
4	Create Interview and Req Gathering Plan	3	1.5w	60h	Team
5	Orient Req Defn Prj Participants	4	1d	4h	Prj Mgr
6	Determine Req. Int'f, Sys Function	5	30d	360h	
10	Perform MH Analysis	7FS-2w	4w	160h	IE/Consultant
11	Determine Computing & Peripheral HW/SW	6	3d	20h	Team
12	Analyze WMS Req and Defn Data	11	3w	50h	Team
13	Publish WMS Req Report	12	1w	24h	Team

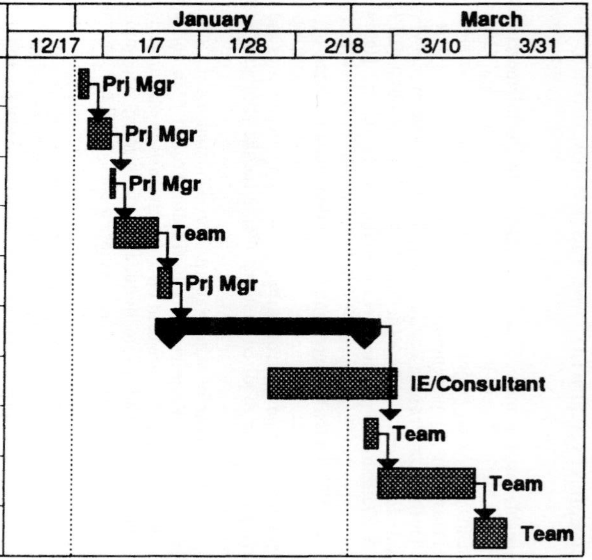

Timeline header: January | March — 12/17, 1/7, 1/28, 2/18, 3/10, 3/31

FIGURE 6-2 WMS Requirements Definition subphase preliminary schedule. In addition to defining the project tasks, task prerequisites, task durations, and task efforts, this schedule also indicates the task dependencies, as illustrated by the arrows connecting the tasks, and the required resources, as listed to the right of each scheduled task.

Subtask number	Subtask name	Subtask description	Prerequisite tasks	Skills required to complete	Estimated work-hours	Estimated work-days	Personnel assigned
7	Interviews	Conduct interviews according to the interview schedule	Tasks 1–5	PM, SA	160	25	TWB LPM
8	Process documentation	Document current warehouse process on flow process charts	Tasks 1–5	PM, SA	160	25	TWB LPM
9	Summarize requirements	Consolidate and summarize requirements information by category within major functional areas	Subtasks 7 and 8	PM, SA	40	5	TWB LPM

PM = project management, SA = system analyst.

FIGURE 6-3a Example task description from the WMS Requirements Definition Project Work Plan.

ID	Task Name	Pre-Req	Duration	Effort	Schedule (resource)
1	Identify Prj Leaders and Responsibilities		2d	6h	Prj Mgr
2	Plan and Organize Req Defn Proj	1	3d	24h	Prj Mgr
3	Conduct Planning Session	2	1d	4h	Prj Mgr
4	Create Interview and Req Gathering Plan	3	1.5w	60h	Team
5	Orient Req Defn Prj Participants	4	1d	4h	Prj Mgr
6	Determine Req. Int'f, Sys Function	5	30d	360h	
7	Conduct Interviews		5w	160h	Team
8	Document Current Processes		5w	160h	Team
9	Summarize Requirements	7,8	1w	40h	Team
10	Perform MH Analysis	7FS-2w	4w	160h	IE/Consultant
11	Determine Computing & Peripheral HW/SW	6	3d	20h	Team
12	Analyze WMS Req and Defn Data	11	3w	50h	Team
13	Publish WMS Req Report	12	1w	24h	Team

Timeline: December (12/17) – January (1/7, 1/28) – February (2/18) – March (3/10, 3/31)

FIGURE 6-3b Example task schedule from the WMS Requirements Definition Project Work Plan.

The project plan's *Project Assumptions* section contains a list of the assumptions that guided the formulation of the estimates and schedules. The integrity of a project plan, and often the success of a project, is predicated on the validity of project assumptions. The following example typifies information normally found in the Project Assumptions section for a WMS Requirements Definition Project Plan:

The logic utilized in the development of this WMS Requirements Definition Project Plan is predicated on the following assumptions:

1. The WMS Requirements Definition subphase is to be approved, personnel are to be selected, and work is to begin by [date].
2. Nine functions including receiving and dock management, put-away, location management, cycle count, order processing, picking, replenishment, packing, and shipping are to be addressed during the WMS Requirements Definition project.
3. Required host interfaces are to be defined during the Requirements Definition project.
4. User interviews are to be the vehicle employed to gather and define the WMS system requirements.
5. Project personnel and users are to be available according to the WMS Requirements Definition Project Work Plan dated [date].
6. Requested changes to the WMS Requirements Definition Project Plan and/or deviations from these assumptions are to be processed using the defined change control procedure.

The final section of the WMS Requirements Definition Project Plan is the *Completion Criteria*. A clear concise understanding of what constitutes project completion, as well as who is required to approve intermediate and final deliverables resulting from project execution, is essential to ensure that the project's objectives are ultimately accomplished. Each task identified in the project plan has completion criteria associated with it. The overall WMS Requirements Definition Project Plan Completion Criteria section might contain a statement similar to this:

Approval of the WMS Requirements Report by the project manager, project administrator, and technical leader constitutes the completion of the WMS Requirements Definition Project.

Completion of the essential WMS Requirements Definition Project and Work Plans signifies a monumental step in the arduous planning process. With these project objectives documents serving as guidelines, the following additional organizational and planning subtasks are feasible:

1. Create a Warehouse Management System Requirements Report outline including the following sections:
   a. Executive Overview
   b. Functional Requirements by Area
      (1) Receiving and dock management
      (2) Put-away
      (3) Location management
      (4) Cycle count
      (5) Order processing
      (6) Picking
      (7) Replenishment
      (8) Packing
      (9) Shipping

    c. Material Handling Analysis Results
    d. Interface Requirements
       (1) Host
       (2) System
    e. Hardware and Software Requirements
       (1) Computing hardware and software
       (2) Peripheral hardware and software
       (3) Material handling hardware and software
    f. Warehouse Management System Recommendations and Strategy
2. Arrange for suitable office space and clerical support for the project team
3. Create WMS Requirements Definition project team orientation packages which include the following items:
    a. WMS Requirements Definition Project Plan
    b. Task assignment sheets from the WMS Requirements Definition Work Plan
    c. Warehouse Management System Requirements Report outline
    d. Change control procedure

Task 3

Once the WMS Requirements Definition project planning is concluded, the project manager may proceed with Task 3, "Conduct a Planning Session." The main objectives of the planning session are to orient the WMS Requirements Definition project team to the project's goals, structure, and control mechanisms and to build team camaraderie. In addition to distributing the team orientation packages, a review of project documentation, such as the WMS Requirements Definition Project Plan, the WMS Requirements Definition Work Plan, the WMS Requirements Report outline, and the change control procedure, is necessary.

Task 4

With the project team in place and oriented, the final project organizing steps may be performed. By completing Task 4, "Create an Interview and Requirements Gathering Plan," the project team is ensured that interviewing and data gathering techniques to be utilized during the WMS Requirements Definition subphase are consistent and effective. The subtasks involved with Task 4 include the following:

1. *Determine present knowledge.* It is important for the project team to ascertain what has been previously documented with respect to the current warehouse operating procedures or systems, and/or to the requirements for future warehouse procedures or systems. Availability of the following key documents must be explored:
    a. Process documents for current warehouse procedures
    b. Training materials for current warehouse procedures
    c. User's guide for present warehouse system software, if applicable
    d. Previous requirements gathering efforts
2. *Specify required information.* Once the level of current knowledge is quantified, the project team can specify what information is required. Relevant information consists of, but is not limited to, current warehousing processes and operating procedures including inputs, outputs, and reports; problems with current processes; the time required to complete current processes; and specifications, process improvements, or requirements for a future warehouse management system.

Example interview questions by functional area, which may be utilized to help gather pertinent information, are as follows:

a. Receiving

    (1) How are receipts scheduled?

    (2) What is the process for receiving merchandise?

    (3) What paperwork or forms are utilized in the receiving process?

    (4) How are the quantity and quality of the received merchandise verified?

    (5) What additional information (lot and/or serial numbers, weight, and so forth) is collected during the receiving process?

    (6) What method is utilized to initialize the identification and tracking process for the merchandise being received?

    (7) How are returns handled?

    (8) How is the availability of the required personnel and equipment determined?

    (9) Once the receiving process is complete, what is done with the information gathered?

    (10) What are the current problems associated with completing a receipt?

    (11) What level of process improvement can be obtained through the elimination of a problem identified by the preceding question?

    (12) How can the receiving process be improved?

b. Put-away and location management

    (1) What types of storage locations are available for put-away?

    (2) What criteria are currently used in the selection of a storage location? Are there additional criteria that can be considered?

    (3) What paperwork or forms are utilized in the put-away process?

    (4) If the location identified for put-away is not available or adequate, how is another location selected?

    (5) How is the availability of the required personnel and equipment determined?

    (6) Once the put-away is complete, what is done with the information gathered?

    (7) What are the current problems associated with completing put-away tasks?

    (8) What level of process improvement can be obtained through the elimination of a problem identified by the preceding question?

    (9) How can the put-away process be improved?

c. Cycle count

    (1) Is inventory accuracy a problem in the warehouse?

    (2) What method is utilized to verify inventory accuracy?

    (3) Are physical inventories performed? At what intervals?

    (4) How are inventory discrepancies resolved?

    (5) How can inventory accuracy be improved?

d. Order processing and picking

    (1) What steps occur during the processing of an order?

    (2) How are orders reviewed, maintained, and, if necessary, combined?

    (3) How is product allocation completed?

    (4) What criteria are utilized to determine what orders are to be processed and in what sequence the selected orders are to be filled?

    (5) How is the availability of the required personnel and equipment determined?

    (6) What picking methods (radio frequency pallet, carton, or piece, pick lists, labels, etc.) are used?

    (7) What picking documentation (pick lists, labels, etc.) must be created to support the picking methods?

(8) What control data (serial numbers, lot numbers, etc.) is recorded during the picking process?

(9) Is pick mechanization (pick-to-light, ASRS, etc.) utilized? If so, how effectively?

(10) If the location identified for picking does not contain adequate stock to complete the pick, what happens? Does this situation occur often?

(11) How are picking locations replenished?

(12) Once picking is complete, what is done with the information gathered?

(13) What are the current problems associated with completing order processing and picking?

(14) What level of process improvement can be obtained through the elimination of a problem identified by the preceding question?

(15) How can order processing and picking be improved?

e. Packing

    (1) How is the packing function accomplished?

    (2) What documentation (packing lists, shipping labels, manifests, etc.) is created? Are the documents easy to edit and reprint?

    (3) How is order checking completed?

    (4) Does order consolidation normally occur during the packing operation?

    (5) Are interfaces to scales, manifesting systems, and printers utilized?

    (6) How is the availability of the required personnel determined?

    (7) Once packing is complete, what is done with the information gathered?

    (8) What are the current problems associated with the packing function?

    (9) What level of process improvement can be obtained through the elimination of a problem identified by the preceding question?

    (10) How can packing be improved?

f. Shipping

    (1) What carrier types (truckload, LTL, parcel) are required?

    (2) Is order checking a necessary function?

    (3) What documentation (packing lists, shipping labels, bills of lading, manifests, etc.) is created? Are the documents easy to edit and reprint?

    (4) How is the availability of the required personnel and equipment determined?

    (5) What controls are in place to ensure orders are placed in the shipping lanes in the proper route and stop sequence?

    (6) How are order cancellations resolved?

    (7) How are trailers loaded and, if required, unloaded?

    (8) What are the current dock management and carrier scheduling requirements?

    (9) Once shipping is complete, what is done with the information gathered?

    (10) What are the current problems associated with the shipping function?

    (11) What level of process improvement can be obtained through the elimination of a problem identified by the preceding question?

    (12) How can shipping be improved?

g. General

    (1) What level of reporting is required by function?

    (2) How is productivity tracked?

3. *Identify people to interview.* Once the level of information required is established, and interview questions that may be utilized to collect the information are formulated, the project administrator is in a position to identify people in each functional area who can provide this information.

4. *Develop an interview schedule.* After the proper people have been identified to be interviewed, the project manager and project administrator are responsible for developing an interview schedule. The normal progression of inventory through the standard warehouse functions serves as the best guide in the establishment of an effective requirements gathering interview schedule.

5. *Create interview packages.*   Although interviewing has repeatedly proved to be the most beneficial method for requirements gathering and definition, it is almost always somewhat unsettling for the interviewee. To ease the interviewee's discomfort and to attempt to improve the overall effectiveness of the interview process, interview packages can be created. These interview packages normally contain the following items:

   a. An interview request letter
   b. An overview of the Warehouse Management System Requirements Definition project scope
   c. An outline of the warehouse functionality to be discussed and/or a list of questions to be asked during the interview
   d. A list of the material the interviewee is requested to prepare and bring to the interview

Task 5

Completion of the interview schedule and packages permits the project team to proceed with Task 5, "Orient WMS Requirements Definition Project Participants." Since end user and management involvement and support are the keys to the project's success, it is important to plan and conduct the orientation meeting in a worthwhile and upbeat fashion. At the meeting, the following essential agenda items are to be covered at a minimum:

1. Introduce the WMS Requirements Definition project team.
2. Explain the reason for conducting the orientation meeting.
3. Describe the WMS Requirements Definition project scope.
4. Specify the purpose of the WMS Requirements Definition project.
5. Review the interview process.
6. Distribute the interview packages.

Some suggestions for increasing user involvement include initiating a contest to name the project and create a logo, and/or if the name and logo have previously been selected, presenting the end users, IS representatives, and management team with a small token such as a key ring, T-shirt, mug, or beverage hugger, with the WMS project name and logo illustrated on it.

Task 6

Once the WMS Requirements Definition project participants are oriented, the project team may begin Task 6, "Determine Warehouse Requirements, Interfaces, and System Function." The primary objective of this task is to collect and document in a clear and concise manner the information necessary to specify the requirements for the warehouse management system.

The subtasks involved with the determination of the warehouse management system requirements include the following:

1. *Conduct interviews according to the interview schedule.*   During each interview session, the interviewer is to review material requested by the project team and

supplied by the interviewee, ask relevant questions about the interviewee's functional area, and document on an interview summary sheet similar to that shown in Fig. 6-4 all answers, problems, inputs, outputs, interfaces, and requirements expressed by the interviewee.

2. *Document current warehouse processes on flow process charts (FPC).* The current process for each warehouse function described by the interviewee is to be documented in a flow process chart[2] as illustrated in Fig. 6-5. Information required to complete the flow process charts is as follows:

a. *#.*   Sequence number of a process step.

b. *Description.*   A narrative explanation of a process step.

c. *Resource or title.*   The person involved in a process step.

d. *Required.*   An indication of whether or not a process step is required in the new warehouse management system.

e. *Quantity.*   The number of times per shift/day during both average and peak periods that a process step is performed.

f. *Qty (U/M).*   The unit of measure (i.e., receipts, picks, etc.) associated with a specified quantity.

g. *Time.*   The time required to complete a process step.

h. *Material handling automation.*   A description of the material handling equipment or automation that is utilized to complete a process step.

i. *Automated picking.*   Mechanized picking equipment such as pick-to-light, carousels, and so forth.

j. *Conveyor.*   Mechanical device used to move merchandise from one location to another. Types of conveyor include belt, gravity, and roller.

k. *Container build.*   Manual or mechanized (i.e., palletizer, robot, and so forth) method used to build pallets or containers.

l. *Fork truck.*   Electric- or gas-powered vehicle with "forks" that allow it to pick up pallets or skids.

m. *Sortation.*   The process of grouping cartons based on predetermined criteria. Sorting technology includes automated scanning, programmable logic controllers (PLCs), and so forth.

n. *Activity.*   A classification or grouping of process steps.

o. *Operation.*   An operation transpires when an item is changed and/or prepared for another activity. An operation also occurs when information is transferred or when planning or calculating happens.

p. *Transportation.*   Transportation is the movement of an item or person from one location to another.

q. *Delay.*   A delay ensues when an item or person is forced to wait for the next process step.

r. *Inspection.*   An inspection is the result of an item being examined for identification or verified as to quality or quantity.

s. *Process.*   A qualitative description or classification of an activity or process step.

t. *Reports and documentation.*   Paper output created as a result of an activity or process step.

u. *Decision.*   The application of rules and/or logic by a person or system to determine the appropriate action.

v. *Manual operation.*   An unautomated procedure that is manual in nature and usually requires paperwork.

w. *Automated operation.*   An automated procedure that involves interaction with a system to accomplish a process step or activity.

Interview Summary Sheet

Date:

Functional Area:

Interviewer:

Interviewee:

Responsibility:

Functional Requirements:

    1.
    2.
    3.

Information Requirements:

    Input   1.
               2.
               3.

    Output   1.
               2.
               3.

Interfaces Used: 1.
               2.
               3.

Problems:

    1.
    2.
    3.

Recommendations:

FIGURE 6-4   Interview summary sheet.

Flow Process Chart

Function Chart #- Page # -

	Activity		Present	Summary Proposed	Savings		Material Handling Automation			Qty	Process
○	Operation						AP	Automated Picking	⊡		Report/Documentation
⊃	Transportation						C	Conveyor	◇		Decision
▽	Delay						CB	Container Build	▽		Manual Operation
▭	Inspection						FT	Fork Truck	▭		Automated Operation
		Total					S	Sortation			

#	Description	Resource Title	Req'd	Quantity Avg	Peak	Qty U/M	Time	Mat'l Handling Present Future	○ ⊃ ▽ ▭	Remarks
	Current Process									
1										
2										
3										
4										
5										
6										
7										
8										
9										
10										
11										
12										
13										
14										
15										
16										
17										
18										
19										
20										
21										
22										
23										
24										
25										
26										
27										
28										
29										

FIGURE 6-5 Flow process chart.

The main reasons for documenting the current process are to provide a logical starting point for understanding the requirements for the future warehouse management system and to permit the project team to begin devising solutions and alternatives for proposed processes.

3. *Consolidate and summarize requirements information by category within major functional areas.* The interview information collected for each major functional area listed in the outline of the Warehouse Management System Requirements Report is to be organized and condensed into a format that facilitates analysis. One way to accomplish this subtask is to sort the data within each major function into established categories such as the following:

a. All flow process charts for a major function
    (1) For example, receiving function
        (a) FPC 1.   Receiving by PO
        (b) FPC 2.   Quality control sampling
        (c) FPC 3.   Returns processing
b. All input for a selected major function
    (1) For example, receiving function
        (a) Pro bill
        (b) Bill of lading
        (c) Purchase order
c. All output for a selected major function
    (1) For example, receiving function
        (a) Receiving worksheet
        (b) Daily receiving report
        (c) O/S/D (over/short/damage) report
        (d) KeyRec log
d. All required interfaces for a selected major function
    (1) For example, receiving function
        (a) Purchase order header and detail
        (b) Receiving summary
e. All problems for a selected major function
    (1) For example, receiving function
        (a) No PO available for item to be received
        (b) Difficult to resolve discrepancies

Task 7

As the warehouse requirements, interfaces, and system function information are being gathered and documented, Task 7, "Perform a Material Handling Analysis," may also be occurring.

The main reason for undertaking this task is to attempt to evaluate and determine the benefits to be realized by the incorporation of material handling automation in the requirements for the future warehouse management system. If a better understanding of possible material handling automation options is deemed necessary before proceeding, a review of Chap. 4, Material Handling Considerations, and/or the addition of a project team member or consultant with expertise in material handling automation may be warranted.

The subtasks associated with the completion of the material handling analysis include the following:

1. *Conduct interviews as discussed in the Task 6 section.* While the current processes are being reviewed, the interviewer responsible for the material handling analysis can either create flow process charts emphasizing the current material handling automation utilized by the warehouse or complete the material handling analysis columns provided on the flow process charts.

2. *Perform a mathematical analysis of the current material handling process.* The most beneficial tool in the analysis of material handling automation requirements and the justification for future systems is rooted in the gathering and statistical application of current warehouse process data. A complete analysis may incorporate the following:

ABC COMPANY

VOLUME ANALYSIS FOR CARTONS SHIPPED

	ACTUAL PEAK DAY	AVERAGE DAY FOR PEAK WEEK	AVERAGE DAY FOR MONTH	PEAK DAY % VS. MO. AVG.	PEAK WK % VS. MO. AVG.	AVERAGE DAY FOR YEAR	PEAK WK % VS. YRLY AVG
MAY 94	28056	26028	19425	44%	34%	25119	4%
APR 94	31344	28889	23414	34%	23%	25119	15%
MAR 94	33879	28676	23432	45%	22%	25119	14%
FEB 94	46842	28521	23999	95%	19%	25119	14%
JAN 94	30490	24465	21990	39%	11%	25119	-3%
DEC 93	25086	22658	17884	40%	27%	25119	-10%
NOV 93	NO DATA	22112	20231		9%	25119	-12%
OCT 93	NO DATA	39536	30566		29%	25119	57%
SEP 93	NO DATA	34196	31620		8%	25119	36%
AUG 93	NO DATA	46232	33977		36%	25119	84%
JUL 93	NO DATA	40806	24897		64%	25119	62%
JUN 93	NO DATA	34660	24142		44%	25119	38%
MAY 93	NO DATA	25831	18919		37%	25119	3%

SUMMARY	
PEAK DAY	46842
AVERAGE DAY	25119
AVERAGE–PEAK DAY % VS. MO. AVG.	49%
AVERAGE–PEAK WEEK % VS. MO. AVG.	28%
AVERAGE–PEAK WEEK % VS. YRLY. AVG.	23%

FIGURE 6-6 Volume analysis.

- Volume analysis (Fig. 6-6)
- Velocity profile (Fig. 6-7)
- Order profile (Fig. 6-8)
- Truck activity summary (Fig. 6-9)
- Rack analysis (Fig. 6-10)
- Unit hours, head count, and cost summary (Fig. 6-11)

3. *Establish material handling system (MHS) requirements.* Based on the information collected and analyzed in subtasks 1 and 2, a preliminary list of base or benchmark requirements for the material handling system can be documented.

4. *Review warehouse tactical and strategic plans.* Before proceeding with the determination of new material handling automation processes and concepts, it is important to review the tactical and strategic 5- to 10-year plans for the business and the warehouse facility. Plan items of particular importance in the definition of material handling system requirements include, but are not limited to, future volume projections, facility changes, and/or new services to be offered (for example, piece picking).

ABC COMPANY

VELOCITY PROFILE

CARTON SHIPPING
10% OF SKU'S = 57% OF SHIPS
20% OF SKU'S = 72% OF SHIPS
30% OF SKU'S = 82% OF SHIPS
40% OF SKU'S = 89% OF SHIPS
50% OF SKU'S = 94% OF SHIPS
60% OF SKU'S = 97% OF SHIPS
70% OF SKU'S = 97% OF SHIPS
80% OF SKU'S = 98% OF SHIPS
90% OF SKU'S = 99% OF SHIPS
100% OF SKU'S = 100% OF SHIPS

FIGURE 6-7   Velocity profile.

ABC Company
Order Profile
Daily Average for 6/1 through 9/30

Lines per order	Orders per day	Percentage of orders	Items per day	Percentage of items	Average item per order
1	4554	73.9	4701	30.9	1.03
2	808	13.1	1696	11.2	2.10
3	266	4.3	853	5.6	3.21
4	127	2.1	552	3.6	4.35
5	77	1.3	414	2.7	5.38
6	50	0.8	331	2.2	6.62
7	36	0.6	285	1.9	7.92
8	28	0.5	238	1.6	8.50
9	23	0.4	247	1.6	10.74
10	20	0.3	228	1.5	11.40
11+	171	2.8	5648	37.2	33.03
Totals	6160	100.0	15,193	100.0	2.47

FIGURE 6-8   Order profile.

ABC Company
High-Level Platform Truck Activity

1993 week ending	Inventory, pallets	Inventory verify, pallets	Combine pallets	Retrieve short cartons	Return to mfg., pallets	Fix wrecked pallets	Returns to storage, pallets
10/3	813	330	2929	5251	1764	2082	714
9/26	367	107	2797	4928	1823	1895	376
9/19	609	72	3310	3381	1863	2240	840
9/05			2576	5303	668	2007	595
8/29			3378	5183	650	1842	455
8/22			2459	3726	1064	1573	599
8/15			2445	2724	2338	1651	524
8/08	217	217	2602	4410	2587	1627	317
8/01			2458	5091	1051	1453	798
7/25	101	101	1364	3073	920	1228	484
7/18	741		1354	3720	1300	1377	420
7/04	182		1505	3561	349	1484	431
6/27	261	22	1763	3276	1696	2231	98
6/20			1074	2174	1271	1434	171
6/13	100	27	1355	1866	532	1384	287
6/06			724	2655	1383	1026	249
5/30	97	22	1588	1265	1817	1301	244
5/23	80		2361	813	1758	1615	46
5/16	433	51	1847	764	953	1272	409
5/09			1711	1548	314	1652	259
5/02			2687	2902	559	1440	80
4/25			2772	1853	393	1233	339
4/18	165	29	1464	1584	652	1183	48
4/11			1858	670	894	1311	88
4/04	163	478	2324	1623	407	1611	172
Total	4329	1456	52,705	73,344	29,006	39,152	9043

FIGURE 6-9   Truck activity summary.

It is important to remember that a fundamental principle of material handling system design is to always design for future volumes and plans.

5. *Develop new process requirements and concepts for the MHS.*   This subtask begins with the creation and quantification of future (based on future volume information) requirements for storage, flow rates, capacities, and workload. By comparing the current process analysis as determined in subtasks 1 and 2 to the future process analysis, a list of of material handling system requirements and high-level flow charts of proposed material handling processes and alternatives can be developed and documented.

6. *Perform an economic analysis of the proposed process concepts and alternatives.*   To compare the alternatives and justify the ultimate proposed process concepts and requirements further, a labor cost savings analysis, return on investment (ROI) analysis, and sensitivity analysis are recommended. (Examples of these analyses are shown in Chap. 8, Justifications and Benefits.) Results gleaned from these analyses provide the financial perspective, which when considered with the technical issues, allows for the final determination of the proposed process concepts and the material handling system requirements.

ABC Company

Rack Analysis–Current

	NORTH SIDE				
AISLE	# OF RACKS 3 PAL	# OF RACKS 5 PAL	# OF PICK FRONTS	# OF STOR. LOCS	TOTAL LOCS
A	26	12	138	414	552
B	34	13	162	648	810
C	42	14	186	578	764
D	44	14	189	593	782
E	46	14	192	608	800
F	46	13	187	593	780
G	45	12	182	572	754
H	44	12	182	566	748
I	45	12	182	572	754
J	46	12	182	578	760
K	46	12	182	578	760
L	46	12	182	578	760
M	47	13	190	602	792
N	47	14	198	620	818
O	46	14	198	614	812
P	47	14	198	620	818
SUBTOTAL	697	207	2930	9334	12264

	SOUTH SIDE				
AISLE	# OF RACKS 3 PAL	# OF RACKS 5 PAL	# OF PICK FRONTS	# OF STOR. LOCS	TOTAL LOCS
J	91		258	824	1082
K	92		258	830	1088
L	92		258	830	1088
M	92		258	830	1088
N	92		258	830	1088
P	92		258	830	1088
Q	92		258	830	1088
R	92		258	830	1088
S	92		258	830	1088
T	92		258	830	1088
U	92		258	830	1088
SUBTOTAL	1011	0	2838	9124	11962
TOTALS	1708	207	5768	18458	24226

FIGURE 6-10   Rack analysis.

7. *Prepare a list of required material handling hardware and software.*   The proposed process concepts and the material handling system requirements determined in subtask 6 have associated material handling hardware and software requirements that can now be determined and documented.

ABC COMPANY
CURRENT PROCESS-UNIT HOURS/HEAD COUNT/COSTS

	1994	1995	1996	1997	1998	1999
CARTONS SHIPPED	(REFLECTS 10% ANNUAL GROWTH)					
ANNUAL (M)	6.90	7.59	8.35	9.18	10.10	11.11
DAILY TOTAL CARTONS	27165	29882	32870	36157	39773	43750
HOURS/CARTON						
RECEIVING	0.0020	0.0020	0.0020	0.0020	0.0020	0.0020
RETURNS	0.0011	0.0011	0.0011	0.0011	0.0011	0.0011
REC'VG CHECKER	0.0010	0.0010	0.0010	0.0010	0.0010	0.0010
PUT-AWAY	0.0045	0.0045	0.0045	0.0045	0.0045	0.0045
REPLENISH	0.0035	0.0035	0.0035	0.0035	0.0035	0.0035
PICKING	0.0128	0.0128	0.0128	0.0128	0.0128	0.0128
SHIPPING CHECKER	0.0010	0.0010	0.0010	0.0010	0.0010	0.0010
SHIPPING/LOADING	0.0073	0.0073	0.0073	0.0073	0.0073	0.0073
CYCLE COUNT	0.0006	0.0006	0.0006	0.0006	0.0006	0.0006
LOCATOR CLERK	0.0011	0.0011	0.0011	0.0011	0.0011	0.0011
LABEL PRINT	0.0008	0.0008	0.0008	0.0008	0.0008	0.0008
AVERAGE HOURS/CARTON	0.0357	0.0357	0.0357	0.0357	0.0357	0.0357
ADJUSTED HEADCOUNT						
RECEIVING	6.8	7.5	8.2	9.0	9.9	10.9
RETURNS	3.7	4.1	4.5	5.0	5.5	6.0
REC'VG CHECKER	3.4	3.7	4.1	4.5	5.0	5.5
PUT-AWAY	15.3	16.8	18.5	20.3	22.4	24.6
REPLENISH	11.9	13.1	14.4	15.8	17.4	19.1
PICKING	43.5	47.8	52.6	57.9	63.6	70.0
SHIPPING CHECKER	3.4	3.7	4.1	4.5	5.0	5.5
SHIPPING/LOADING	24.8	27.3	30.0	33.0	36.3	39.9
CYCLE COUNT	2.0	2.2	2.5	2.7	3.0	3.3
LOCATOR CLERK	3.7	4.1	4.5	5.0	5.5	6.0
LABEL PRINT	2.7	3.0	3.3	3.6	4.0	4.4
TOTAL ADJUSTED HEADCOUNT	121.2	133.3	146.7	161.4	177.5	195.2
CARTONS/EMPLOYEE/HOUR	28.0	28.0	28.0	28.0	28.0	28.0
ANNUAL LAB. COST (K$)	$21.0	$21.6	$22.3	$22.9	$23.6	$24.3
RECEIVING	$143	$162	$183	$207	$235	$266
RETURNS	$78	$89	$101	$114	$129	$146
REC'VG CHECKER	$71	$81	$92	$104	$118	$133
PUT-AWAY	$321	$364	$412	$467	$529	$599
REPLENISH	$250	$283	$320	$363	$411	$466
PICKING	$913	$1,034	$1,172	$1,328	$1,504	$1,704
SHIPPING CHECKER	$71	$81	$92	$104	$118	$133
SHIPPING/LOADING	$521	$590	$668	$757	$858	$972
CYCLE COUNT	$43	$48	$55	$62	$71	$80
LOCATOR CLERK	$78	$89	$101	$114	$129	$146
LABEL PRINT	$57	$65	$73	$83	$94	$107
TOTAL ANNUAL LABOR COST	$2,546	$2,884	$3,268	$3,703	$4,195	$4,753

FIGURE 6-11 Current process—unit hours, head count, and cost summary.

Task 8

With the warehouse system and material handling requirements documented, the project team can progress to Task 8, "Determine Preliminary Computing and Peripheral Hardware and Software Requirements." The objectives of this task are to evaluate the current computing environment, to quantify the level of equipment and technology currently utilized in the warehouse, and to develop a computing and peripheral hardware and software strategy that allows the WMS requirements to be met. Important aspects to remember during the preliminary hardware and software determination process are to review the strategic 5- to 10-year business plans and to apply the strategic information to the hardware and software requirements without limiting future system evaluation efforts. For example, if radio frequency technology is a requirement, do not specify a particular vendor unless capital investment with that vendor has already occurred.

The subtasks involved with the determination of preliminary computing and peripheral hardware and software requirements include the following:

1. *Review Chap. 3, Technology Considerations.* An understanding of available technology including the advantages and disadvantages of each is a prerequiste to computing (i.e., computing platform, data base, and so forth) and peripheral (i.e., radio frequency terminals, scanners, and so forth) hardware and software Requirements Definition. By reviewing Chap. 3 and calling on previous experiences, the project team can begin to form an initial strategy for the future warehouse management system. Examples of strategic elements to consider are the following:
   a. Computing platform
      (1) Mainframe computer versus midrange computer versus microcomputer
      (2) Host-based versus distributed versus client/server
      (3) IBM AS/400 versus HP/9000, IBM RISC/6000 versus DEC VAX, etc.
      (4) OS/400 versus UNIX/AIX versus Windows NT, etc.
   b. Data base
      (1) Relational versus hierarchical versus object-oriented
   c. Data base access
      (1) Open data base connectivity (ODBC) compliant or not
   d. Connectivity
      (1) System network architecture (SNA) versus transmission control protocol/Internet protocol (TCP/IP)
   e. Data collection
      (1) Paper versus portable terminals
      (2) Batch versus real-time terminals
      (3) Spread versus narrow-band radio frequency
   f. Input/output devices
      (1) Nonprogrammable terminals versus workstations
      (2) Laser versus dot matrix printers
2. *Collect system sizing data.* Utilizing data gathering sheets similar in nature to those illustrated in the indicated figures allows for the collection and consolidation of relevant information such as:
      (1) Resource deployment totals (Fig. 6-12)
      (2) Material handling and information-based equipment totals (Fig. 6-13)
      (3) Warehouse processing volumes (Fig. 6-14)
      (4) Physical warehouse layout information (Fig. 6-15)
      (5) Computing environment summary (Fig. 6-16)

Area	Shift					
	1		2		3	
	Avg.	Peak	Avg.	Peak	Avg.	Peak
Receiving						
Detailed receiving						
Staging, inbound						
Returns						
Put-away						
Quality control						
Replenishment						
Order processing						
Picking						
Packing						
Checking						
Staging, outbound						
Loading or shipping						
Administrative—clerk						
Supervisor(s)						
Drivers						
Other (please specify)						

FIGURE 6-12   Resource deployment totals. List the total average and peak number of warehouse employees by area and shift. If employees work in more than one warehouse area, use decimal fractions (i.e., 0.75 receiving, 0.25 staging).

The statistical and logistical information required to complete the work sheets can be obtained from a variety of sources including requested material supplied by interviewees during the interview process, quantity details as documented on the flow process charts, analyses completed as a result of Task 7 ("Perform a Material Handling Analysis"), and reports containing current warehouse operation figures and statistics.

3. *Derive a computing and peripheral hardware and software strategy.*   The project team can now relate the technology knowledge base gained in subtask 1 to the statistical data acquired in subtask 2 to define the general requirements and overall computing and peripheral hardware and software strategy to be used with the future warehouse management system. Examples of this data application and strategy development may possibly consist of one of the following scenarios:

a. *Example 1:* The warehouse that the WMS is being considered for is located near a busy airport, where radio data transmission within the spread spectrum frequencies is sure to be unreliable due to interference. In this case, the project team is likely to surmise that the narrow-band radio frequency technology with its FCC-licensed dedicated frequencies is a better strategic choice.

Area	Equipment	Information-based†
Receiving		
Returns		
Put-away		
Picking		
Packing		
Loading or shipping		
Other (please specify)		

*For example: Put-away operation—12 fork trucks and 150 feet of conveyor. Indicate if the same equipment is used by two areas of the warehouse (i.e., the same fork trucks are used by put-away and picking operations).
†For example: the number of RF terminals; number of hand-held scanners; number of fixed scanners. Please include brand name and model number.

FIGURE 6-13   Material handling and information-based equipment totals. List the types and quantity of material handling and information-based equipment used in each area of the warehouse.* Only include the equipment that is to be utilized with the new system.

    b. *Example 2:* During the WMS Requirements Definition project, it becomes obvious that the management of daily warehouse activities requires a significant amount of data manipulation. Additionally, to simplify and enhance these routine planning functions, graphical display is desired. Based on these requirements, the project team might conclude that a client/server system with its graphical user interface (GUI) capabilities, intelligent workstations, and inherent distributed data processing methodology is the logical alternative.

4. *Prepare a list of the computing and peripheral hardware and software requirements.*   This subtask entails the documenting of the computing and peripheral hardware and software requirements as determined in subtask 3.

Receiving		
	Average	Peak
Number receiving docks used		
Number of trailers per day		
Number of POs received per day		
Average number of lines per PO		
Number of cartons received per day		
Average number of cartons/movable unit		
Total lines received per day (at receipt or to be palletized):		
% palletized, 1 SKU per pallet		
% palletized, 2+ SKUs per pallets		
% not palletized		
Cross-docks (flow-through)		
Number of QC SKUs per day		
Number of QC cartons per day		
Returns		
	Average	Peak
Number of returns per day		
Number of lines per return		
Number of cartons per return		

FIGURE 6-14   Warehouse processing volumes. Attempt to obtain both average and peak warehouse processing volumes. If time is a restraint, obtain peak information first.

Task 9

Upon completion of the data gathering and analyses required in Tasks 6 to 8, the project team is positioned to proceed to Task 9, "Analyze the Warehouse Management System Requirements Definition Data." Since the requirement information has been derived from a variety of sources, this task is essential to allow the overall project goal of providing a complete and understandable description of the warehouse management system requirements to be met.

To analyze the WMS Requirements Definition data properly, the following subtasks are recommended:

1. *Consolidate and summarize requirements information by category within major functional areas.* Requirement information such as the following, resulting from the culmination of previous tasks, must be organized and consolidated by major functional area (for example, receiving function). By summarizing the information in this manner it not only permits analysis to be performed, but it also translates the data into a format that is conducive for the generation of the Warehouse Management System Requirements Report.

Put-Away Operation		
Method of put-away	Percent of each type	
Pallet		
Carton (other than on pallets)		
Eaches (other than in cartons)		
	Average	Peak
Number of pallets put away per day		
Number of cartons put away per day		
Number of eaches put away per day		
Replenishment		
Method of replenishment	Percent of each type	
Pallet		
Carton (other than on pallet)		
Eaches (other than in cartons)		
	Average	Peak
Number of pallets replenished per day		
Number of cartons replenished per day		
Number of eaches replenished per day		
Carton or Pallet Picking		
	Average	Peak
Number of orders per day		
Average number of lines per order		
Number of full pallets picked per day		
Number of cartons picked per day		
Eaches Picking		
	Average	Peak
Number of orders per day		
Average number of lines per order		
Number of eaches picked per day		

FIGURE 6-14   (*Continued*)

   a. Flow process charts
   b. Input summary
   c. Output summary
   d. Interface requirements
   e. Problem summary

Bulk Picking		
	Average	Peak
Number of orders per day		
Average number of lines per order		
Packing and Checking		
	Average	Peak
Number of orders per day		
Average number of lines per order		
Number of cartons packed per day		
Number of pallets created per day		
Shipping		
	Average	Peak
Number of trailers shipped per day LTL % Full %		
Number of parcel orders shipped per day		
Number of lines per parcel order		
Number of customer pickups per day		
Number of shipping docks used*		
Number of orders shipped per day		
Total cartons shipped per day		
Total pallets shipped per day		
Number of documents printed per order		
Type of documents printed per order		
Cycle Count		
	Average	Peak
Number of locations counted per day		

*Note: Indicate if the same dock doors are used for shipping and receiving.

FIGURE 6-14   (*Continued*)

  f. Material handling proposed process concepts and requirements
  g. Material handling hardware and software requirements
  h. Computing and peripheral hardware and software strategy
2. *Analyze the requirements information.*   The steps associated with this subtask
   include the following:
   a. Review requirements data and problem areas.
   b. Determine relative importance of requirements (minor to major).
   c. Streamline current processes.

Inventory Information

Regulatory compliances: oil, paint, batteries, chemicals? ____

Environmental requirements: Frozen ____; Chilled ____; Expensive ____

Other: Controlled substances ____; Dangerous ____; Office supplies ____

Disk Sizing Information	
Total number of storage locations	
Total number of SKUs	
Average number of movable units in DC	
Number of serialized SKUs	
Average number of serialized SKUs in DC	
Number of days of history to be stored	

Sizing and Material Handling Profile Information	
Projected annual growth—assume current year is year 0:	
Year 1 (19)	%
Year 2 (19)	%
Year 3 (19)	%
Year 4 (19)	%
Year 5 (19)	%
Total number of SKUs (PN)	
% conveyable	% nonconveyable
SKU profile (relationship of SKUs to ship volumes):	
10% of SKUs = ____ % of ships or revenue	
20% of SKUs = ____ % of ships or revenue	
30% of SKUs = ____ % of ships or revenue	
40% of SKUs = ____ % of ships or revenue	
50% of SKUs = ____ % of ships or revenue	
60% of SKUs = ____ % of ships or revenue	
70% of SKUs = ____ % of ships or revenue	
80% of SKUs = ____ % of ships or revenue	
90% of SKUs = ____ % of ships or revenue	
100% of SKUs = ____ % of ships or revenue	

FIGURE 6-14   (*Continued*)

Physical Warehouse Layout Information		
Total square footage of warehouse		
Number of warehouse buildings		
Description of physical separation of warehouse buildings (cement walls, 1 mile, etc.)		
Are warehouse blueprints or a layout available?		
Number of receiving dock doors		
Number of shipping dock doors		
Are the same doors used for receiving and shipping?		
Total number of customer pickup locations		
Location Information		
Types and total number of stocking locations	Type of location*	Total number of locations
	Floor bulk	
	Drive-in rack	
	Pallet rack	
	Shelving	
	Outside	
	Remote	

*These are only examples of possible location types.

FIGURE 6-15   Physical warehouse layout information.

Computing Environment Summary	
Host-computer machine (brand and model)	
Host-computer operating system	
Host-computer preferred communications method	
File structure description	
File types	
File access method	
Preferred connectivity protocol	

FIGURE 6-16   Computing environment summary.

Flow Process Chart

Receiving by P.O.   Chart #- 2   Page # - 1

Summary

Activity	Present	Proposed	Savings
○ Operation	18	15	3
▭ Transportation	4	3	1
▽ Delay		5	2	3
▢ Inspection		2	2	0
Total	29	22	7

Material Handling Automation		Qty	Process
AP	Automated Picking	0	Report/Documentation
C	Conveyor	1	Decision
CB	Container Build	9	Manual Operation
FT	Fork Truck	12	Automated Operation
S	Sortation		

#	Description	Resource Title	Req'd	Quantity Avg	Peak	Qty U/M	Time (Min.)	Mat'l Handling Present	Future	○ ◻ ▽ ▢	Remarks
	Proposed Process										
1	Call warehouse	Carrier	opt.	10	20	D	---				D=Delivery/Truck
2	Add Rev Log & assign conf. #	RevClerk	yes	10	20	RLL	1.0				RLL=Rcv Log Line
3	Drive to warehouse	Trucker	yes	10	20	D	---				
4	Inspect load	SecGuard	yes	10	20	D	3.0				
5	Update Rcv Log with arrival time	SecGuard	yes	10	20	RLL	0.5				
6	Assign dock door	RevClerk	yes	10	20	D	0.3				
7	Move load to rev door	Trucker	yes	10	20	D	3.0				
8	Record seal number(s)	Receiver	yes	10	20	S	0.5				S=Seal
9	Unload trailer	Receiver	yes	10	20	D	30.0	FT			
10	Record BOL quantities	Receiver	yes	8000	16000	Car	0.05	CB			Car=Carton
11	Close trailer	Receiver	yes	10	20	D	1.0				
12	Reconcile BOL quantities	RevClerk	yes	4	8	D	2.0				Exception
13	Open box to verify pack	Receiver	yes	150	250	Car	0.8				
14	Count merchandise	Receiver	yes	150	250	Car	0.3				
15	Inspect a box of each SKU	Receiver	yes	150	250	Car	2.5				
16	Identify damage by inventory class	Receiver	yes	15	25	Dis	0.5				Dis=Discrepancy
17	Assign movable unit to inventory	Receiver	yes	300	1050	Plt	0.5				Plt=Pallet
18	Reserve storage location for inventory	N/A	yes	300	1050	Plt	0.5				
19	Complete receiving load	Receiver	yes	10	20	D	1.0				
20	Verify/Adjust overages	Supervisor	yes	30	105	Plt	5.0				Exception
21	Adjust remaining Rcv Log discrepancies	Supervisor	yes	15	25	RLL	3.0				Exception
22	Close Rev Log, & create history	N/A	yes	10	20	RLL	1.0				

FIGURE 6-17   Flow process chart for the proposed receiving by purchase-order process.

d. Brainstorm new concepts and identify proposed and alternative processing methods. (An alternative process normally satisfies the same objective in a slightly less advantageous way.)

e. Develop and document proposed and alternative process flows on flow process charts (see Fig. 6-17).

f. Validate that the proposed process flows meet the stated requirements, and eliminate problem areas. Proper flow process chart utilization, including comparisons of each FPC's activity and process; present, proposed, and savings summaries; and verification that requirements identified in current process steps are met in the proposed process, provides a practical methodology for validation. Examples of conclusions drawn through this validation methodology are as follows:

(1) *Employs fewer manual processes.* A comparison of the current (Fig. 6-18) versus the proposed (Fig. 6-17) FPCs for the receiving process illustrates that the proposed process reduces the number of manual operations by 50 percent.

(2) *Incorporates more automated operations and decision logic, therefore resulting in more consistent operations and fewer errors.* Again, a com-

Flow Process Chart

Receiving by PO Chart #- 1 Page # - 1

	Summary				Material Handling Automation		Qty	Process
Activity	Present	Proposed	Savings		AP	Automated Picking	7	Report/Documentation
○ Operation	18				C	Conveyor	2	Decision
⇨ Transportation	4				CB	Container Build	18	Manual Operation
▽ Delay	5				FT	Fork Truck	2	Automated Operation
☐ Inspection	2				S	Sortation		
Total	29							

#	Description	Resource Title	Req'd	Quantity Avg	Peak	Qty U/M	Time (Min.)	Mat'l Handling Present	Future	○ ⇨ ▽ ☐	Remarks
	Current Process										
1	Call warehouse	Carrier		10	20	D	---				D=Delivery/Truck
2	Schedule appointment	RcvClerk	yes	10	20	D	1.0				
3	Drive to warehouse	Trucker	yes	10	20	D	---				
4	Check-in at rcv. office	Trucker		10	20	D	0.5				
5	Inspect seals	SecGuard	yes	10	20	S	5.0				S=Seal
6	Update log	SecGuard	yes	10	20	D	0.5				
7	Deliver Pro Bill or BOL to office	Trucker	yes	10	20	D	---				
8	Create blind receiver	RcvClerk		4	8	D	3.0				Exception
9	Print PO	RcvClerk		4	8	D	0.5				Exception
10	Create Receiving Worksheet	RcvClerk		10	20	D	2.5				
11	Assign Receiver	RcvClerk	yes	10	20	D	1.0				
12	Unload trailer	Receiver	yes	10	20	D	30.0	FT			
13	Open box to verify pack	Receiver	yes	150	250	Car	0.8				Car=Carton
14	Count merchandise	Receiver	yes	8000	16000	Car	0.05	CB			
15	Complete Receiving Worksheet	Receiver		10	20	D	20.0				
16	Inspect a box of each SKU	Receiver	yes	150	250	Car	2.5				
17	Log discrepancies	Receiver	yes	30	50	Dis	3.0				Dis=Discrepancy
18	Report to receiving supervisor	Receiver		30	50	Dis	5.0				Exception
19	Resolve discrepancies	Suprvisr	yes	30	50	Dis	7.0				Exception
20	Apply pallet ID label	Receiver	yes	300	1050	Plt	0.3				Plt=Pallet
21	Open boxes logged on report	Receiver		30	50	Car	1.2				
22	Return receiving worksheets to office	Receiver		10	20	D	3.0				
23	Compare actual to P/Ls and POs	RcvClerk		10	20	D	4.3				
24	Determine stocking locations	RcvClerk	yes	300	1050	Plt	2.0				
25	Assign location to pallets	RcvClerk	yes	300	1050	Plt	0.5				
26	Create put-away labels	RcvClerk		300	1050	Plt	1.0				
27	Create daily receiving report	RcvClerk	yes	3	3	Rep	10.0				Rep=Report
28	Create O/S/D report	RcvClerk	yes	3	3	Rep	8.0				
29	Create KeyRec	RcvClerk	yes	10	20	Rep	4.0				

FIGURE 6-18 Flow process chart for the current receiving by purchase-order process.

parison of the current versus proposed FPCs for the receiving process not only depicts a 50 percent reduction in the number of manual operations, but it also shows a 250 percent increase in the number of automated operations.

(3) *Minimizes exception processing.* The number of activities that may require exception processing decreased from four in the current FPC receiving process to three in the proposed FPC receiving process.

(4) *Meets defined requirements.* The requirements described and denoted in the current FPC process for the receiving process are all met in the proposed FPC process for receiving.

3. *Create a weighted matrix of the WMS.* The relative importance factors assigned to the requirements and the proposed process flows validated in subtask 2 can be combined with the requirement information consolidated in subtask 1 to create a

Functional Area	Req. #	Requirement Description	Weighting Factor	Comments
Receiving	1.0	Receiving by PO	3	
	1.1	Schedule deliveries into the warehouse from carriers and vendors.	3	
	1.2	Provide a receiving log function to enter receipt information. Associate the appropriate POs with the receiving log.	3	
	1.3	Assign a confirmation number to the scheduled appointment/receipt.	2	
	1.4	Include the capability to update the receiving log with the carrier arrival time.	2	

Functional Area	Req. #	Requirement Description	Weighting Factor	Comments
Receiving	1.0	Receiving by PO (cont.)		
	1.5	Assign dock doors.	3	
	1.6	Record seal numbers.	1	
	1.7	Provide a radio frequency (RF) trailer unload function to perform BOL level receiving and to expedite the release of dock doors and carriers.	3	
	1.8	Allow for a RF detailed receiving function based on the information in the receiving log.	3	
	1.9	Display SKU receiving information including pallet configuration details.	3	
	1.10	Record SKU information including serial number, lot number, and expiration date.	3	
	1.11	Allow for and direct quality control sampling.	3	

FIGURE 6-19 WMS Requirements Definition weighted matrix. Weighting factors: 1, not a required function; 2, required function that may be accomplished through alternate processing; 3, required function that is to be performed as described.

weighted requirements matrix as shown in Fig. 6-19. This weighted matrix is an invaluable tool for future system evaluation efforts, only if the requirements information and documented process steps included in the matrix are not viewed as fixed or rigid, but rather as a guide to convey the requirements that must be met by the new warehouse management system. It is important to remember that flexibility in the manner in which requirements are satisfied is the key to to maximizing

Functional Area	Req. #	Requirement Description	Weighting Factor	Comments
Receiving	1.0	Receiving by PO (cont.)		
	1.12	Assign and record discrete inventory tracking IDs for each unit created.	3	
	1.13	Provide for exception processing including the dispositioning of damaged merchandise.	3	
	1.14	Allow for the reconciliation of quantity discrepancies.	3	
	1.15	Create movable unit records.	3	
	1.16	Call put-away to reserve storage locations for the inventory.	1	
	1.17	Provide a means to close the receipt (receiving log) and update inventory summary and receiving files.	3	
	1.18	Capture vendor and carrier compliance data.	2	
Receiving	2.0	--continue requirements--		

FIGURE 6-19   (*Continued*)

the benefits reaped from the Requirements Definition project as well as any future system evaluation efforts.

Task 10

Once the requirements data is consolidated, analyzed, and verified, the project team can prepare to conclude the Warehouse Management System Requirements Definition subphase by completing Task 10, "Publish the Warehouse Management System Requirements Report." The WMS Requirements Report, which summarizes the system requirements, goals, and objectives, is to contain the following sections at a minimum:

- Executive Overview
- Functional Requirements by Area
  Receiving and dock management
  Put-away
  Location management
  Cycle count
  Order processing
  Picking
  Replenishment
  Packing

Shipping

Security

- Material Handling Analysis
- Interface Requirements

Host

System

- Hardware and Software Requirements

Computing hardware and software

Peripheral hardware and software

Material handling hardware and software

- Warehouse Management System Recommendations and Strategy

Many of the figures referenced in the example WMS Requirements Report are illustrated (with the tasks they are deliverables of) earlier in this chapter. To eliminate redundancy, figures are not reexhibited; however, the example WMS Requirements Report does include the original figure number and a page reference at the appropriate position within the report format.

The WMS Requirements Report *Executive Overview* briefly describes the purpose of the report, lists major goals to be attained through the implementation of a warehouse management system, and overviews the remaining sections. An example of a WMS Requirements Report Executive Overview is as follows:

To remain competitive in today's business environment, our company's warehouse operations must be continually honed in an endeavor to reduce costs and improve the quality of customer service. Fully realizing that this effort to continue evolving and managing our distribution operations requires a sophsticated system, we have completed the first step, the Requirements Definition, in the implementation of a warehouse management system (WMS). The requirements for this comprehensive and flexible WMS that is expected to automate our current manual operations, while it centrally controls the diverse technologies and equipment in our warehouse, are documented in this Warehouse Management System Requirements Report.

The main objectives to be attained through the implementation of a WMS include:

1. Improving customer service through the reduction of order cycle times and better shipping accuracy
2. Improving productivity by reducing product handling, optimizing travel time of storage and retrieval equipment, and optimizing warehouse tasks
3. Increasing inventory accuracy through real-time activity confirmation and the associated reduction of paperwork required for current warehouse operating procedures
4. Maximizing facility utilization through location maintenance and rewarehousing
5. Reducing the number of errors and the frequency of exception processing through the use of automated operations and system applied decision rules and logic

The requirements identified in this report result from a review of current warehouse procedures, installed system software, tactical and strategic business plans, interviews with personnel from each warehouse functional area, and technical analyses of data and information. Information contained in this report is to be used in the evaluation of solution options including third-party or internally developed warehouse management systems.

Subsequent sections of this document as listed below provide a more detailed view of the report contents.

Section 1: Executive Overview

Section 2: Functional Requirements by Area
Section 3: Material Handling Analysis
Section 4: Interface Requirements
Section 5: Hardware and Software Requirements
Section 6: Warehouse Management System Recommendations and Strategy

The *Functional Requirements by Area* section of the WMS Requirements Report provides, for each warehouse major functional area, descriptions of the requirements in both a narrative and matrix format, a summary of the inputs and outputs, and a synopsis of relative problems. Additionally, diagrams depicting the process flows for each major function are a useful tool to demonstrate system requirements. The following paragraphs illustrate an abbreviated example of a WMS Requirements Report Functional Requirements by Area section:

The premise of the receiving process is to implement a warehouse management system (hereafter referred to as WMS) that allows real-time control capabilities of items entering the facility. The proposed flow outlined in Fig. 6-20 and the receiving requirements listed in the WMS Requirements Definition weighted matrix shown in Fig. 6-19 detail the fundamental requirements of a design that incorporates many of the features envisioned by the personnel of our company.

The current receiving process is initiated by the vendor or carrier scheduling an appointment with the receiving office. Upon arrival, carriers register at the receiving office and present the receiving documentation consisting of either the PRO bill or bill of lading. After security checks the load and verifies the seals are intact, the merchandise to be received can be unloaded.

All items entering the warehouse are received against an authorization such as a purchase order (PO) that gates their legitimate receipt or acceptance into the warehouse. The WMS receiving function is to receive items electronically against the appropriate authorization, identify and reconcile deviations, update inventory records, and initiate the tracking of items through the warehouse.

Input Summary

The input* and its associated information as illustrated in Fig. 6-21 are required for the receiving function. (*This is an abbreviated example. Many more inputs may be required for the receiving process.)

Output Summary

The output* and its associated information as shown in Fig. 6-22 are required for the receiving function. (*This is an abbreviated example. Many more outputs may be required for the receiving process.)

Problem Summary for Receiving

A synopsis of major problems commonly encountered during the receiving process is as follows:

1. Receiving discrepancies are difficult to track and resolve.
2. Receiving authorizations are not available for all receipts.
3. A significant amount of paperwork and manual data entry is required to track receipts.
4. Twenty-five percent of the merchandise received is not accurately identified and routed.

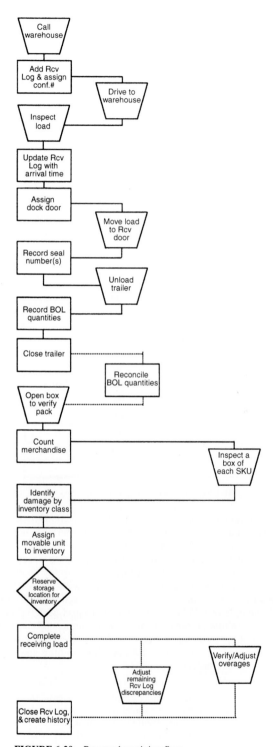

FIGURE 6-20 Proposed receiving flow.

Input	Information Required
Purchase Order (PO)	PO Number
	Host or DC Origination Indicator
	Status
	Carrier
	Vendor
	PO Line Number
	SKU
	Quantity Ordered
	Quantity Received

FIGURE 6-21   Receiving input summary.

Output	Information Required
Movable Unit	Unit Identification
	Unit Status
	Unit Location
	SKU(s)
	Quantity of SKU(s)
	Quality of SKU(s)
	FIFO Date
	Weight
	Velocity
	Environment

FIGURE 6-22   Receiving output summary.

The WMS Requirements Report *Material Handling Analysis* section contains the results of the mathematical analyses of the current material handling processes, the requirements and high-level flow charts relating to the proposed material handling concepts and alternatives, and the justifications for the proposed material handling concepts derived from the labor savings, ROI, and sensitivity analyses. To emphasize the value of integrating material handling automation with the WMS, this section may also list key benefits to be realized with the implementation of the proposed system. An abbreviated example of a WMS Requirements Report Material Handling Analysis section is as follows (although previous WMS Requirements Report example sections encompassed receiving, the Material Handling Analysis section cites a picking example; the change is based on the fact that a material handling analysis of the picking function gives a better example):

Determining the level of material handling automation that provides the greatest advantage and return on investment when integrated with a warehouse management system is the catalyst that motivated the project team to complete a material handling analysis. By gathering and analyzing current and proposed processes and data, evaluating material handling equipment currently utilized, and reviewing warehouse tactical and strategic plans, new process requirements and concepts have been developed for a material handling system (MHS). The results of the material handling analysis are detailed in this section of the WMS Requirements Report.

Current Process Analysis

In order to evaluate and recommend a new material handling system, it is necessary to have a detailed understanding of the current process. To gain this knowledge current processes have been reviewed through interviews with key personnel, and methods and performance data has been collected from current warehouse reports. The results of the analysis are as follows:

- Volume analysis (Fig. 6-6)
- Velocity profile (Fig. 6-7)
- Order profile (Fig. 6-8)
- Truck activity summary (Fig. 6-9)
- Rack analysis (Fig. 6-10)
- Unit hours, head count, and cost summary (Fig. 6-11)

Design Year Statistics

Based on the principle that material handling systems are to be designed for future volumes and plans, projected design year requirements have been calculated and summarized below (see Fig. 6-23).

Proposed Material Handling Process Concept and Alternatives

By utilizing current process analysis data and the design year statistics information in conjunction with observations, analysis, and estimates, the following major areas have been identified as candidates for the implementation of material handling automation.

Proposed Process Requirements and Concepts

- *Label picking and sortation conveyors for high-velocity cartons is the recommended process* for picking cartons. Customer-required ship address labels, which also serve as pick lists, are applied to each carton as it is picked. After the label is applied, the carton is to be placed on a powered conveyor that transports the carton to a downstream sortation conveyor responsible for routing the carton to the correct shipping lane. Order sequence integrity is to be maintained by controlling the batch size to a prescribed maximum per trailer loaded.
- *Pick/pack to order using a pick-to-light system is the recommended process* for pick/pack. With the pick-to-light system, picking instructions for items associated with a particular order are provided. The order items are picked into a labeled shipping carton. When the carton is full, it is sent via powered conveyor to the sortation conveyor, which then routes the carton to the correct shipping lane.

Conceptual diagrams of the carton picking and pick/pack processes are illustrated in Figs. 6-24 and 6-25.

Alternative Process Requirements and Concepts

- *Carton picking with radio frequency technology is an alternative process* for carton picking. This process is similar to the process delineated for label picking and sorta-

ABC COMPANY

DESIGN YEAR STATISTICS

AVERAGE CARTON VOLUMES–15%/YR GROWTH (1999)	68485
PEAK MONTHLY CARTON VOLUMES–20% OVER AVG. (1999)	82182
SKU GROWTH (CURRENT IS 425)	625
VELOCITY PROFILE	SKUS–20% SHIP–82%
TURN RATE	FAST – 18 SLOW – 3.4
PRODUCTION PLAN	MFG–65% VENDOR–35%
PLANNED SHIFTS	FULL–1ST FULL–2ND PARTIAL–3RD
CARTONS PER PALLET	64
LABOR RATE: 3%/YR INFLATION PLUS 30% BENEFITS (1999)	$11.96/HR

DAILY CARTON VOLUME DETAIL		
	AVG DAY	PEAK MONTH
1994	27523	33028
1995	33028	39634
1996	39633	47560
1997	47559	57071
1998	57071	68485
1999	68485	82182

FIGURE 6-23   Design year statistics.

tion conveyors for high-velocity cartons with the exception that the each carton label is scanned after it is applied. The addition of RF technology provides enhanced accuracy due to its real-time nature. Since, in this particular case, customer ship labels must be applied regardless of the picking method selected, the improved accuracy normally gained through the use of RF technology is negated by the dramatic increase in the number of work-hours and system transactions resulting from a proposed RF implementation.

- *Pick/pack batch with RF technology is an alternative process* for pick/pack picking. In this process items are picked into a tote with a batch (bulk) picking process. Subsequent downstream sortation is used to sort the items by order and to ensure that the items are placed in the correct shipping container.

CARTON PICKING MODULE

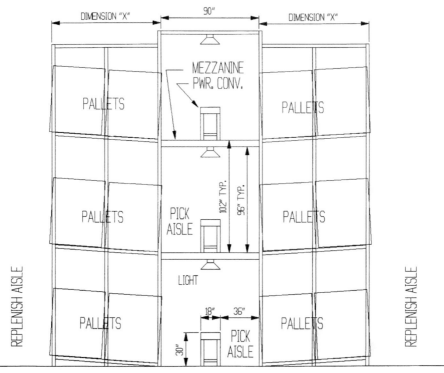

FIGURE 6-24   Conceptual diagram of the carton picking process.

Potential Benefits from Material Handling Automation

The proposed process concepts and requirements have been justified through a series of analyses including:

- Labor cost analysis (Fig. 8-6)
- Return on investment analysis (Fig. 8-17)
- Sensitivity analysis (Fig. 8-18)

Based on the technical design considerations and the results of the above analyses, it can be concluded that the following significant potential benefits may be realized through the integration of material handling automation and a warehouse management system:

- Reduced picking labor due to the use of flow racks, radio frequency devices, and pick-to-light systems
- Reduced cycle time to process orders
- Maximized facility utilization through the efficient use of warehouse space
- Improved organization and disciplined material movement and handling

The *Interface Requirements* section of the WMS Requirements Report defines the interfaces required to allow efficient data communication between the warehouse

PICK/PACK MODULE

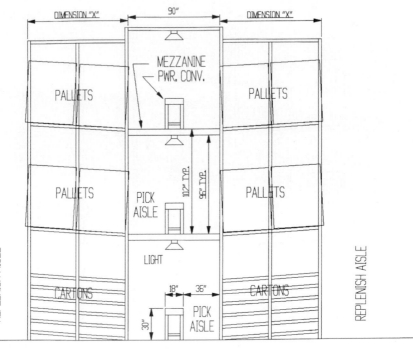

FIGURE 6-25 Conceptual diagram of the pick/pack process.

management system and the host or other systems (for example, a pick-to-light system). A narrative description and preliminary interface information requirements are normally included for each interface specified. The following paragraphs illustrate an abbreviated example of a WMS Requirements Report Interface Requirements section:

> Since the data base for our real-time distributed warehouse management system requires new and separate master files as well as periodic updates from the current host data base, host interfaces with the ability to transfer pertinent data must be defined. Additionally, due to the fact that separate systems (for example, pick-to-light and parcel manifesting) are to be operated in conjunction with the WMS, system interfaces must also be identified. Based on these two hypotheses, the following list of WMS host-computer and system interfaces has been formulated. (This is an abbreviated example.)
>
> - *Purchase order.* Purchase orders that are created on the host system during the purchasing process are downloaded by the host, and include information relevant to pending deliveries to the warehouse. Data to be transferred across the interface includes PO number, status, carrier, vendor, PO line number, SKU, and quantity ordered.
> - *Receiving summary.* Receiving information including purchase order number, vendor, SKUs, quantities, and quality is sent to the host computer to update inventory summary data bases. Data to be transferred across the interface includes delivery number, authorization number, SKU, quantity received, inventory quality, and vendor.

The WMS Requirements Report *Hardware and Software Requirements* section outlines the computing, peripheral, and material handling hardware and software strategy necessary to fulfill the requirements for the future warehouse management system. Information supplied in this section consists of a system sizing summary and a strategic computing, peripheral, and material handling hardware and software requirements list. It is important to reiterate that where applicable, the computing, peripheral, and material handling hardware and software requirements are to remain general, thereby enhancing available options during the system evaluation process. An abbreviated example of a WMS Requirements Report Hardware and Software Requirements section is as follows:

To define the requirements for our warehouse management system effectively, a strategic direction for the ultimate selection of computing, peripheral, and material handling hardware and software has to be identified. As a means of developing this hardware and software strategy, evaluations of the current computing environment as shown in Fig. 6-26 and quantifications of the level of equipment and technology currently utilized in the warehouse as illustrated in Fig. 6-27 have occurred. Additionally, the physical warehouse layout information (Fig. 6-28) and summaries of resource deployment totals (Fig. 6-29) and warehouse processing volumes (Fig. 6-30) have also been analyzed. The application of the current data versus the WMS requirements has resulted in the following list of WMS computing, peripheral, and material handling requirements.

Hardware and Software Requirements

- Support a midrange computer or microcomputer computing platform.
- Select or develop an open UNIX system with a built-in fourth-generation language (4GL) and relational data base independence (possibly a client/server environment utilizing graphical user interfaces).
- Ensure that data base access is open data base connectivity (ODBC) compliant.
- Allow for access to host interfaces via structured query language (SQL) and/or TCP/IP.
- Support separately or in combination narrow-band and spread-spectrum radio frequency terminals.
- Support a mixture of nonprogrammable terminals and intelligent workstations.
- Provide for printing capability through the utilization of intelligent printer data streams (IPDS).

Computing Environment Summary	
Host-computer machine (brand and model)	IBM 3090
Host-computer operating system	CICS
Host-computer preferred communications method	DDM
File structure description	Interface files are expected to be 50–150 bytes in length and contain 10–30 fields. Information is to be conveyed in one file (i.e., SKU master), or in an attempt to eliminate redundancy, two files (i.e., PO header and detail).
File types	VSAM
File access method	VTAM
Preferred connectivity protocol	X.25

FIGURE 6-26   Current computing environment summary.

Area	Equipment	Information-based
Receiving	4 hand trucks	2 CRTs (IBM 3274)
	4 fork trucks	1 printer (IBM 4224)
Returns	1 hand truck	1 PC (Apple)
Put-away	6 fork trucks (shared with Replenishment)	
Picking	3 cherry pickers	2 bar code printers (Zebra Z90)
	1 fork truck	
	10 tuggers with 30 trailer carts	
Packing	100 ft of gravity conveyor	2 printers (IBM 4224)
	2 scales	1 bar code printer (Zebra Z90)
Loading or shipping	4 hand trucks	2 CRTs (IBM 3274)
	4 fork trucks	2 printers (IBM 4224)

FIGURE 6-27   Current material handling and information-based equipment summary. This is a summary of the types and quantity of material handling and information-based equipment used in each area of the warehouse. The list only includes the equipment that is to be utilized with the new system.

- Support material handling automation equipment such as conveyor systems, pick-to-light systems, pallet and carton flow racks, and sorters, and, if applicable, their associated interface requirements.

The *Warehouse Management System Recommendations and Strategy* section of the WMS Requirements Report describes the project team's recommended plan for proceeding with the implementation of a WMS. Many system options including the evaluation of third-party vendor solutions or an internal development feasibility study are available. Owing to the fact that the remainder of this chapter deals with system evaluation efforts, an example of the Warehouse Management System Recommendations and Strategy section is not provided.

Physical Warehouse Layout Information	
Total square footage of warehouse	350,000 ft^2
Number of warehouse buildings	2
Description of physical separation of warehouse buildings (cement walls, 1 mile, etc.)	The warehouse buildings are 0.25 miles apart. A county road separates the two buildings.
Are warehouse blueprints or a layout available?	Yes
Number of receiving dock doors	4
Number of shipping dock doors	4
Are the same doors used for receiving and shipping?	No
Total number of customer pickup locations	1

Location Information		
Types and total number of stocking locations	Type of location	Total number of locations
	Floor bulk	300
	Drive-in rack	1000
	Pallet rack	3000
	Shelving	700
	Outside	0
	Remote	0

FIGURE 6-28   Current physical warehouse layout information.

EVALUATION OF SYSTEM OPTIONS

With the warehouse management system goals, objectives, and requirements documented as a result of the Requirements Definition study, an evaluation of potential system options can be made. System options are normally catagorized into two groups, external systems such as those offered by third-party WMS vendors, and internal systems developed by the Information Systems department.

EXTERNAL SYSTEM OPTIONS AND EVALUATION

Properly selected and installed packaged software can have many advantages over internal or custom-developed systems including lower costs, additional or improved functionality, and compressed installation schedules. Conversely, the wrong choice

	Shift					
	1		2		3	
Area	Avg.	Peak	Avg.	Peak	Avg.	Peak
Receiving—clerks	2	3	2	3		
Receiving—receiving	4	8	4	8		
Returns	1	2	1	2		
Put-away	4	6	4	6		
Quality control	1	2	1	2		
Replenishment			3	5	4	6
Order processing	1	2	1	2		
Picking			8	13	4	7
Packing			4	8	2	3
Shipping—clerk	1	2	2	3	1	2
Loading or shipping			4	8	2	6
Administrative—clerk	2	2	2	2	2	2
Supervisor(s)	4	4	4	4	2	2

FIGURE 6-29　Current resource deployment totals. This is a summary of the total average and peak number of warehouse employees by area and shift.

can result in disillusioned users and longer implementation times as modifications are generated to comply with user requirements. Based on these two assumptions, it becomes clear that evaluating third-party WMS vendors and their respective products can prove to be a harrowing experience without a defined evaluation process in place.

As is the case with any legitimate process, there are a series of tasks and subtasks associated with evaluating and selecting the appropriate third-party WMS vendor and product. These tasks are:

Task 1　　Document the WMS Goals, Objectives, and Requirements
Task 2　　Establish Selection Criteria
Task 3　　Prepare a List of Potential WMS Vendors
Task 4　　Execute the Request for Information Process
Task 5　　Request a Vendor Demonstration and Presentation
Task 6　　Perform an Installed Site Visit
Task 7　　Conduct the Request for Proposal Process
Task 8　　Select a WMS Vendor

Task 1

The first task, "Document the WMS Goals, Objectives, and Requirements," is necessary to identify clearly what is to be accomplished though the implementation of a

Receiving		
	Average	Peak
Number receiving docks used	4	8
Number of trailers per day	10	20
Number of POs received per day	50	60
Average number of lines per PO	5	5
Number of cartons received per day	8000	16,000
Average number of cartons/movable unit	20	20
Total lines received per day (at receipt or to be palletized):		
% palletized, 1 SKU per pallet	80	80
% palletized, 2+ SKUs per pallets	15	15
% not palletized	5	5
Cross-docks (flow-through)	0	0
Number of QC SKUs per day	300	600
Number of QC cartons per day	300	600
Returns		
	Average	Peak
Number of returns per day	25	40
Number of lines per return	1	2
Number of cartons per return	3	5

FIGURE 6-30   Current warehouse processing volumes. This is a summary of the average and peak warehouse processing volumes.

warehouse management system. At this point, it is important to remember that detailed descriptions of the WMS goals, objectives, and requirements are especially critical, in that this strategic information is to be submitted to potential third-party WMS vendors for the purpose of mapping and evaluation.

By assuming that a Requirements Definition study as defined earlier in this chapter has previously been completed, it is inferred that the goals, objectives, and requirements relevant to the future WMS are already contained in the Requirements Report resulting from the study. As illustrated later in this process, the documented requirements information can prove to be an invaluable resource during the creation of RFIs and/or RFPs.

Task 2

To realize the most benefit from a third-party WMS vendor, it is vital that Task 2, "Establish Selection Criteria," be performed in conjunction with the documentation of the WMS goals, objectives, and requirements. Although it may not be intuitively obvious, the selection criteria determined in this task influence or are required for the com-

Put-Away Operation		
Method of put-away	Percent of each type	
Pallet		60
Carton (other than on pallets)		25
Eaches (other than in cartons)		15
	Average	Peak
Number of pallets put away per day	300	1050
Number of cartons put away per day	75	265
Number of eaches put away per day	45	155
Replenishment		
Method of replenishment	Percent of each type	
Pallet		75
Carton (other than on pallet)		25
Eaches (other than in cartons)		0
	Average	Peak
Number of pallets replenished per day	250	950
Number of cartons replenished per day	100	300
Number of eaches replenished per day	0	0
Carton or Pallet Picking		
	Average	Peak
Number of orders per day	30	60
Average number of lines per order	80	120
Number of full pallets picked per day	0	0
Number of cartons picked per day	6000	24,000
Eaches Picking		
	Average	Peak
Number of orders per day	30	60
Average number of lines per order	400	425
Number of eaches picked per day	60,000	125,000

FIGURE 6-30   (*Continued*)

pletion of all the remaining tasks in the evaluation of external systems. Examples of possible selection criteria to be considered are as follows:

1. *Perceived fit of the third-party packaged software to the defined WMS fundamental and functional requirements.*   The mapping of requirements (including workable alternative processes) needs to indicate a match of 80 to 90 percent.

Bulk Picking		
	Average	Peak
Number of orders per day	0	0
Average number of lines per order	0	0
Packing and Checking		
	Average	Peak
Number of orders per day	30	60
Average number of lines per order	400	425
Number of cartons packed per day	1200	2500
Number of pallets created per day	0	0
Shipping		
	Average	Peak
Number of trailers shipped per day	8	25
LTL % Full %	LTL, 50%; Full, 50%	LTL, 50%; Full, 50%
Number of parcel orders shipped per day	10	10
Number of lines per parcel order	3	5
Number of customer pickups per day	0	0
Number of shipping docks used*	4	8
Number of orders shipped per day	30	60
Total cartons shipped per day	7200	26,500
Total pallets shipped per day	0	0
Number of documents printed per order	2	2
Type of documents printed per order	BOL, PL	BOL, PL
Cycle Count		
	Average	Peak
Number of locations counted per day	200	200

*Note: Indicate if the same dock doors are used for shipping and receiving.

FIGURE 6-30   (*Continued*)

2. *Installation services available (for example, software personalization and installation, training, and so forth).*   The level of installation services offered by vendors varies dramatically; it is therefore wise to determine a minimum level of required services.

3. *Product stability.*   Through market research, customer references, and possibly the vendor demonstration, determine how stable the vendor's packaged WMS product is. Information such as year of WMS product release and number of installations provides a perspective on product stability.

Inventory Information

Regulatory compliances: oil, paint, batteries, chemicals? No

Environmental requirements: Frozen No ; Chilled No ; Expensive Yes

Other: Controlled substances No ; Dangerous No ; Office supplies Yes

Disk Sizing Information	
Total number of storage locations	5,000
Total number of SKUs	2,500
Average number of movable units in DC	25,000
Number of serialized SKUs	200
Average number of serialized SKUs in DC	2,000
Number of days of history to be stored	30

FIGURE 6-30 *(Continued)*

4. *Customer references.* Always check vendor references, especially within the same industry niche, very thoroughly. References normally tend to be surprisingly objective. An independent verification of a vendor's credentials by a consultant or other source may also prove to be a worthwhile activity.

5. *Vendor characteristics.* Since a partnership formed with a selected WMS software vendor is sure to span several years, it is important to ensure that a chemistry exists between any potential vendors and the selecting company.

6. *Financial stability.* It is good business practice to evaluate a vendor's financial stability by obtaining financial statements and performing a business credit check.

7. *Customer service (warranty support, 24-hour hot line, and so forth).* As with installation services, the level of customer service required must be determined.

8. *Long-term product plans.* It is essential to access how a vendor's long-term product plans affect the selecting company. Questions such as "What enhancements are planned for the proposed WMS product?" and "How is new information technology incorporated?" must be considered.

9. *Solution cost.* All aspects of cost including the software license fee, installation services, equipment, ongoing maintenance, and so forth must be calculated and compared to the selecting company's budget.

Task 3

With the defined selection criteria serving as a guideline, Task 3, "Prepare a List of Potential WMS Vendors," can occur. Since the recent emphasis placed on the benefits gained through efficient warehouse operations has resulted in a proliferation of software vendors touting warehouse management systems, a myriad of resources has also become available to aid in the selection process. Identification of potential third-party WMS vendors is possible through:

1. Reviews of trade publications such as *Automatic I.D. News* and *Modern Materials Handling*

2. Memberships in associations such as the Council of Logistics Management (CLM) and the Material Handling Institute (MHI)
3. Attendance at trade shows geared toward logistics such as PROMAT (sponsored by the Material Handling Institute) and Scan-Tech
4. Consultants specializing in warehouse automation
5. Technology (radio frequency, computer equipment) vendors

Task 4

Once the prerequiste information including the WMS vendor selection criteria, list of potential WMS vendors, and the WMS requirements has been established, Task 4, "Execute the Request for Information Process," is possible. The objective of this task is to describe the processing steps associated with effective RFI utilization. When properly created and executed, an RFI is a powerful tool in the collection of the detailed WMS vendor information required to reduce a field of 10 to 25 potential vendors to a manageable group of two to three highly qualified candidates.

The subtasks associated with Task 4 include the following.

1. Establish an RFI strategy. The creation of an RFI strategic plan that describes the company's motivation for issuing an RFI and outlines the guidelines, steps, and schedule to be adhered to is the first step toward the successful completion of the RFI process. If assistance is required in the creation of the WMS and MHS RFIs and their associated strategies, the addition of a consultant with expertise in logistics and/or material handling may be necessary.

2. Create a WMS RFI. By integrating the goals, objectives, and requirements information contained in the WMS Requirements Report with the defined customer selection criteria and the RFI strategy, a WMS RFI can be created. A complete RFI normally contains the following sections at a minimum:

- Introduction
- Overview
- Vendor Response Guidelines
- Background
- Fundamental System Requirements
- Functional Requirements
- Hardware and Software Requirements
- Implementation Approach and System Support
- Vendor Information and References
- Vendor Pricing Information

Before presenting examples of each section, a review of the following general RFI preparation guidelines that may increase the quality and quantity of the responses gleaned through the RFI process is appropriate.

1. Utilize a simple reponse format that permits vendors to respond on the actual RFI form or on a diskette included with the RFI. Additionally, whenever practical, allow the vendors to respond to requirements with a yes/no or number rating system response and provide minimum guidelines for narrative inquiries.

2. Enable vendors to perform mapping and evaluation activities by providing an adequate level of information.

3. Use common warehouse industry terms to describe requirements, processes, and operations.

The RFI's *Introduction* briefly introduces the company, and describes the company's warehousing and distribution operations. Normally, a list of facilities to be affected by a WMS implementation, as well as a sampling of statistical data supporting the need for a WMS, is also included. An example of a WMS RFI Introduction section is detailed below.

Introduction

Generic Company, one of the largest retail distributors of parts in the United States, is a consumer product distribution leader and innovator. The principal goal of Generic Company is to develop the people, organization, and systems necessary to provide the fastest, most efficient, and accurate retail distribution services to its customers.

At the present time, Generic Company operates two warehouses located in Anywhere, North Carolina, and Somewhere, New York. In addition to these two warehouses, Generic Company is planning to build a state-of-the-art facility in Everywhere, Iowa. Information required for the warehouse operations is available from Generic Company's host computer located at the corporate headquarters in Charlotte, North Carolina.

Generic Company's need for a warehouse management system is reenforced by its current requirement of supplying parts to 500 stores nationwide. In simple terms, 500 stores basically relate to up to 500 store replenishment orders from its warehouse facilities each day. Additionally, Generic Company plans to continue the number and size of stores served by its warehouses by averaging a yearly growth rate of 7 percent.

As to be expected in any retail environment, Generic Company experiences definite seasonal peaks. The peak for receiving parts occurs in August and September. The busiest shipping period is November and December.

The *Overview* section of the RFI defines the company's strategic objectives in terms of its vendor expectations relative to packaged software and services. Additionally, it summarizes the RFI response process and associated vendor selection criteria. An example of a WMS RFI Overview section follows.

Section 1: Overview

Generic Company's strategic objective is to implement an on-line, real-time warehouse management system to support all its physical distribution processes, inventory and location management requirements, and special customer services. Pursuant to this goal, Generic Company is interested in contracting a vendor with a proven warehouse management system package and the necessary skills to:

- Design any necessary customizations and interfaces
- Implement and test modifications and interfaces
- Integrate the necessary computing and peripheral hardware and software
- Incorporate material handling automation (MHA) as specified by Generic Company's MHA consultant
- Install the software
- Perform appropriate testing including acceptance test support
- Provide accurate documentation
- Train users and system support personnel
- Provide ongoing maintenance support

Generic Company initiated the process of selecting the right warehouse management system by conducting a Requirements Definition study. The requirements identified during the study resulted from a review of current warehouse procedures, installed system software, and tactical and strategic business plans, interviews with personnel from each warehouse functional area, and technical analyses of data and information. Information collected during the study has been utilized to create this RFI.

Once a vendor submits a RFI response, the response will be evaluated based on the following criteria:

- Perceived fit of the vendor's solution to Generic Company's fundamental and functional requirements
- Vendor's ability to fulfill the detailed service requirements listed above

Based on the results of this RFI process, it is our intention to narrow the prospective vendor list to two or three. The two or three highest ranked vendors will then be asked to present a detailed presentation and provide for an off-site visit to a warehouse or a distribution center where the proposed system is installed. Upon completion of the presentations and site visits, a formal request for proposal may be issued.

The RFI's *Vendor Reponse Guidelines* section contains the vendor's instructions for preparing an RFI response. By utilizing this section both to permit key assumptions and guidelines to be described and to control the vendor response mechanism, an environment for equitable and expedient WMS vendor evaluation and selection can be fostered. An example WMS RFI Vendor Response Guidelines section is illustrated below.

Section 2: Vendor Response Guidelines

The vendor response guidelines section provides the instructions for preparation that are to be used in responding to this RFI. In an attempt to minimize the effort required to prepare a response, a diskette with the RFI sections has been included with this package. Perspective vendors are invited to complete and return their responses on the diskette provided. (As a precaution, a hard copy of the vendor's RFI response is also required.)

Additional RFI response guidelines are as follows:

1. For a vendor to be considered in the RFI process, three hard copies of the RFI response (or one hard copy and the diskette containing the response) must be submitted to the warehouse manager by 5 P.M. EST on January 31, 1997. Copies of the response are to be delivered to Mr. Warehouse Manager, Generic Company, 1234 Warehouse Drive, Charlotte, NC 28223.
2. The cost of preparing the RFI response is the sole responsibility of the vendor.
3. All information supplied in this RFI and submitted by vendors in response to this RFI is considered confidential.
4. It is the responsibility of the vendor to inquire about and clarify any requirement of the RFI that is not understood. Questions may be addressed to Warehouse Manager, (000) 000-0000 (telephone), (111) 111-1111 (fax).
5. Each section within the RFI contains instructions for proper completion. Vendors responding must follow the instructions to be considered.
6. The vendor solution rating system detailed below is to be utilized in Secs. 4 and 5 where identified Generic Company requirements are to be mapped against the vendor's system capabilities.
   a. Ratings are to be entered in the left margin beside each and every requirement in the designated sections.
   b. If necessary, comments clarifying a vendor's response may be entered after each requirement.
   c. The format for the vendor's assessment of requirements is X/Y/Z with the following permissible values for X, Y, and Z:
      (1) Permissible values for X are the following:
         1 = Does not satisfy the requirement
         2 = Partially satisfies the requirement
         3 = Closely or alternatively satisfies the requirement
         4 = Completely satisfies the requirement
      (2) Permissible values for Y are the following:
         A = Customization required—design changes to data base or addition of new functions and features

B = Minor change required—minor changes to data base or displays

C = System enhancement required—addition of requirement increases the value of base package and improves its marketability (i.e., add at no cost)

D = Process change required—specified requirement is met in an alternative way, thereby possibly requiring a process change for the customer

E = Nonstrategic direction for system—the addition of this requirement to the base system is not perceived as strategic and/or is cost-prohibitive

(3) Permissible values for Z are the following:

C = Comment to clarify response included

N = No comment required to clarify response

7. All responding vendors are to be notified by February 28, 1997, as to whether they have qualified as finalists.

The *Background* section of the RFI describes the WMS scope, specifies the company's motivation for issuing a WMS RFI, and delineates the warehouse functions to be supported by a proposed system. Key benefits and objectives the company intends to realize through the implementation of a WMS are also a valuable addition to this section. An example of a WMS RFI Background section follows.

Section 3: Background

In today's business environment, Generic Company has become increasingly aware of the importance of efficient warehouse operations. With greater customer demands for products and services on a more timely basis at lower costs, business trends indicate that our warehouses have become the focal point of our distribution network.

Based on customer needs, our company's recent and planned growth, new industry requirements, and our internal Requirements Definition study results, the apparent solution is a warehouse management system designed to improve inventory control, manage warehouse space, and increase employee productivity. In addition to improving internal warehouse operations, the warehouse management system is slated to be the required bridge between the various host-level systems such as production scheduling, purchasing, logistics planning, and order management systems, allowing us to provide the proper customer order response time critical in today's competitive market.

With a real-time accurate assessment of available inventory, labor, and equipment it is envisioned that the WMS can issue and manage tasks and monitor performance in each of the following functional areas:

- Receiving and dock management
- Put-away
- Location management
- Cycle count
- Order processing
- Picking
- Replenishment
- Packing
- Shipping

The main objectives to be attained through the implementation of a WMS include:

1. Improving customer service through the reduction of order cycle times and better shipping accuracy
2. Improving productivity by reducing product handling, optimizing travel time of storage and retrieval equipment, and optimizing warehouse tasks
3. Increasing inventory accuracy through real-time activity confirmation and the associated reduction of paperwork required for current warehouse operating procedures
4. Maximizing facility utilization through location maintenance and rewarehousing
5. Reducing the number of errors and the frequency of exception processing through the use of automated operations and system-applied decision rules and logic

The RFI's *Fundamental System Requirements* section depicts the general system requirements. Areas covered in this section include requirements related to processing environments (i.e., distributed, host-based), information technology (i.e., EDI, RF, bar coding), access and control of information (i.e., security, reliability, ease of use), and solution elements (i.e., real-time application, host interface capabilities). An example of a WMS RFI Fundamental System Requirements section is detailed below.

Section 4: Fundamental System Requirements

Generic Company has determined that the general system requirements listed in this section are fundamental to its future operations, and therefore, to a perspective warehouse management system.

When responding to this section, vendors must utilize the vendor solution rating system outlined in Sec. 2, Vendor Response Guidelines, to indicate their system's capabilities relative to Generic Company's fundamental requirements.

4.1. *Distributed processing.* Since timely data is essential to effectively control the warehouse inventory and perform the order fulfillment process, a separate computer located at the warehouse is a fundamental system requirement. The use of a distributed computer running an interactive real-time application enables the warehouse to run independent of the availability and response time of the host business system.

4.2. *Real-time application software.* With the assumption that the warehouse is to operate separately from the host business system, it is mandatory that the warehouse management system is a real-time on-line application. The WMS is to track inventory and direct work flow within the warehouse in a manner that allows the system to respond to normal as well as unexpected events and changes.

4.3. *Host interface capabilities.* In an effort to reduce the amount of data that is duplicated while allowing the WMS to run independently, host interfaces with the capability to perform real-time and batch updates of information required for host and warehouse operations are necessary. Host interfaces are to be accessible via SQL or TCP/IP.

4.4. *Data integrity.* In physical distribution, information flow precedes product movement. To that end, communication procedures relating to system processors, networks, and peripherals are to ensure file integrity in the event of transmission problems. Additionally, vital business data is to be backed up daily, and recovery procedures are to be available to perform selective or full recovery depending on the type of failure experienced.

4.5. *Security.* A security function that allows management control of employee authorization through the use of a distinct and valid ID and password is mandatory. Generic Company's preference is that access and update authority is granted at the transaction and data base field levels.

4.6. *Capability of being audited.* All transactions and data base updates must be audited easily. Information to be collected includes employee, date, time, type of activity, location, items affected, and so forth.

4.7. *Labor productivity data and measurements.* The WMS must be capable of providing user-defined reports that utilize time-stamped system transactions to create labor productivity data. The ability to apply defined measurements to the productivity data is considered to be helpful.

4.8. *Ease of use.* The following ease-of-use features, which facilitate prompt problem determination and change management, as well as simplifying training, are considered to be highly desirable: fast path functions, multiple sessions, context-sensitive help facilities, graphical user interfaces (GUI), user-maintained tables for setting parameters, and ad hoc query and report generation capabilities.

4.9. *Bar code technology.* The WMS is to utilize bar codes and bar code scanning devices to ensure that inventory is accurately identified and tracked within the warehouse. Preferred bar code symbologies include type 3 of 9 or Interleaved 2 of 5.

4.10. *Radio frequency technology.* Implementation of an interactive WMS and bar code technology allows reliable real-time data collection to be performed through the use of radio frequency terminals with attached scanners. It is Generic Company's intent to limit or eliminate the use of paper in its daily warehouse processing.

The *Functional Requirements* section of the RFI outlines the basic warehouse management system requirements deemed necessary to meet the business processes in each functional area. Additionally, to provide perspective to the requirements' functional relevance, a process overview is also included. An example of a WMS RFI Functional Requirements section follows.

Section 5: High-Level Functional Requirements

By completing a Requirements Definition study for a WMS, Generic Company has discovered that the business processes being performed in each functional area of the warehouse can be improved, and that through its disciplined approach to implemention, it has been able to identify the high-level functional requirements described in narrative and bulleted format in this section.

When responding to this section, vendors must utilize the vendor solution rating system outlined in Sec. 2, titled Vendor Response Guidelines, to indicate their system's capabilities relative to Generic Company's high-level requirements listed by function immediately after the process overview for each warehouse functional area.

5.1. Receiving

 a. Process overview. The receiving process is initiated by the vendors and/or carriers scheduling an appointment with the receiving office. Upon arrival, carriers register at the receiving office and present the receiving documentation consisting of either the PRO bill or bill of lading. After security checks the load and verifies that the seals are intact, the merchandise to be received can be unloaded. All items entering the warehouse are received against an authorization such as a purchase order (PO) that gates their legitimate receipt or acceptance into the warehouse.

 b. Receiving requirements

 5.1.1 Capture carrier compliance information.

 5.1.2 Download receipt information including ASNs, POs, advanced packing list, and return authorizations from the host.

 5.1.3 Reconcile receiving data against the appropriate authorization.

 5.1.4 Provide the capability to receive unexpected receipts.

 5.1.5 Direct the pallet building process by displaying pallet configuration information.

 5.1.6 Record item information such as serial number, lot number, and expiration date, as appropriate.

 5.1.7 Allow for and direct quality control sampling.

 5.1.8 Assign and record discrete inventory tracking IDs for each unit created.

 5.1.9 Provide for exception processing including the dispositioning of damaged merchandise and the reconciliation of quantity discrepancies.

 5.1.10 Capture vendor compliance data.

 5.1.11 Update inventory summary and receiving files.

 5.1.12 Record transaction information for audit and productivity tracking purposes.

5.2. Put-away

 a. Process overview. The put-away process provides control of the movement of merchandise from the receiving areas to the appropriate storage locations in the warehouse. A storage location may be a primary pick, reserve, overflow, or any other defined location.

 b. Put-away requirements

 5.2.1 Determine preferred and alternate put-away locations for all inventory units based on predefined criteria and logic.

 5.2.2 Update inventory records to reflect a change in location for all inventory units moved during put-away operations.

 5.2.3 Prevent inventory units from being placed in locations for which they are not authorized.

5.2.4 Maintain an audit trail of all inventory units put away for problem resolution, audit purposes, and labor productivity tracking.

5.2.5 Bypass put-away for inventory allocated to orders due for shipment.

5.2.6 Perform dynamic rewarehousing whenever possible to maximize the use of warehouse space.

The RFI's *Hardware and Software Requirements* section delineates the computing, peripheral, and material handling hardware and software strategy requisite to satisfy the requirements for a WMS. Information supplied in this section consists of a system sizing summary and a strategic hardware and software requirements list. An example of a WMS RFI Hardware and Software Requirements section is shown below.

Section 6: Hardware and Software Requirements

As a result of conducting a WMS Requirements Definition study, Generic Company has been able to set a strategy for the computing, peripheral, and material handling hardware and software that must ultimately be integrated with the WMS to create a total solution. However, in an effort to allow itself the most latitude in the WMS vendor selection process, the following requirements list purposely indicates a strategic direction without specifying particular brands or vendors.

1. Hardware and software requirements list
   a. Support a midrange or microcomputer computing platform.
   b. Select or develop an open UNIX system with a built-in 4GL and relational data base independence (possibly a client/server environment utilizing graphical user interfaces).
   c. Ensure data base access is open data base connectivity (ODBC) compliant.
   d. Allow for access to host interfaces via SQL and/or TCP/IP.
   e. Support separately or in combination narrow-band and spread-spectrum radio frequency terminals.
   f. Support a mixture of nonprogrammable terminals and intelligent workstations.
   g. Provide for printing capability through the utilization of intelligent printer data streams (IPDS).
   h. Support material handling automation equipment such as conveyor systems, pick-to-light systems, pallet and carton flow racks, and sorters, and, if applicable, their associated interface requirements. (An independent material handling automation consultant is currently being utilized by Generic Company.)
2. Pilot warehouse sizing information
   a. Computing environment summary (Fig. 6-26)
   b. Equipment summary (Fig. 6-27)
   c. Physical warehouse layout information (Fig. 6-28)
   d. Resource deployment totals (Fig. 6-29)
   e. Warehouse processing volumes (Fig. 6-30)

Based on the strategic hardware and software requirements and the provided pilot warehouse sizing information including warehouse processing volumes, present equipment summary, host configuration detail, resource deployment, and physical warehouse layout, vendors are requested to submit preliminary estimates of

1. The processor size and disk access and storage device (DASD) required for the computing platform selected
2. The number and type of terminals and associated peripheral devices (i.e., controllers, scanners, etc.) required for the radio frequency technology selected
3. The number and type of all other major peripheral devices required (i.e., printers, in-line scanners, etc.) to implement the proposed WMS

The *Implementation Approach and System Support* section of the RFI is a vehicle to determine the type and level of vendor support provided before, during, and after a WMS implementation. Topics typically covered in this section include standard implementation plans, testing, training, documentation, warranty, and ongoing maintenance. An example WMS RFI Implementation Approach and System Support section is as follows.

Section 7: Implementation Approach and System Support

A proven WMS implementation approach in combination with proper and adequate installation and post-warranty system support is of paramount importance to Generic Company. Vendors are asked to respond to this section by including a brief narrative describing their approach to the major implementation and system support topics listed. Underneath each major topic, points of special interest to Generic Company are provided to serve as minimum response guidelines.

7.1. Implementation plan
   a. Describe the standard implementation approach including descriptions of phases and major tasks.
   b. Delineate the proposed project organization and responsibilities for both the vendor and Generic Company.
   c. Denote the normal time frame associated with installing the base WMS package.
7.2. Testing
   a. List and describe conventional testing practices (i.e., unit, functional, and acceptance) and standards (i.e., documented test plans and data and automated software test programs) utilized.
   b. Indicate whether acceptance test support including a data base and direction in the creation of an adequate test plan are available.
7.3. Training
   a. List and describe (i.e., course titles, duration, media, type, and targeted participants) for all training services and materials included with the base package.
   b. Indicate whether training is performed on-site or at the vendor location.
   c. Specify whether or not a training data base with sample training data is provided.
7.4. Documentation
   a. List and briefly describe the normal content of all documentation supplied with the base package.
   b. Indicate the level of and the navigation through the help text supplied.
   c. Describe how documentation is updated with each new release or any software fixes.
7.5. Warranty
   a. Explain what the warranty encompasses and how long the warranty period is.
   b. Describe user and technical support (hot-line and on-site—include hours available and typical response time to be expected) provided.
7.6. Ongoing maintenance
   a. Describe user and technical support (i.e., hot-line and on-site—include hours available and typical response time to be expected) available.
   b. Indicate the correct procedure to use to report problems with the software.
   c. Depict the procedure used to incorporate software fixes and new releases to a package that is installed at a customer site.
   d. Indicate whether or not a user group is sponsored for the WMS package.

The RFI's *Vendor Information and References* section is utilized to extract pertinent vendor data such as financial stability, product stability, resource skills and potential availability, and references. Since a partnership formed with a selected WMS vendor is destined to cover several years and to cost hundreds of thousands of dollars, it is essential that vendor profile information be examined in conjunction with functional requirement synchronization, thereby resulting in the optimum vendor and product selection. An example of a WMS RFI Vendor Information and References section is detailed below.

Section 8: Vendor Information and References

This section of the RFI is designed to provide each vendor with the opportunity to profile its company, describe its product planning strategies, and to detail successful installations of its proposed WMS product. Vendors are asked to respond to this section by entering a reponse to the right of the inquiry, attaching a separate sheet, or including a brief narrative, as appropriate.

1. Vendor profile
   a. Company name:
   b. Sales representative name and telephone number:
   c. Technical support contact name and telephone number:
   d. Headquarter address:
   e. Telephone number:
   f. FAX number:
   g. Year company was established:
   h. Parent or holding company, if appropriate:
   i. Number of professionals employed in the following areas:
      (1) Product development:
      (2) Installation and support:
      (3) Sales and marketing:
   j. Annual revenue for
      (1) 1995:
      (2) 1994:
      (3) 1993:
      (4) 1992:
   k. Describe business relationships with hardware and/or software vendors whose products are required to implement the proposed WMS software.
2. Package history
   a. Year of proposed WMS product release:
   b. First installation date for proposed WMS software:
   c. Number of package installations for proposed WMS software:
   d. Current release number and date:
   e. Provide a narrative that lists and describes the modules included in the base WMS software package release being proposed.
   f. Next scheduled release number and date:
   g. Describe how new technologies and functions are integrated into the base WMS software package.
3. *Vendor References.* Three customer references that may be contacted by Generic Company are to be listed below.
   a. Reference 1: company name, application name and release level installed, technical environment (hardware, operating system), contact information (contact name, title, telephone number), contact instructions.

The *Vendor Pricing Information* section of the RFI provides a means for gathering initial budgetary pricing information for each proposed WMS solution. Even though an RFI is not a request for proposal, the preliminary financial implications submitted by each vendor are an integral factor that must be considered in the overall vendor evaluation process. An example WMS RFI Vendor Pricing Information section follows:

Section 9: Vendor Pricing Information

Although this is not a request for proposal, some consideration is to be given to the financial implications for each vendor solution proposed. In order to allow financial comparisons of competing vendor solutions to be performed, a minimum level of cost information is being requested in this section.

Vendors are asked to respond to this section by entering a reponse to the right of the inquiry, attaching a separate sheet, or including a brief narrative, as appropriate.

1. List the license fees by module for modules required.
2. Explain how (by CPU, warehouse, etc.) the number of license fees needed is determined.
3. Indicate whether or not an enterprise license is available.
4. Describe any standard license fee discounting policy.
5. List and describe common additional installation services and/or costs not provided

with the licensing of the software, but required during implementation (e.g., radio frequency site survey).

6. Provide hourly rates and estimated hours (or fixed prices) for all services typically required during implementation.
7. Indicate the price associated with annual support and maintenance.
8. Include a list price preliminary estimate for the processor size and DASD required for the computing platform selected.
9. Provide the list prices for the terminals and associated peripheral devices (controllers, scanners, etc.) specified in the Hardware and Software Requirements section of this RFI.
10. Provide the list prices for all other major peripheral devices (printers, in-line scanners, etc.) specified in the Hardware and Software Requirements section of this RFI.

3. Create a MHS RFI.   Since a MHS is envisioned as a subset of the total WMS solution, a considerable amount of the information conveyed or required in a WMS RFI must also be included in a MHS RFI. Based on this assumption, the example WMS RFI detailed in subtask 2 is to serve as a model for the preparation of a MHS RFI. [It is important to note that the type of MHS requirements defined and/or vendors targeted may warrant an RFI strategy that encompasses multiple versions (entire MHS versus individually defined subsystems) of the MHS RFI. For example, a MHS consultant who utilizes subcontractors can probably interpret the requirements and quote on all aspects of a system, while a MHS vendor focused on a specific product or technology, such as a pick-to-light system, may only receive details about and bid on a portion of the MHS.]

The modifications by section required to convert the WMS RFI to a MHS RFI are as follows.

Introduction

• Utilize as written.

Overview

• Change, as appropriate, the vendor required skill list to exhibit skills more germane to a material handling vendor.

Vendor Response Guidelines

• Amend the vendor solution rating system, so that MHS vendors need only respond with permissible values of X.

Background

• Rewrite this section with an emphasis on the MHS role in relation to the total WMS solution.
• Include a modified list of the key benefits and objectives to be gained through the MHS.

Fundamental System Requirements: Update this section with a list of general system requirements that are more pertinent to a MHS. Examples of MHS fundamental requirements may include the following:

• *Equipment type.*   List the type of material handling equipment required to support the project.
• *Volume rates.*   Specify the rate and time criteria the material handling equipment needs to meet.
• *Product characteristics.*   Define the weight, size, and unit load of all products to be handled by the listed equipment. A matching of product to equipment type is also helpful.

- *Codes and regulations.* Iterate that all proposed equipment must comply with applicable federal, state, and local laws and regulations. Codes such as the National Electrical Code (NEC), Occupational Health and Safety Act (OSHA), Underwriters Laboratories (UL), and American National Standards Institute (ANSI) are most often cited.

Functional Requirements

- Within this section include narrative and high-level flow charts of the proposed MHS process, bulleted MHS requirements, conceptual process diagrams and warehouse layouts, and preliminary equipment sketches (both top and side views).

Hardware and Software Requirements

- Incorporate the list of material handling hardware and software requirements formulated as a result of the material handling analysis completed during the WMS Requirements Definition study.

Implementation Approach and System Support

- Keep the same topics (testing, training, documentation, and so forth), but adjust the minimum response guidelines as necessary.

Vendor Information and References

- Remove inquiries relating to package history.
- Solicit information and experience relevant to equipment to be installed.

Vendor Pricing Information

- Eliminate references to software licensing.
- Request pricing details for the specified material handling hardware and software.

4. Submit RFIs. Once the WMS and MHS RFIs have been successfully created, they are ready for submission to the potential vendors identified earlier in this process.

5. Evaluate RFI responses. Organization and consistency are the keys to creating a productive environment for the evaluation of RFI responses. While the proposals are out to the vendors, the evaluation process can commence with the establishment of the RFI response evaluation team. This team, which is to be comprised of the key personnel required for a successful WMS implementation, is to be staffed at an adequate, yet minimum level. (A team member representing every department affected by the WMS implementation is not necessary and/or conducive to an efficient and meaningful evaluation process.)

Once named, the RFI evaluation response team can continue the evaluation process by reviewing the original RFI strategy. Areas of particular importance that must be focused on include:

1. Designated RFI process goal of selecting two or three highly qualified WMS vendors
2. Established selection criteria
3. Defined RFI response evaluation techniques

With the RFI strategy understood, the RFI evaluation response team can continue the evaluation process by creating a weighted matrix, similar to that shown in Fig. 6-31. Guidelines to consider during the development of the response matrix are as follows:

1. List all categories as defined by the major RFI sections on which the vendors are to be rated.

RFI Major Catagories	Req. #	Requirement Description	Weighting Factor	Vendor Evaluation		
				Vendor A	Vendor B	Vendor C
Fundamental System Requirements	4.0		25			
	4.1	Distributed Processing				
	4.2	Real-Time Application Software				
	4.3	Host Interface Capabilities				
	4.4	Data Integrity				
	4.5	Security				
	4.6	Auditability				
	4.7	Labor Productivity Data/Measurements				
	4.8	Ease-of-Use				
	4.9	Bar Code Technology				
	4.10	Radio Frequency Technology				

RFI Major Catagories	Req. #	Requirement Description	Weighting Factor	Vendor Evaluation		
				Vendor A	Vendor B	Vendor C
High-Level Functional Requirements	5.0		25			
	5.1	Receiving				
	5.1.1	Capture carrier compliance information.				
	5.1.2	Download receipt information from the host.				
	5.1...	--continue requirements--				
	5.2	Put-away				
	5.2.1	Determine preferred and alternate put-away locations for all inventory units based on predefined criteria and logic.				
	5.2.2	Update inventory records to reflect a change in location for all inventory units moved during put-away operations.				
	5.2...	--continue requirements--				

FIGURE 6-31   RFI evaluation matrix.

RFI Major Catagories	Req. #	Requirement Description	Weighting Factor	Vendor Evaluation		
				Vendor A	Vendor B	Vendor C
Hardware and Software Requirements	6.0		10			
	6.1	Midrange or Microcomputer Platform				
	6.2	UNIX-Based Operating System				
	6.3	ODBC Compliant System				
	6.4...	--continue requirements--				
Implementation Approach & System Support	7.0		20			
	7.1	Implementation Plan				
	7.2	Testing				
	7.3	Training				
	7.4...	--continue requirements--				

RFI Major Catagories	Req. #	Requirement Description	Weighting Factor	Vendor Evaluation		
				Vendor A	Vendor B	Vendor C
Vendor Information and References	8.0		10			
	8.1	Vendor Profile				
	8.2	Package History				
	8.3	Vendor References				
Vendor Pricing	9.0		10			
	9.1	Software License				
	9.2	Installation Services and Costs				
	9.3	Annual Support and Maintenance				
	9.4...	--continue requirements--				
Totals						

FIGURE 6-31   (*Continued*)

2. Assign a weighting factor at the category level to ensure that the relative importance of each category within the evaluation process is appropriately designated. The selection criteria defined in Task 2 is normally cited as the most influential element in the determination of weighting factors.

3. List under each category the individual requirements associated with each major RFI section.

4. Stipulate weighting factors to the individual requirements within each category, as appropriate. Although the addition of weighted individual requirements complicates the evaluation process, it may be deemed necessary based on specific business needs.

After the prerequisite steps have been accomplished and vendor RFI responses have been obtained, the team can begin evaluating the RFIs. While performing the evaluation process, the team must realize that the ultimate success of the process is predicated on the team's ability to reach a consensus for each matrix item by vendor. Once the scores for the matrix items are determined and tallied and the weighting factors are applied, the list of WMS vendor finalists can be prepared.

Task 5

With the pool of potential WMS vendors narrowed to two or three qualified candidates, Task 5, "Request a Vendor Demonstration and Presentation," can be conducted. The objectives of this task are to allow the vendor to demonstrate its product and capabilities while showcasing its supporting organization and to permit the selecting company to validate and analyze the vendor more thoroughly with respect to previously established selection criteria. Some aspects to be particularly cognizant of and/or to be openly requesting information on are as follows:

1. The supporting vendor organization (beyond the sales force)
2. The project team that is earmarked for "you," the selecting company
3. The vendor's level of understanding and experience relative to the selecting company's business environment
4. The degree of commitment exhibited by the vendor's executives
5. Examples of deliverables, such as documentation, training scripts, and so forth
6. Product planning material for the proposed product
7. Vendor characteristics—especially chemistry between the vendor and the selecting company

Task 6

After the vendor demonstrations and presentations are completed, the next logical step in the WMS vendor evaluation process is Task 6, "Perform an Installed Site Visit." This visit is not to be viewed as a sales activity, but rather as an opportunity to garner information relevant to implementation issues encountered by the installed customer. Areas to focus on during the visit include:

1. The installed site customer's overall satisfaction with the vendor's WMS
2. Critical success factors cited by the customer and/or vendor
3. The composition and associated commitment required from the customer project team
4. The vendor's adherence to the price and schedule quoted to the installed customer
5. The level of support supplied to the installed site customer during and after the WMS implementation
6. The overall morale of the installed site customer's employees

7. The installed site customer's rating of the training and documentation delivered with the system

8. Whether or not the proper level of technology and material handling automation has been applied with the WMS

9. Reliability of the WMS and associated hardware and software

10. What the installed site customer might have done differently

Task 7

If a decision on a third-party WMS vendor still cannot be rendered after the RFI evaluation process, vendor demonstrations and presentations, and the installed site visits, then Task 7, "Conduct the RFP Process," can be employed to aid the selection process. Because an RFP response entails a significant amount of vendor effort in terms of time and personnel to evaluate and complete, it is wise to only issue RFPs to WMS vendors who have been both identified as qualified finalists and notified of their status.

The objective of this task is to describe the processing steps related to effective RFP creation and utilization. In pursuit of this objective, it is important to remember that RFIs and RFPs only differ in the level (not type) of detailed information conveyed and required in the latter. For example, an RFI mandates that a vendor must respond to high-level requirements, while an RFP instructs a vendor to respond to the same high-level requirements and the individual process steps associated with each of the requirements.

Since the subtasks for RFP processing are almost identical to those documented for RFIs, the subsequent list of subtasks and their explanations only highlights the deviations between the RFI and RFP processing steps.

1. *Establish an RFP strategy* (same as RFI).

2. *Create a WMS RFP.* With the WMS RFI detailed in Task 4 serving as a model, the following modifications by section are needed to convert a WMS RFI to a WMS RFP.

    a. Introduction

        (1) Utilize as written.

    b. Overview

        (1) Update this section to indicate that the goal of the RFP process is to select the WMS vendor who is to receive the implementation contract.

        (2) Incorporate additional selection criteria that are to be considered in meeting this goal, as appropriate.

        (3) Indicate the steps (Requirements Definition study and RFI) that have occurred prior to the issuance of this RFP.

    c. Vendor Response Guidelines

        (1) Revise the vendor solution rating system and critical RFP response dates, as required.

    d. Background

        (1) Utilize as written.

    e. Fundamental System Requirements

        (1) For each fundamental requirement receiving an affirmative response, instruct the vendors to explain specifically how the requirement is met by their system.

    f. Functional Requirements

        (1) Add the weighted matrix (minus the weighting factors) created during the Requirements Definition study. See Fig. 6-32.

        (2) Require the vendors to respond to both the individual requirements and their associated process steps.

Permissible values for X:

1 = Does not satisfy the requirement.
2 = Partially satisfies the requirement.
3 = Closely or alternatively satisfies the requirement.
4 = Completely satisfies the requirement.

Permissible values for Y:

A = Customization required—design changes to data base or addition of new functions and features.
B = Minor change required—minor changes to data base or displays.
C = System enhancement required—addition of requirement increases value of base package and improves its marketability (i.e., add at no cost).
D = Process change required—specified requirement is met in an alternate way, thereby possibly requiring a process change for the customer.
E = Nonstrategic direction for system—the addition of this requirement to the base system is not perceived as strategic and/or is cost-prohibitive.

Permissible values for Z:

C = Comment to clarify response included.
N = No comment required to clarify response.

Functional Area	Req. #	Requirement Description	Vendor Response		
			X	Y	Z (Comments)
Receiving	1.0	Receiving by PO			
	1.1	Schedule deliveries into the warehouse from carriers and vendors.			
	1.2	Provide a receiving log function to enter receipt information. Associate the appropriate POs with the receiving log.			
	1.3	Assign a confirmation number to the scheduled appointment/receipt.			
	1.4	Include the capability to update the receiving log with the carrier arrival time.			
	1.5	Assign dock doors.			
	1.6	Record seal numbers.			

FIGURE 6-32　RFP functional requirements matrix.

　　(3) Expand the functional process overviews in an effort to provide the vendors with additional process definition.
　g. Hardware and Software Requirements
　　(1) Indicate if any of the hardware and software requirements have migrated from a strategic scenario to a specific technology, brand, language, and so forth. (For example, if it has been decided that narrow-band RF equipment manufactured by vendor X is the selecting company's preference, then

Functional Area	Req. #	Requirement Description	Vendor Response		
			X	Y	Z (Comments)
Receiving	1.0	Receiving by PO (cont.)			
	1.7	Provide a radio frequency (RF) trailer unload function to perform BOL level receiving and to expedite the release of dock doors and carriers.			
	1.8	Allow for a RF detailed receiving function based on the information in the receiving log.			
	1.9	Display SKU receiving information including pallet configuration details.			
	1.10	Record SKU information including serial number, lot number, and expiration date.			
	1.11	Allow for and direct quality control sampling.			

Functional Area	Req. #	Requirement Description	Vendor Response		
			X	Y	Z (Comments)
Receiving	1.0	Receiving by PO (cont.)			
	1.12	Assign and record discrete inventory tracking IDs for each unit created.			
	1.13	Provide for exception processing including the dispositioning of damaged merchandise.			
	1.14	Allow for the reconciliation of quantity discrepancies.			
	1.15	Create movable unit records.			
	1.16	Call put-away to reserve storage locations for the inventory.			
	1.17	Provide a means to close the receipt (receiving log) and update inventory summary and receiving files.			
	1.18	Capture vendor and carrier compliance data.			
Receiving	2.0	--continue requirements--			

FIGURE 6-32 *(Continued)*

specify this preference in the RFP.)

   (2) Ask for more sound estimates of the amounts and types of computing and peripheral hardware and software required for proper implementation of the proposed WMS.

h. Implementation Approach and System Support

   (1) Keep the same major topics, but request that more detailed information be provided. (For example, indicate that a preliminary project implementation schedule must be included with the RFP response.)

i. Vendor Profile

   (1) Utilize as is. (This section can be omitted, if the vendors previously supplied their profile information in an RFI response.)

j. Vendor Pricing Information

   (1) Instruct the vendors to prepare budgetary pricing numbers that include appropriate discount schedules for the software license fees and computing and peripheral hardware and software.

   (2) Ask the vendors to calculate budgetary pricing estimates for installation services (i.e., training, documentation, project management, and so forth), software modifications (based on the results gleaned from completing the requirements matrix), and integration services (i.e., computing and peripheral hardware and software installation).

   (3) Specify whether the implementation contract fee is to be a fixed price or billed hourly.

3. *Create an MHS RFP.* The information required to prepare an MHS RFP or proposal specification is not normally available until the conclusion of the detailed design segment. Additional information on the MHS's specifications is available in Chap. 9, Design Phases.

4. *Submit RFPs.* Once the WMS RFP has been successfully created, it is ready for submission to any identified vendor finalists.

5. *Evaluate RFP responses.* The same basic strategy used to evaluate RFI responses can be used to appraise RFP responses. The one exception to the evaluation matrix format might be the addition of an intangible category. This intangible category can be utilized to allow abstract elements such as the chemistry between a vendor and the selecting company to be factored into the RFP response evaluation.

Task 8

Comparison of the selection criteria to the information acquired from the RFI and RFP evaluations, the vendor presentations and demonstrations, and installed site visits usually produces WMS vendor evaluation results in a format that permits Task 8, "Select a WMS Vendor," to be completed. If no clear choice of a WMS vendor is forthcoming, or if most of the vendor requirements and process mapping results indicate a custom system is needed, then an internal WMS feasibility study is probably warranted.

INTERNAL SYSTEM OPTIONS AND EVALUATION

An internal feasibility study (IFS) entails an evaluation of all factors including resources, skills, and budgets required to produce a warehouse management system "in house." Because of the inherent complexity of a WMS, companies normally only turn to the possibility of developing an internal system if a highly customized system is required, and their Information Systems department has the appropriate skill mix to give credence to the estimating, planning, and developing efforts. The deliverables resulting from an IFS are a preliminary project plan, software and hardware sizing estimates, a preliminary work plan, and a financial prospectus.

The following tasks are associated with an internal feasibility study:

Task 1   Perform an Evaluation of External Systems

Task 2   Analyze Staffing Capabilities

Task 3   Create a Preliminary WMS Design and Implementation Project Plan

Task 4   Estimate the Software Development Effort

Task 5   Develop Quote Packages for Computing, Peripheral, and Material Handling Hardware and Software

Task 6   Prepare a Preliminary WMS Design and Implementation Work Plan

Task 7   Conduct a Financial Evaluation

Task 1

The internal feasibility study process begins with the execution of Task 1, "Perform an Evaluation of External Systems." While initiating internal system-sizing efforts by evaluating third-party warehouse management systems may appear contradictory, it provides both pertinent information relevant to the necessity of performing an IFS and unforeseen benefits.

First, as has been stated previously, poor results gained from vendor requirement and process mapping attempts may indicate that a custom system is needed. This validation is essential, since it is commonly accepted that unless a system is very custom, it costs more to conduct an internal system development effort.

A primary benefit of completing an external system evaluation is the expertise and experience that the third-party vendors bring into the picture. Since legitimate vendors are continually refining their view of the warehouse and improving their arsenal of proven warehouse processes, they may be able to suggest alternative methods for meeting requirements. Hopefully, these identified alternative methods in conjunction with reasonable software modifications enable the selecting company to utilize a more cost-effective WMS package. Additionally, a vendor's technical knowledge and ability to size required computing, peripheral, and material handling hardware and software requirements is normally invaluable to the selecting company.

A secondary benefit of an external system evaluation is the significant amount of free consulting that is realized as a by-product. Vendors who estimate, plan, and implement warehouse management systems on a daily basis are an excellent source for IFS information such as:

• Warehouse process modifications
• Project staffing requirements
• Project planning
• Hardware and software recommendations

Task 2

Prior to attempting to perform the required internal feasibility study planning and estimating tasks, it is vital that Task 2, "Analyze Staffing Capabilities," be completed. At this point, it is appropriate to emphasize that proper staffing is crucial to achieving valid WMS project estimates, designs, and implementation plans. In the absence of adequate staffing, a realistic IFS cannot be accomplished.

Items to consider while analyzing the staffing capabilities include the following:

1. Whether or not the appropriate skill mix (i.e., project managers, system integrators, structured or object-oriented programmers) is resident within the company
2. Knowledge level of staff (i.e., warehouse systems background, in particular)
3. Staff experience level (i.e., association with and/or the completion of projects of this complexity before)
4. Availability of the required resources
5. Normal mode of programming (i.e., fix mode rather than coding from scratch)
6. Other projects currently in the plan (i.e., what priority does this project have in the overall workload plan)

Task 3

Once the staffing issues have been evaluated and resolved, and a project manager for the internal feasibility study has been named, Task 3, "Create a Preliminary WMS Design and Implementation Project Plan," is possible. In order to estimate properly the duration, effort, and complexity affiliated with the internal development of a WMS, a preliminary project plan that includes all required tasks and subtasks and their associated descriptions, as well as a skills summary necessary to perform the defined tasks and subtasks, must be prepared. Example WMS Project Plans and/or project schedules for external or detailed design and implementation or installation that can be utilized in the fulfillment of this task are detailed in Chap. 9, Design Phases, and Chap. 10, Project Management.

Task 4

After a WMS Preliminary Project Plan has been established, the estimating activities are initiated with Task 4, "Estimate the Software Development Effort." To ensure that consistent results are obtained, a standard software sizing methodology such as estimating matrices can be devised. These matrices normally incorporate predefined complexity and size factors that are systematically applied to functional requirement specifications to yield estimates for individual task durations for each function to be coded. The estimated task durations, which include the effort to design, code, perform screen updates, document, and unit test each function, are used to prepare the preliminary WMS project schedule and work plan. Example estimating matrices and their associated sizing assumptions are illustrated in Figs. 6-33 to 6-35.

Complexity	Definition (for structured programming)
Simple	Screen-based: Simple calculations; may contain simple (1 or 2) screens Logic-based: Straight line, no subroutine program
Moderate	Screen-based: Moderate calculations; may contain simple to complex (3 or 4) screens Logic-based: Single program with up to 5 subroutines
Complex	Screen-based: Complex calculations; contains complex (>4) screens Logic-based: Multiple programs and/or more than 5 subroutines

FIGURE 6-33   Software sizing estimating matrix—complexity definitions.

Size	Definition (for structured programming)
Small	Uses few files, one to three file updates, and one program
Medium	Uses two to four file updates; creates work files, data areas, and data queues; requires one-dimensional arrays and/or simple exception handling
Large	Uses many files, file updates, and programs; requires complex exception handling and/or multidimensional arrays

FIGURE 6-34   Software sizing estimating matrix—size definitions.

	Estimated* work-days (d) per project team member		
	Task complexity		
	Simple	Moderate	Complex
Task size			
Small	5 d	7 d	10 d
Medium	9 d	15 d	25 d
Large	15 d	20 d	40 d

*Estimates based on a structured programming approach.

FIGURE 6-35   Software sizing estimating matrix—duration.

Task 5

Technological and financial information pertinent to third-party hardware and software required for proper WMS implementation is collected through the execution of Task 5, "Develop Quote Packages for Computing, Peripheral, and Material Handling Hardware and Software." The quote packages, which are crafted from information contained in previously created RFIs and RFPs, are comprised of the following:

- Hardware and software requirements
- Statistical warehouse sizing information such as warehouse processing volumes, resource deployment figures, equipment summaries, and so forth
- A brief narrative of the hardware and/or software proposed usage
- Projected design year requirements
- Conceptual drawings, as appropriate

Third-party vendors are requested to provide the following information in their responses:

- Pricing information
- Technical specifications for proposed hardware and/or software
- Installation procedures
- Testing approach
- Technical support strategy
- Training methods
- Standard documentation

- Warranty support
- Maintenance support

If assistance is required in the formulation of the hardware and software quote packages, it may be necessary to retain consultants with expertise in logistics and/or material handling to furnish any necessary guidance.

Task 6

The information gleaned from the completion of the software development sizing effort and the third-party vendor hardware and software quotation responses can be combined with the tasks and subtasks outlined in the preliminary project plan to permit Task 6, "Prepare a Preliminary WMS Design and Implementation Work Plan," to be achieved. Creation of the work plan, which consists of the preliminary project schedule and duration and effort estimates for personnel and other required resources, is significantly simplified by devising estimating matrices for individual project elements (i.e., testing, training, and designing) and by utilizing the work plan development steps described in the Estimated Schedule and Associated Resource Requirements section of this chapter.

Task 7

Task 7, "Conduct a Financial Evaluation," is accomplished through the proper application of cost information to the capital and expense items and to the defined task and subtask criteria detailed in the preliminary project and work plans. The objectives of this task are to estimate accurately the cost associated with the design and implementation of a WMS, and to determine the ramifications of the estimated cost with respect to the company budget.

The cost-estimating process is based on the assumption that total project cost is the sum of the component costs for each major project category. Major project categories include:

- Software development (i.e., design, code, and test)
- Implementation (i.e., project management, training, and documentation generation)
- Computing hardware and software
- Peripheral hardware and software
- Material handling hardware and software

Component costs that are calculated and summed for each category are as follows:

- Direct labor
- Indirect labor
- Capital Expense
- Expense items

Examples of direct and indirect labor calculations are depicted in Chap. 8, Justification and Benefits.

Once the total project cost estimate has been calculated, it is compared to the company budget to ascertain whether the internal development of the WMS is feasible.

REFERENCES

1. *Managing the Requirements Definition Subphase,* Science Research Associates, Chicago, 1987.
2. Kadota, Takeji, and Salvendy, Gavriel, "Charting Techniques," in *Handbook of Industrial Engineering,* Wiley-Interscience, New York, 1982, pp. 3.3.1–3.3.6.

CHAPTER 7
SIMULATION

INTRODUCTION

One of the most powerful tools to assist in the design and operation of distribution systems is computer simulation. Automated distribution systems are highly integrated systems consisting of many interdependent functions and interactions. Simulation models designed to analyze these interdependencies provide valuable insight into system operations and performance. These models can include all system design parameters such as product range, material handling equipment, operational and maintenance personnel, storage facilities, operational logic, and expansion opportunities.

Experimentation with models is quicker, easier, and less expensive than experimenting with actual operating facilities. With good simulation analysis, a system design is validated as operationally feasible before substantial capital investment is committed. The initial capital costs associated with many distribution automation projects have increased the need for this analytical tool to help determine which design best meets company objectives. Examples where simulation is used within a distribution automation project include the following:

- Justification of new capital expenditures
- Design sensitivity to product mix and volume changes
- Determination of system throughput
- Bottleneck recognition
- Experimenting with scheduling practices
- Validating material handling improvements

Companies are under constant pressure to balance the tradeoffs between high facility utilization, in-process inventory levels, and ship dates to customers. Simulation is the only tool that can balance the associated waiting time costs for customers, products, equipment, and labor. While it is important to keep customer waiting times to a minimum to prevent buying from the competition, inefficiencies in production and inventory storage costs can also cripple a company. With a simulation, the waiting time, production, and storage cost can be balanced and compared to the added value of products being sold to determine profits.

The ingredients for a successful simulation project include having a well-defined set of objectives, using a team approach to the project, following good simulation methodology, obtaining accurate data to use in the model, and using the results effectively to improve the system design. Simulation projects can become lengthy and nonproductive if they are not based on a sound strategy and clear direction. This chapter provides an outline for achieving a successful simulation project.

OVERVIEW

Simulation has emerged as one of the most appropriate modeling tools in the design and operation of production systems. In its broadest sense, computer simulation is the process of designing a mathematical logical model of a real system and experimenting with this model on a computer.[1] When a simulation model is executed in the computer, simulated operations occur as they would in an actual distribution facility, but at an accelerated rate. Statistics are automatically collected by the model to report on equipment and personnel utilization, bottleneck operations, in-process inventory levels, and facility throughput.

Simulation modeling and analysis provide the following key benefits:

- System design is tested before implementation.

- Model building and experimentation increase understanding of the real system.

- Cost is minimized when changes to layout or equipment are tested before the capital investment.

- System performance measures under different scenarios are generated and can be compared.

- System capability is demonstrated.

- An additional tool is provided for training personnel.

Proper planning and good simulation methodology are essential to achieving these benefits. Eight time-tested steps for successfully performing a simulation study, as shown in Fig. 7-1, are described in this chapter. A DC automation project example is also presented to assist in the detail review and definition for each of these steps.

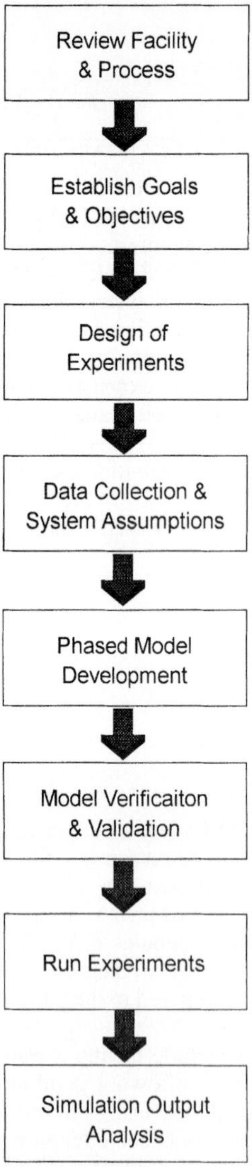

SYSTEM MODELS

When a system is subject to study, there are basically two ways to perform the study. One can experiment with the actual system or experiment with a model of the system. Experimentation with the actual system, in most cases, is not possible due to the cost and logistics of such a proce-

FIGURE 7-1 Simulation steps.

dure. A more acceptable method is to create and study a system model. There are two basic types of system models: physical and mathematical.

Physical Models

A physical model is a mock-up or re-creation of the actual system. Examples of physical models include cockpit simulators, wind tunnels, miniature models, and gravity tanks. Owing to the dynamic interactions involved, physical models are difficult and costly options for modeling most distribution systems.

Mathematical Models

Mathematical models are used to predict or determine the quantitative relationships of a system. Linear programming, physics, and geometry are examples of mathematical models. A mathematical model is any model that uses calculations, formulas, or algorithms to predict the behavior of a system. Mathematical models have been proven to be very accurate as long as the model correctly defines the system. There are basically two types of mathematical models: analytical and simulation.

Analytical Models. Analytical models are derived from a known formula to obtain the solution. The solution may be difficult to obtain and may involve time and computing resources. Analytical models are abundant in advanced math, physics, and engineering courses. Queuing models, linear programming, and regression analysis are examples of analytical models used in production system analysis. Analytical models address specific classes of problems that meet certain assumptions and conditions.

Simulation Models. Many systems contain a number of dynamic, interacting algorithms that make analytical solutions virtually impossible. In these cases, simulation is used to predict the system and its performance. Depending on the number of replications used to run the model and the model validity, a reasonable estimation can be achieved of the system behavior. Simulation is often the only way to model complex production systems and their wide array of variability. Simulation itself does not provide a single solution as with an analytical model. But simulation does provide the decision maker with a tool for analyzing a system under various configurations and weighing the results in order to select the design configuration that best satisfies the performance measurements.

SIMULATION BENEFITS

Distribution automation projects typically involve process reengineering and associated warehouse layout and material handling modifications. In some cases, the project's scope may involve a new warehouse being built. There is significant investment associated with fitting or refitting a warehouse, which can skyrocket if it is discovered after implementation that the proposed design does not meet the company's needs. Owing to the significant investment and permanence of the decisions being made, the project team should consider the use of simulation analysis in achieving the following benefits.

Test before Implementation

The project team can develop a proposed system model and analyze and verify different design scenarios before a system is recommended and implemented. The project risk is reduced since these tests can estimate throughput capabilities with a given amount of personnel and equipment. This provides verification that the proposed system will meet the required goals and objectives.

In the following example, a simulation study is performed to test a proposed layout designed for a distribution center (DC) that must maximize space utilization to avoid building expansion costs. The current and proposed DC layouts are shown in Fig. 7-2. The proposal is to change the current layout from wide aisle storage racks with fork trucks to narrow aisle storage racks with man-aboard wire-guided vehicles. This change requires new wire-guided vehicles that operate at different travel and lift rates than the current fork trucks. In addition, these vehicles are unable to pass each other due to the narrow aisles created by installing additional storage racks.

Simulation is used to answers questions such as:

- Will this strategy meet the desired throughput?
- How many wire-guided vehicles are required?
- Which put-away or picking strategy minimizes labor and equipment: pick and pass (operators dedicated to a group of aisles and passing orders to the next group of aisles to complete order picking), or complete order picking by one operator?
- Which put-away or pick strategy minimizes order pick cycle time and vehicle interference in the aisles?

Provide System Understanding

Although the primary objective in developing the simulation model is to assess the performance of a system for a prescribed set of parameters, another inherent benefit is the valuable insight gained just by going through the model creation process. Since detailed information on the system parameters, or design specifications, is required to build the simulation model, the modeling process forces the data collection discipline which is required to design and implement a distribution automation project successfully. Information obtained for the model includes put-away, picking, and shipping process times, product arrival times, equipment capacities, conveyor lengths and speed, and queue sizes.

This data collection process can uncover scenarios that need to be considered in the system design that might otherwise have been overlooked. For example, to simulate the model input for the proposed narrow aisle wire-guided vehicle system in Fig. 7-2, a detailed analysis of customer orders is required. What is the distribution of the number of line items on an order and the quantity required per line? This helps determine the number of pick locations the operator must pick from before the pallet is full and must be dropped in the shipping area. How does this order data fluctuate during the month and the year? The last three days of every month may require the DC to pick and ship over 50 to 60 percent of its monthly orders. Holiday seasons may require an increased number of larger orders.

Getting into this order detail can provide eye-opening results. The fluctuations in order demand can significantly impact the system's load. Remember, the proposed system must be able to manage peak, not just average, distribution periods.

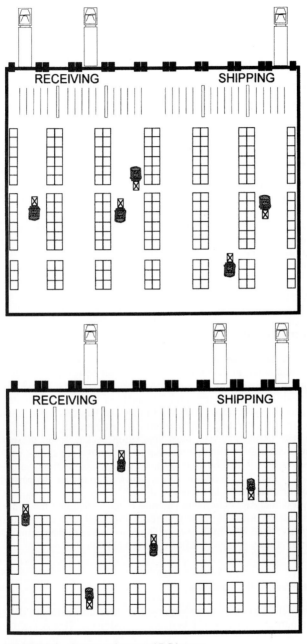

FIGURE 7-2 Current and proposed DC layouts.

In addition, simulation modeling can be used to test and understand the impacts of certain software designs in the warehouse management system (WMS). This is especially useful when additional code is required; the logic can be tested before the investment in code development. For example, in the proposed narrow aisle wire-guided system in Fig. 7-2, the model can help determine what effect the picking logic will have on system performance. How are system throughput and operator productivity impacted if the software combines multiple orders and directs the operator to pick all the quantity at one time versus the software directing the operator to pick one order at a time?

Minimize Costs

With simulation models, changes are made to the equipment capacities and quantities to analyze equipment cost and utilization versus the system throughput. This analysis assists the team in deciding the minimum initial equipment investment required to support the desired throughput objectives and estimating when additional equipment investment is required. In the proposed layout in Fig. 7-2, each wire-guided vehicle represents a significant capital investment. The simulation model is designed to change the number of wire-guided vehicles easily and review the results. The model can also be modified to analyze the impacts to the system with various equipment downtime rates.

Show Performance Measures

The outputs of a simulation model provide performance measurements that are compared to the automation project's goals. The relevant performance measurements are determined prior to conducting the simulation and are analyzed across the various test runs. Typical simulation outputs include machine and labor utilizations, system throughput, product cycle times, and queue times and lengths. Some simulation packages also incorporate operating costs and profit measurements.

To assist with output analysis, most simulation packages can present the performance measurements in graphical form such as bar charts, pie charts, and X-Y graphs. For example, a bar chart may be used to show average picking time per design alternative, as shown in Fig. 7-3. In addition, confidence intervals for the performance measurements can be created, as shown in Fig. 7-4.

In simulation analysis, the system parameters are varied and the generated performance results are recorded. After output results are analyzed, system parameters are adjusted until a refined system develops. The objective is to determine the most suitable configuration for a system to achieve the stated goals.

Demonstrate System Capability

Simulation models, especially those with two- or three-dimensional (3D) graphics and animation support, provide a superb system demonstration. This is useful for presentations to team members and more importantly to the executive management who must approve the automation project. An animated simulation presentation that visually demonstrates the efficient flow of product through the proposed system design is invaluable in increasing buy-in and support for the project. Figure 7-5 shows a 3D distribution system simulation model.

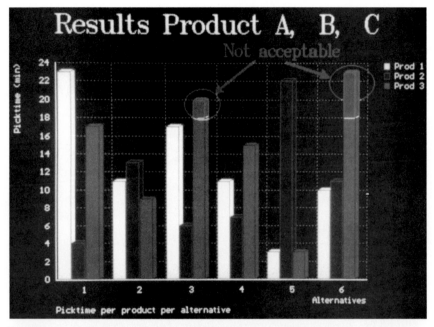

FIGURE 7-3 Simulation model—bar-chart results. (*Courtesy of F&H Simulations.*)

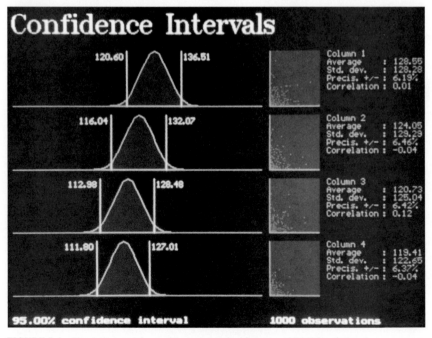

FIGURE 7-4 Simulation model—confidence intervals. (*Courtesy of F&H Simulations.*)

FIGURE 7-5   3D simulation model overview. (*Courtesy of F&H Simulations.*)

Train Personnel

If the simulation package has animation and graphics support, it can be included as part of the training class for the DC personnel. The graphical display of new DC layout changes, equipment additions, or changes in operating procedures helps personnel visualize how the new system will look and operate. This is a great supplement to the standard training documentation and classes.

Simulations are also used as an ongoing education and analysis tool after the system is implemented to train new personnel who will be supporting the system, such as industrial engineers and system analysts. Experimenting with the simulation model enables new personnel to gain years of operational experience.

STEPS IN SIMULATION MODELING

The time required to perform a simulation study will vary depending upon the complexity of the system, but the steps involved are always the same. There are eight steps to completing a simulation study:

1. Review of facilities and processes
2. Establishment of goals and objectives
3. Design of experiments
4. Data collection and system assumptions
5. Phased model development

6. Model verification and validation

7. Run experiments

8. Simulation output analysis

An in-depth look at each of these steps follows. An example is presented of an actual simulation study completed as part of a distribution automation project design. As each step is reviewed, reference is made back to the automation project example to help illustrate how a simulation study evolves.

Automation Project Example

The proposed automation project includes a warehouse management system and a conveyor transportation and sortation system. The major area for change is a new automated picking and shipping process. A simulation model is required to validate the proposed system and throughput rates. Figure 7-6 shows the current warehouse layout and flow, with the numbers referencing the following process steps:

1. The process begins with a printed order ticket distributed to the picking operators from the DC office.

2. The pickers travel through the warehouse on rider pallet trucks and pick the required carton quantities for an order.

3. Pickers stack these cartons on pallets and deliver the pallets to the appropriate shipping lane.

4. At the shipping lane, operators apply shipping labels to the cartons and then transport and stack the cartons onto the truck.

Current inefficiencies have been identified with this process. A carton is handled multiple times, increasing the opportunity for error or damage. In addition, the picking operator must make a trip through the warehouse to complete each individual order. In order to decrease the cycle time for processing orders and to minimize potential picking errors, an enhanced process is sought to minimize the number of times products are handled and the picker travel distance.

The proposed WMS software for the automation project will group orders together into a "wave" based on a truck-loading sequence. A wave is a batch of orders that are picked together as a bulk pick. This means that the order quantity for SKU "A" on all the orders in a wave is summed together. The orders going into each batch are based on a reverse truck-stop sequence, i.e., the first wave picked is the last truck stop since it is loaded into the front of the truck. For each wave, picking operators receive actual shipping labels printed in the pick location sequence, whereas in the current process the operators receive the individual printed order tickets, and pick one order at a time.

Figure 7-7 depicts the layout and flow for the proposed conveyor transportation and sortation system. The following process description refers to the numbered steps on this figure.

1. There are three picking modules with belt take-away conveyors. The pickers receive the actual shipping labels in pick location sequence. They walk to each location, apply the shipping labels, and place the cartons on the conveyor belt, repeating this process for all locations.

2. The picking conveyors provide carton queue space and transport the cartons to a sortation conveyor.

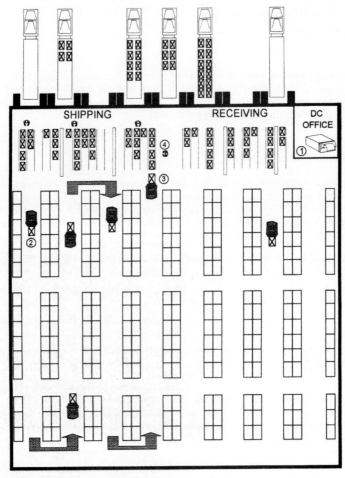

FIGURE 7-6   Current DC layout.

3. A bar code scanner reads the shipping label and identifies the assigned truck and corresponding shipping lane.

4. Cartons are diverted to the appropriate shipping lane.

5. Shipping operators load the cartons directly onto a truck since the shipping labels have already been applied at the pick location.

Step 1: Review of Facilities and Processes

Any simulation study must begin with a complete review of the facilities and the processes for the proposed simulation model. This review may be a tour of an existing facility, or it may be a detailed review of the design documentation for a proposed facility. In either case, the project team must have a complete understanding of the

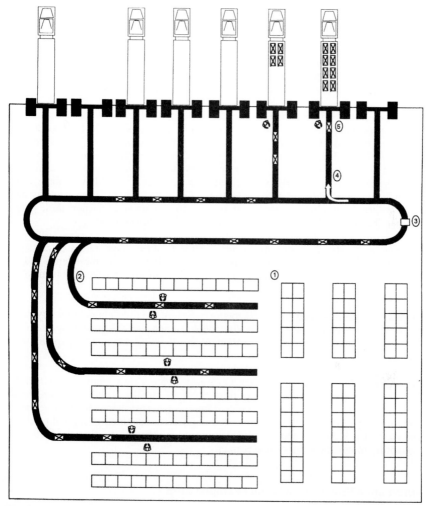

FIGURE 7-7   Proposed conveyor transportation and sortation layout.

operational characteristics of the system. Several things need to be accomplished in this review as discussed below.

Tour the Existing Facility.   A complete tour of the facility is the starting point for any simulation model. Even if the persons developing the simulation work in this facility, it is recommended that they schedule a complete walkthrough or tour and think about how they would model each aspect of the system. When thinking about how a system will be modeled, many questions as to why and how each operation is performed will arise that may not have been previously considered as significant.

List System Components.   During the tour, list all the system components: machines, storage locations and drop zones for inventory, workstations, tool locations, and anything else that is used in the process. Product components and parts for kits should also be listed.

List System Resources. System resources include conveyors, fork trucks, automated guided vehicles (AGVs), maintenance personnel, machine operators, overhead cranes, robots, or any other resource that is used in distribution to handle products. Record resource capacities and ask about any scheduled or unscheduled downtimes.

Note Operational Characteristics. As each system element is reviewed, make notes of the operational characteristics. The following questions should be answered: Is there any special processing logic? In what order are products processed by a resource: FIFO, LIFO, or some other queue discipline? Are all the shipping operators assigned to specific lanes or are floater operators moving from lane to lane as product accumulates? Are the fork trucks dedicated to specific operations? For example, a turret truck may be used to put away and retrieve full pallets within aisles, whereas standard fork trucks with faster speeds may be used to pick up and transport these pallets between the end of each aisle and the receiving and shipping docks.

List System Terminology and Acronyms. Most facilities have names or acronyms that are used to describe equipment or parts. Create a dictionary of these terms and use them in the simulation to make the model familiar to all who are involved in the process.

Draw or Obtain a System Layout Drawing. Before the tour, obtain a facility drawing, or while on the tour draw the facility and make notes on the drawing for each element in the system. After the tour is complete, this drawing acts as a "trip to the floor" whenever a facility review is needed.

Get to Know the System in Detail. Obtain or create a flow chart that defines the flow for every part, carton, or pallet throughout the facility. This flow chart should highlight decision points where operational logic is required. For example, at a quality operation, rejected products may proceed to a different operation than good products. The more that is known about the processes and system capabilities, the higher the probability of modeling the system correctly.

Automation Project Simulation Example: Step 1

With reference to the proposed conveyor transportation and sortation example, the following information is noted by touring the current DC, as well as examining the design document and proposed layout drawings.

System Components. System components include the picking and shipping locations, shipping labels, carton inventory, and bar code scanner. The bar code scanner read rate is 99.85 percent.

System Resources. System resources include the picking operators, belt take-away conveyors for picking, sortation conveyor, gravity conveyors as shipping lanes, and shipping operators.

Note Operational Characteristics. Special processing logic is required for the conveyors, as outlined in the following:

Wave changeover. Since waves must be loaded onto the trucks in reverse truck-stop sequence, orders from different waves cannot get mixed together on the sortation conveyor. The end-of-wave tote is loaded onto the belt take-away conveyor by the picking operator to indicate a new wave is being started. Cartons from the new wave are not allowed on the sortation conveyor until the end-of-wave totes from all three belt take-away conveyors are on the shipping lanes.

Slug load. Cartons will be released from the picking conveyor onto the sortation conveyor in batches. A slug load refers to the number of cartons in each batch.

List System Terminology and Acronyms

Shipping label. A peel-off label that contains ship-to information, picking location, and a UCC128 bar code for shipping lane number.

Belt take-away conveyor. A belt conveyor that runs in the middle of the pick module where picking operators place labeled cartons for transportation to the shipping lane.

Wave. A group of orders processed through the system in a batch.

Slug load. The number of cartons released together from the picking conveyor onto the sortation conveyor.

Draw and Label System Layout. A CAD drawing of the proposed layout is obtained that shows measurements for the different conveyor lengths.

Construct System Flow Chart. Figure 7-8 depicts a flow chart for the proposed process.

Step 2: Establishment of Goals and Objectives

Once the team knows in detail how the system operates, goals and objectives must be established to determine the model focus. The goals and objectives determine how the model is defined and what aspects of the system are modeled. This is important in defining the level of detail to be included in the model development, the performance measures, and the output analysis. Questions to be asked at this point are as follows.

What Is the Primary Reason for the Simulation Model? Why are you simulating in the first place?

What Is the Scope of the Model? The project scope defines how much detail the model should have. It is important to try to keep the simulation model simple. The more variables that are added, the more complex the model creation becomes. In addition, as more variables are added, the interaction between all the variables becomes more difficult to determine, requiring many more experimentation runs. Verify that the variables included are critical to the project success. The key is to focus on the problems to be resolved and not make the model too complex to understand. Remember that identifying what does not need to be included may be just as important as what is included.

What Questions Will the Simulation Answer? Determine the critical areas or concerns that the project team has about the proposed system and make sure the simulation will answer these concerns. Also review what decisions will be made based on the

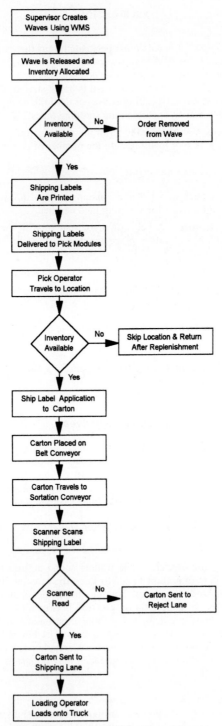

FIGURE 7-8 Proposed process flow chart.

questions the simulation will answer. Make sure the questions can be answered realistically through simulation. Questions such as "how many times does the picking operator stop to talk with friends?" are not valid.

What Kind of Changes Need to Be Made for the Study? Define what options or changes, if any, might be added to the current system. What are the impacts of each change?

What Issues Need to Be Investigated? Based on the system components, available resources, and the scope of the study, determine any specific issues that need to be addressed. Contact with other departments may be required to obtain this detail. Utilization percentages, effects of downtime, and bottlenecks are some examples.

What Are the Measures of Performance? What specific performance measures should the simulation calculate? Again, thought must be given to the critical measurements in order to keep the model simple. Throughput, overall processing time, volume versus resource utilization, and system delays are some valid measures of performance.

Automation Project Simulation Example: Step 2

In reference to the conveyor transportation and sortation proposal, the following are noted.

Primary Reason for Simulation. The simulation objective is to ensure the system design meets the design-year peak picking volume. (The conveyor lengths must be sufficient to keep inventory flowing to the shipping lanes, given the wave and slug-load logic designed in the WMS.)

Define the Model Scope. The model scope begins with picking operators holding the shipping labels for each wave. The ending point is the carton loaded on the truck. To reduce the model complexity, the simulation will not include any analysis on the replenishment of the forward-picking locations (i.e., the wait time associated with delayed replenishment and the amount of buffer storage in the pick location). In addition, the model will not address the no-read cartons through the conveyor scanner that get sent to the reject lane. The team decided that the percentage occurrence of both these events does not significantly impact the system throughput performance.

Define the Questions That the Simulation Will Answer

- Can this new layout concept meet our peak shipping volumes?
- Can eight waves be completed in an 8-hour shift?
- What is the percentage of time the conveyor system is full, thereby stopping the picking operators from placing cartons on the belt take-away conveyor?
- What effect does varying the slug load have on conveyor throughput and utilization?
- Will any additional picking and shipping operators be required?

Define the Addition or Changes to the System. One change or option that must be considered is another more expensive type of conveyor system. The operating speeds and downtime percentages are the main differences between the conveyor systems.

Define the Issues to Be Investigated. Since applying the shipping label in the pick module is a new operation, the industrial engineering group must supply a time standard to be used in the simulation model. Also, the shipping supervisor must analyze and define the number of waves processed during a shift.

What Are the Performance Measures?

- Throughput capacity in cartons
- Time to pick all waves
- Pick stop percentage (due to full conveyor)
- Average carton cycle time (i.e., average time in system)
- Picking and shipping operator utilization

Step 3: Design of Experiments

The design of experiments is where tests are established and designed to provide the information that is desired from the simulation goals and objectives. Different simulation scenarios are defined that experiment with the system configuration.

Define the Base Scenario.   The base scenario is the basic system configuration and is used to validate the model. If the simulation objective is to make modifications to the current layout, then the base scenario should be the current system. If a completely new layout and process is undergoing simulation analysis, then the base scenario is based on the initial proposed system design. All other scenarios are compared to the base.

Define the Changes That Need to Be Investigated.   These changes correspond directly to the changes defined in the goals and objectives. New equipment may be added, operation times changed, or different methods of material handling may be used.

Establish the Number of Scenarios—One for Each Change.   If too many variables are changed in a scenario (a single simulation run), it is impossible to determine what change caused what effect. It is advised that each change represents an additional scenario.

Automation Project Simulation Example: Step 3

In reference to our conveyor transportation and sortation example, the following can be noted.

Define the Base Scenario.   The base scenario is the proposed system, with eight waves in an 8-hour shift, and shipping the estimated volumes from the peak period. Refer to Fig. 7-9 for the base-system parameters.

Define the Changes for Investigation.   The following system parameters will be changed and analyzed:

Wave size.   The wave size used in the model is based on the 5-year company growth outlook for peak shipping volumes. The wave size is varied to simulate the effects of underachieving or overachieving this objective.

Slug load.   The amount of cartons released onto the sortation conveyor at one time will be increased and decreased to determine if there is any significant throughput change.

Scenario	Wave size	Slug load	Carton size	Pick rate
Base	4500 cartons	40 cartons	12 × 18 × 24 in	12 cartons per minute
2,3	+40%, − 40%	Base	Base	Base
4,5	Base	+50%, − 50%	Base	Base
6,7	Base, + 20%	Base	+50%	Base
8,9	Base, + 20%	Base	Base	+25%

FIGURE 7-9   Simulation scenarios example.

Carton size.   The carton sizes are based on a particular product growth plan through the proposed design year. If the product mix is different, the effect on conveyor capacity and queue space could be changed. It is decided to increase the volume of the two largest carton sizes by 50 percent to determine the impact.

Picking rates.   This is a new picking method that has never been performed at the DC. Since the rate is a labor standard and not based on actual pick observations, the operation rate will be varied to determine if bottlenecks occur. Since the loading operation is being performed in the current process, the rate used is already an established standard and, therefore, will not be varied.

Establish the Number of Scenarios.   Figure 7-9 shows one set of planned scenarios. Note that each scenario incorporates at least one of the system parameters defined for change in the previous section and that only one parameter is changed in each scenario. After running these scenarios and after more is learned about the system behavior, additional scenarios may be created to define the system sensitivity further to parameters found to have the greatest impact.

Step 4: Data Collection and System Assumptions

After the system has been flow-charted and organized, pertinent information about the system is collected. Operation characteristics, such as operation time, move time, set-up time, and downtime, are collected for each element in the system. There are several sources available for obtaining this data such as computer data bases, maintenance records, industrial engineering standards or time studies, interviews, automatic data collection devices, and equipment specifications.

If data is difficult to obtain, a sensitivity analysis can identify those components that are most important and need to be estimated most accurately. The others can be roughly estimated. The simulation model may be used to perform the sensitivity analysis by running several test runs, varying the questionable piece of data, and noting any significant performance results.

It is important to stay focused on the simulation objective during data collection. For example, in the conveyor transportation and sortation proposal, the primary reason for the simulation is to ensure the conveyor lengths are sufficient to keep inventory flowing. The staffing requirements are secondary since adding more operators can be accomplished easier than adding conveyor length. Therefore, accurate data for the proposed conveyor lengths is more critical than accurate picking rates.

The raw data collected for the simulation must be summarized into a usable form and is typically characterized using probability concepts. Most simulation packages review the important statistical concepts applicable to data collection and analysis and

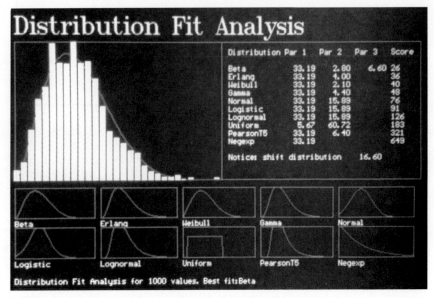

FIGURE 7-10   Distribution fit analysis. (*Courtesy of F&H Simulations.*)

may even provide a method to assist the user in selecting the right statistical distribution for a process based on historical data. The historical data is imported into the simulation tool and the software attempts to identify the best fitting distribution. Figure 7-10 gives an example of a distribution fit analysis.

Some of the most important and most used statistical distributions in simulation modeling are the normal, negative exponential, Erlang, uniform, and empirical. The empirical distribution is user-defined, since not all data can be represented by a theoretical distribution. The density graphs for the other four distributions are shown in Figure 7-11. Following are some examples where each of these distributions may be applied:

Normal.   The picking operation takes 4 minutes on average per order. Due to the variations in the number of items per order, variations in the pick time occur around the average.

Negative exponential.   Trucks arrive at the receiving dock at a certain time interval. It is known that 10 trucks arrive in an hour, and the interarrival time is 6 minutes. However, it is possible that, e.g., two trucks arrive in 5 minutes and that it then takes 20 minutes before the next truck arrives. Every truck arrival is independent of the other truck arrivals.

Erlang.   The maintenance department repairs fork trucks. Sometimes it takes a short time, other times it takes a very long time. On the average it takes 20 minutes.

Uniform.   An operator in a picking operation is assigned to several aisles. A pick from the closest aisle takes 8 minutes and a pick from the farthest aisle takes 11 minutes, but each aisle has the same probability of being selected.

Empirical.   A quality check is made right before the shipping operation. Ninety percent of the orders are correct and sent to the shipping operation. Seven percent

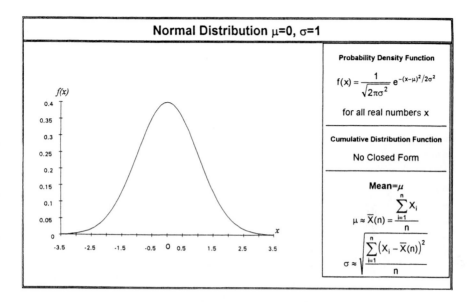

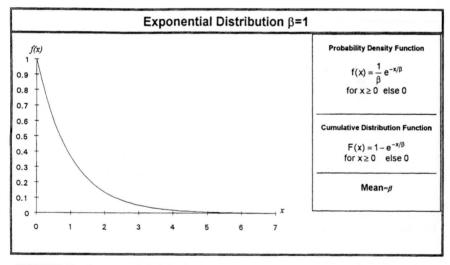

FIGURE 7-11   Common statistical distributions.

have minor problems and are sent to an area for rework. Three percent have back-ordered items and are sent to a holding area.

Data Collection.   The following list includes items that usually require data collection. Unless specifically noted in the objectives, highly detailed data is discouraged. For example, if there is a short setup time for each part at a location, this time can be added to the overall operation time, thus eliminating the need for separately inputting

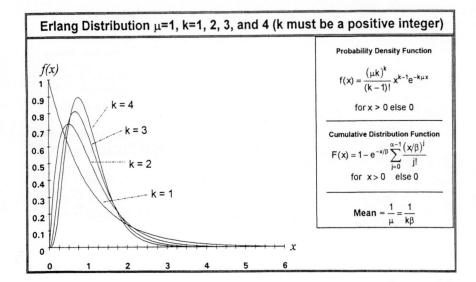

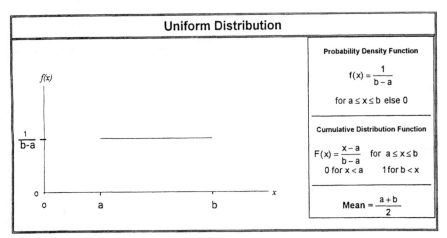

FIGURE 7-11   (*Continued*)

setup time for each part at each location. Using the proper distribution and combining information speeds the running of the model and the time to build the model, and still gives valid results.

Operation rates.   Operation times for each of the different functions included in the process. These times can also be related to the parts, cartons, or pallets used in the process.

Scheduling.   How and when parts, cartons, or pallets arrive at each location.

Downtime.   Scheduled and unscheduled, as well as mean time to repair (MTTR) and mean time between failures (MTBF) for equipment.

Changeover and setup times. Changeover time for different products or part types. Define the shift schedule for all the operators in the system model. Include start and end times as well as breaks, lunch or dinner times, shift crossover, and overtime.

Material handling interfaces. Material handling interfaces as well as detailed information on the material handling devices, for example, transport speeds, conveyor lengths, and fork truck lift speeds.

Operational constraints due to other system components. What constraints if any are placed on a system component as a result of the interaction with other system components. For example, with narrow aisle storage racks, only one fork truck or wire-guided vehicle can enter at a time.

Other important factors. Any other information or data that is required for the system to operate correctly.

Creating an Assumptions Document. An assumptions document contains all the information about the system, by defining the purpose of the simulation and each element in the system. This document becomes the specification for the simulation model. The model is validated by what is written in this document, so it is important not to leave anything out. All involved parties must arrive at a consensus on the system description and any assumptions incorporated due to the unavailability of data.

Sections of the Assumptions Document. The following should be included in the assumptions document. Note that the information gathered in steps 1 to 4 provides input for several sections in this document.

1. *Introduction.* A brief summary of what is to be accomplished in the study, including the model scope and the important points to be made.
2. *System layout.* The CAD or hand drawing of the system to be simulated. System components and resources should be identified on the layout.
3. *Overall objectives.* The goals and objectives for the simulation study.
4. *Issues to be investigated.* The design issues to be addressed by the simulation.
5. *Measures of performance.* The agreed-to performance measurements the simulation model will provide.
6. *Description and assumptions for each component and resource.* A list defining each system component and resource to be simulated along with all the data that has been collected. All the assumptions made about every part of the model are documented as well. It is critical to verify the data collected and assumptions made with the appropriate personnel, including employees and supervisors working in operations and personnel and the industrial engineering group. For proposed system models, data and assumptions should be validated with the design team and potential equipment vendors.
7. *Definition of each experiment to be conducted.* A list outlining all the experiments to be performed with the completed model. By defining exactly what the model should do before it is built, the model will contain everything needed to perform the experiments.

Automation Project Simulation Example: Step 4

Data Collection. In reference to the conveyor transportation and sortation example, the following data is collected and assumptions made:

Operation rates. The industrial engineering department provided these standard rates which will be modeled using a normal distribution:

- *Pick rate.* 12 cartons per minute with a standard deviation of 3.
- *Load rate.* 270 cartons per hour with a standard deviation of 10.

Scheduling

- *Wave size.* The wave size used in the model is based on the 5-year company growth outlook. The average number of cartons released in a wave is 4500. Since there are eight waves planned each 8-hour shift, a negative exponential distribution is used with an interarrival time of 0.8 second (4500 cartons per hour).
- *Shipping lane assignment.* An analysis is completed by randomly sampling current orders from 10 peak volume days, and determining a carrier ratio. Based on this analysis, five lanes (representing two carriers) receive 15 percent more cartons than the remaining lanes. An empirical distribution is used.
- *Slug load.* 40 cartons at a time will be released onto the sortation loop. After 40 cartons have been released, the model will select the conveyor with the current largest queue to release the next slug load.

Downtime. The vendor contacted for an estimate on the proposed conveyor system reports a breakdown, on average, every 500 hours of production that requires 2 hours to repair. The relative infrequency of this downtime is not viewed as an important system component and will not be included in the model.

Changeover and setup times

- *Wave changeover.* To ensure the proper loading sequence on the trucks, a complete wave of orders must be off the sortation conveyor and on the shipping lanes before the next wave of cartons can be released from the picking conveyors onto the sortation conveyor. The simulation must be designed to model this changeover since there is associated system idle time if the three end-of-wave totes, one for each picking conveyor, arrive at the shipping lanes at different times.
- *Shift schedule.* Breaks and meals will be staggered among the picking and shipping operators such that the interruption to product flow is minimized. Therefore, the model will not incorporate employee breaks.
- *Material handling interfaces.* The length of the belt take-away, sortation, and shipping conveyers is needed for the model since this length combined with the carton size defines the available queue space for the in-process inventory. The speed of the conveyors combined with the carton size is needed to determine the system process time.

Other important factors

- *Carton sizes.* The carton size is an important factor in determining adequate queue space on the conveyor system. Based on the design-year criteria and product mix, 40 percent of the cartons are 12 in long, 35 percent of the cartons are 18 in long, and 25 percent of the cartons are 24 in long. Again, an empirical distribution is used.

Step 5: Phased Model Development

The simulation model is the most visible part of the simulation study. It is important to keep in mind that a simulation model is a representation of the system and should be constructed to model the effect of the system. The model development should adhere to the goals and objectives and be completed in phases of increasing complexity. The initial phase in model development is to capture the basic flow and logic of the system. Part movement and operation times are added and verified as the model develops. As soon as the basic model function has been encoded, more detail can be added for each operation, such as equipment downtimes, operator breaks, and queuing disciplines (FIFO, LIFO, etc.), until the desired function is achieved.

Since this book does not cover the modeling techniques used with a particular simulation software or package, it is assumed that the identified personnel, in-house or vendor, have the skills necessary to build a correct model with the selected software. Many simulation packages are designed to allow simple model creation with little or no programming ability. However as the model complexity increases, the need for good programming skills may be important. Figure 7-12 displays a simulation model in development.

Model Documentation. Most simulation models are self-documenting as a result of entering the data required to build the model. Comment lines should also be added wherever possible to explain logic. Items that may seem unimportant at the time may mean a great deal in the future when trying to remember or explain why the model was developed as it was.

Variables, Part Attributes, and Functions. One of the most important parts of building the simulation model is to define and document the use of system variables, part, carton, or pallet attributes, and functions fully. If possible, this definition should be done within the simulation software. The more complex the logic within a model, the more important it becomes to document the use and function of variables and attributes, since they are typically utilized in various code logic.

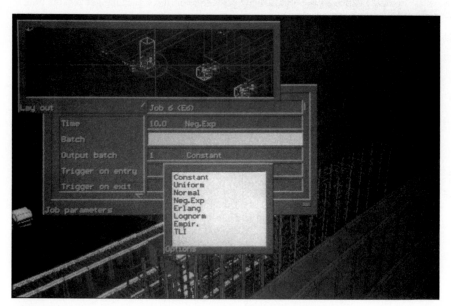

FIGURE 7-12 Simulation model development. (*Courtesy of F&H Simulations.*)

Code Documentation. Any use of subroutines or program files should always be included in the model documentation. It is important that everything used to make the model run is included in the documentation, such as spreadsheet files and any data that is imported into the model.

Automation Project Simulation Example: Step 5

In reference to the conveyor transportation and sortation example, the following attributes, variables, and functions need documentation. The complete documentation on how the attributes and variables are used, and how the functions operate, would be specific to the simulation package chosen for the model development.

Variables

- *Variable 1.* Counts the cartons released in a slug load. Used in the wave changeover function.
- *Variable 2.* Records the begin time the conveyor gets full. Used in the conveyor full function.

Attributes. The cartons are assigned the following attributes:

- *Attribute 1.* Time of creation, used to calculate the average time in the system for each carton.
- *Attribute 2.* Shipping lane assignment (based on an empirical distribution; see Step 4: Data Collection and System Assumptions), used to determine which shipping lane to route the carton.
- *Attribute 3.* End-of-wave tote identifier: whether the item traveling through the system is a carton or the end-of-wave tote. Used in the wave changeover function.

Functions

- *Wave changeover function.* Checks that all three end-of-wave totes are on the shipping lanes before releasing new wave cartons onto the sortation conveyor.
- *Slug-load function.* Determines when the slug load of 40 cartons has been reached and assigns the next belt take-away conveyor that will release cartons onto the sortation conveyor.
- *Conveyor full function.* Used to track the time the picking operation is stopped because all conveyors are full.

Step 6: Model Verification and Validation

Model verification ensures that the input data is correct, and model validation ensures that the model has successfully captured the operational characteristics of the system. To verify the model, check all the data entered into the model and make sure that it is correct as specified in the assumptions document.

A valid model must be seen as closely representing reality. All system components need to be validated to ensure their representation in the model is correct. If there is an actual system that is being modeled, output from the base scenario simulation run can be validated against actual system output. For example, average cycle time output by a model is compared to cycle time data obtained from the actual operation. When the system being modeled does not yet exist, the model is validated against the system assumptions document. In addition, a subjective validation is made by assessing whether the results are realistic in terms of experience with similar systems.

The basic rule of thumb for model validation is that the model behaves as expected. Reviewing the model outputs with several individuals involved with the current and proposed distribution operations is important. If the model is not behaving as expected, validate the assumptions document, and then check the logic and make sure it is working correctly.

The warm-up period should also be adequately defined. This is the time at the beginning of the model run in which statistics are not collected because the system has not yet reached a steady state. For example, in a continuous three-shift picking operation, there is always product in process, whereas at the start of the model, there is no inventory in process. If a sufficient warm-up period is not defined, then throughput results are overstated and utilizations understated. In an operation that starts cold each day, i.e., all in-process products are processed through the system by the end of each day, the warm-up period is not required.

Step 7: Run Experiments

Once the model has been verified and validated, the design of experiments is conducted. The scenarios and scenario replications are run (15 to 30 replications for high reliability), and the output results compared. The simulation software may have a vehicle for predefining the various scenarios and then running the experiments and replications automatically. If not, then the necessary changes must be made to the model for each experiment and the model runs initiated by the user for a sufficient number of replications. As all the experiments are run, each separate output data file should be saved so the results can be compared and confidence intervals for performance parameters can be calculated.

Step 8: Simulation Output Analysis and Presentation

This is the final step in the simulation study, and the one that is most visible. If any of the previous steps have been neglected or performed incorrectly, the results may be invalid. Good output analysis involves correct application of statistical analysis. Since the model represents outputs for a real system, the statistical analysis of the outputs from a simulation is similar to the statistical analysis of the data obtained from an actual system.

Focus on the output analysis should be on the performance measurements defined in step 2 and documented in the assumptions document. Simulation software generates a considerable amount of data that requires careful sifting to sort out the meaningful results. The performance measurements help identify what is important and what is not.

Once the important data has been identified, charts and graphs should be constructed that visually convey this information to management. The way the data is presented is very important when decisions hinge on the simulation study presentation. The team should do a thorough job analyzing the results and presenting final conclusions to the executive committee.

Automation Project Simulation Example: Step 8

Figure 7-13 gives the summarized results from one set of simulation scenarios designed for the conveyor transportation and sortation example. The results indicate

	Scenario number				
	Base	2	4	6	8
Average wave (in cartons)	4500	6300	4500	4500	4500
Pick rate per minute	12	12	12	12	15
Slug load (in cartons)	40	40	60	40	40
Carton length (inches)	12, 18, 24	12, 18, 24	12, 18, 24	12, 27, 36	12, 18, 24
Output results					
Time to pick all waves	8:22	10:43	8:38	9:28	7:48
Output (carton per module)					
Pick module 1	15352	20727	14989	16092	13974
Pick module 2	10424	16321	10347	9068	10727
Pick module 3	10224	13352	10664	10840	11299
Percent of time picking stopped (conveyor full)					
Pick module 1	5	15	4	14	8
Pick module 2	2	8	2	9	4
Pick module 3	0	2	0	3	1
Average time in system (minutes)					
Pick module 1	21	33	20	32	20
Pick module 2	24	36	25	29	23
Pick module 3	26	37	27	30	25

FIGURE 7-13   Simulation results.

that the conveyor system can handle the peak volumes in the design year, but there are areas for throughput and operational improvement. One area of concern is the high percentage of time picking is stopped due to the conveyors being full. The team has recommended two design changes to address this problem, which they will evaluate by modifying the simulation and designing additional scenarios. First, the higher velocity SKUs should be relocated from pick module 1 to pick module 3 to provide additional conveyor queue space since module 3 is farthest from the sortation conveyor. This can be tested in the model by decreasing the carton interarrival time at pick module 1 and increasing the carton interarrival time at module 3.

The second recommendation for reducing the idle pick time caused by the full conveyors is to add shipping operator "floaters." When a shipping conveyor becomes full, the floater assists the loading operation for that lane. To test this change in the model, a function is needed that increases the number of operators at a lane based on the queue sizes in the shipping lane.

Scenario 4 indicates the slug-load size has a negligible effect on the system throughput. However, the 50 percent increase in carton sizes modeled in scenario 6 impacts the system throughput significantly. The team decided the model with the two changes recommended above should also be rerun with this increased carton size to determine if throughput problems persist. The increase in the pick rate in scenario 8 helps the throughput time but also causes more conveyor full conditions. Again, this scenario should be rerun with the two recommended changes incorporated to reevaluate the pick rate impact.

The team must repeat steps 3 to 7 to incorporate and test the high-velocity SKU and floater proposals before final recommendations are made to management.

SELECTING SIMULATION SOFTWARE

Following is a list of different features that are useful for modeling distribution systems and should be considered when selecting a simulation software package or vendor.

Animation

It is extremely helpful when building the model to see a representation of the system being simulated. Model verification and validation are enhanced with the ability to watch the flow of product through the base scenario model. For example, a model validation issue is easily identified if the facility being modeled never has a bottleneck problem in a particular shipping lane, but the base scenario animation shows product queuing up in the modeled shipping lane.

Also, animation is a valuable tool for increasing confidence and buy-in for the recommendations resulting from the simulation analysis. Not everyone is familiar with the robustness of simulation as an analysis tool; however, they are often impressed by the animated system representation. A package that provides true 3D animation as well as 2D representations is preferred.

Graphical Model Building

Graphical model building is essential for those developing the simulation. The graphical approach helps in visualizing all the elements in the system as the model is being built, as well as minimizes the amount of coding required by the modeler. Usability is key to selecting simulation software, and graphical modeling makes simulation much easier to understand and master.

Flexibility

It is important when modeling distribution facilities that the modeling software has the flexibility to handle complex logic and routings. Typically some kind of programming interface should be available in order to handle the complex logic of warehousing systems. The ability to import data from spreadsheets is also important.

Material Handling Capabilities

A modeling software that is strong in conveyors, rack storage, AGVs, fork trucks, and warehouse pickers is required for distribution systems.

Output Analysis Tools

The modeling software should enable "what if" analyses with design of experiment capability. Any feature that summarizes the output data for the user is extremely helpful, such as having the ability to curve-fit data to a distribution. This is particularly helpful in situations where output from one model is used as input to another model. The breadth of performance measures provided are also an important feature. Packages with product cost analysis capability are useful when developing the justification and benefits analysis.

Documentation, Training, Consulting, and Customer Support

The availability of training will jump start the effective use of any simulation package. Good documentation and ongoing customer support are essential for reference after the initial training. Also, consultant availability is important if there are limited skills available within the organization to complete the simulation study in the time frame required.

REFERENCE

1. Pritsker, Alan B., *Introduction to Simulation Using SLAM II,* 4th ed., Systems Publishing/Wiley, West Lafayette, IN, 1995.

CHAPTER 8
BENEFIT ANALYSIS AND JUSTIFICATION

INTRODUCTION

One of the key steps in the successful implementation of a warehouse automation project is a thorough and accurate justification. A justification is the process by which a company decides whether or not the costs of a particular project substantially "justify" the benefits. Management team members must be certain the proposed project will help them achieve the company's goals and objectives and provide a good investment on their capital funds. In addition, even if a company is convinced it must invest in warehouse automation to remain competitive or gain an advantage over its competitors, it still must justify that the proposed solution is the proper course of action.

A warehouse automation solution is a combination of a warehouse management system, material handling equipment, processes, people, layout, and technology. There are many alternatives for each component of the solution and therefore many viable solutions and combinations that may achieve the project goals. Selecting the best alternative requires the completion of a cost-benefit analysis where the company estimates the expected savings and investment costs for each alternative and calculates the return on investment (ROI).

Expected benefits should include savings that are both tangible, such as direct labor savings, and intangible, such as increased revenues due to customer service improvements. Intangible savings are more difficult to define, quantify, and measure than tangible savings, because they are more speculative. However, overreliance on tangible savings in investment decisions leads to (1) too heavy a focus on cost reduction rather than profit improvement; (2) missed opportunities regarding the profitability effects of quality, throughput, and other intangible items; and (3) underestimation of the costs and lost revenues of the "do-nothing" option. Logistics and warehousing professionals are well advised to evaluate carefully the extent to which intangible savings should be used in the benefit analysis and justification process.

OVERVIEW

A benefit analysis and justification is a six-step process which balances the initial and ongoing costs of an automation investment against the expected savings to help identify the "best" alternative. The six steps in this process are briefly described below and then further detailed later in this chapter.

Step 1   Identify the alternatives

Step 2   Determine costs

Step 3   Calculate savings

Step 4   Compare the alternatives

Step 5   Perform supplemental analysis

Step 6   Specify the preferred alternative

In addition to describing each step, examples and sample data collection spreadsheets and ROI spreadsheets are provided in this chapter. For additional information on the subject of economic justification, several texts commonly used by colleges and industry are listed in the Bibliography at the end of this chapter.

Step 1: Identify the Alternatives

As seen throughout this text, there are numerous design and system alternatives to choose from when automating a warehouse system. In order to select the most effective and efficient system, the top three or four proposals should first be identified. This is done by completing a broad or high-level analysis, making sure that all finalists meet the goals and objectives of the company. Then, a more detailed review begins of the finalists, by the company's own resources, if the proper skills are available, or through outside advisors such as management or engineering consultants, or systems and material suppliers.

Step 2: Determine Costs

A major component of any project is the bottom-line expense to the company. Each alternative must be evaluated separately, reviewing both the implementation and operating costs. These costs include material handling and storage equipment, software for the warehouse management system, direct and indirect labor, and other miscellaneous items. Costs are based on the proposed process flow for each alternative and estimated through the project design year. Typically, a contingency factor is added to the cost estimate in an attempt to quantify risk.

Step 3: Calculate Savings

Savings are divided into two groups: tangible and intangible. Tangible savings are those where an actual "out of pocket" dollar savings can be calculated; for example, direct and indirect labor savings, lease or construction cost reductions, or avoidance due to space savings.

There are two types of intangible savings, hard and soft. Hard intangible savings are those that can be calculated once an assumption or estimate of the improvement is made. Examples are percentage of increased sales or market share due to improved customer service, improved fill rates, reduced picking errors, inventory carrying cost savings due to reduced safety stock, increased inventory turn rates, and reduced cycle time. Soft intangible savings are those that have value, but no assumption or estimate has been made to enable a calculation to be done. Examples are employee morale, pride, loyalty, safety, and ergonomics.

Step 4: Compare the Alternatives

In this step, the cost of the investment is compared to the expected savings resulting from the investment through the specified design year of the project. Cash flows are summarized for all costs and savings for each alternative. Many methods for calculating economic performance are available, including present worth, annual worth, future worth, payback period, benefit to cost ratio, and internal rate of return. Financial personnel can assist in defining any assumptions that need to be made, such as the company's desired rate of return (hurdle rate), inflation, tax rates, payback period, and labor rates. The preferred depreciation method for capital investments can also be established by the finance personnel.

Step 5: Perform Supplemental Analysis

The comparison between alternatives using specified economic performance measurements is normally quite straightforward and the alternative that has the most attractive results is recommended. However, since the savings, costs, and interest rate are estimates, a supplementary analysis is recommended to determine the sensitivity of the estimating process to the desirability of the alternatives. This step, as well as the addition of contingency to the cost estimates, should adequately quantify the risk of an automation project.

Step 6: Specify the Preferred Alternative

The final step in conducting a benefits analysis is the selection of the preferred alternative. This selection is usually based on financial and nonfinancial considerations. A summary chart comparing the alternatives is developed to enable an informed decision by management. This summary includes the assumptions, economic results, sensitivity analysis, intangible benefits, and specific risks. The final decision depends on management's reaction to risk, uncertainty, intangible results, and the availability of capital for the project.

ANALYSIS

This section describes the complete six-step process in detail, utilizing examples and charts which will assist in conducting the benefit analysis and justification.

Step 1: Identify the Alternatives

The three major components of the solution are the system design, material handling design, and technology equipment, which are discussed in other chapters of this handbook. A thorough understanding of all these chapters is required to obtain a working knowledge and vocabulary of the components of possible alternative solutions. Since there are many possible warehouse automation solutions that address the project goals, this process step is difficult for anyone unfamiliar with all the various solutions and equipment available. A vital supplement to this step is employing assistance from those who have had prior experience in the warehouse automation process. Warehouse automation is a specialized skill, and experience is critical. Equipment suppliers are a good source of knowledge, but their solutions generally utilize equipment they sell and install. Other viable solutions or alternatives may be overlooked. Independent consultants with experience in warehouse automation are a good option. Since they are not restricted to specific suppliers or equipment, the alternatives they select cover a broader spectrum.

Money spent to develop several good alternatives and evaluate them is money well spent. It should be considered good insurance against a potentially disastrous decision. In most automation projects of any size, the cost of a good study is a small portion of the total investment. In addition, the technical feasibility and risk of each alternative should be carefully considered. The more complex a solution or the more "leading edge" the technology, the higher the risk. Unless these risky solutions offer clear savings advantages, more traditional or conventional solutions should be selected.

One alternative that must be included in the analysis is the "do-nothing" alternative. This alternative is necessary to establish a base to which other alternatives will be compared to during the economic analysis. The do-nothing alternative assumes continuing business as usual, with adjustments for labor, equipment, and space to contain projected growth. If a new process is required because the company plans on offering a new service, a base process must be designed for that service. Usually, the base alternative for a new process is one with minimum automation and capital investment, and the maximum labor requirement.

Step 2: Determine Costs

Cost estimating falls into four primary areas: equipment, software, labor, and others. Several work sheets are included which provide sample formats to help develop an accurate cost listing. These work sheets can be altered as needed to meet the project scope.

Cost estimates are calculated for each year through the design year of the project. One definition of the design year is the last year that the investment for this automation project is intended to meet projected volume growth. Another definition is to use the planning horizon, which is usually 5 years. The majority of capital cost occurs during implementation, but if additional equipment is required to meet increased volumes in the out years, then the associated costs must be included in the analysis.

Typically, a contingency factor is added to the cost estimate in an attempt to quantify the risk. This contingency is added as a percentage of the costs and is based on the level of accuracy and completeness of the estimate, as well as the design stability of the alternative. For example, a proposed alternative with a high degree of complexity and a lengthy implementation plan is likely to experience design and schedule changes before completion. Improvements and oversights may occur which can affect capital and labor cost estimates. The volatility of these estimates requires a larger contingency factor than those of more conventional, short-term alternatives. Contingency factors should be different for different elements of the solution, depending on the confidence in the estimate of the element.

Equipment Cost Estimates. The proposed process flow for an alternative solution dictates the type of technology and equipment required for the job. Proposed process flow charts should be developed for each warehouse functional area. In addition, if layout changes are involved, a proposed DC layout is needed to estimate the new material handling and storage equipment requirements. Refer to Chap. 3, Technology Considerations, and Chap. 4, Material Handling Considerations, for help in developing these specifications. Using the proposed equipment lists and layout requirements, budget estimates can be obtained from selected vendors.

Figure 8-1 depicts a sample equipment cost estimate work sheet that lists equipment required in each functional area. One work sheet is needed for each alternative. Once completed, this work sheet is used as input to the return on investment calculation. Note that the work sheet includes system hardware as well as material handling equipment. A proposed receiving process flow chart is reviewed to show how these work sheets are completed (see Fig. 8-2).

The current receiving process begins with the operator completing receiving forms for each incoming delivery and then entering the required data into a PC to generate SKU labels. To expedite this process and reduce errors, the proposed process begins with the supplier transmitting advance shipping notices (ASNs) via EDI to a warehouse management system (WMS). The receiving operator will then verify incoming merchandise against the ASN by scanning the UPC number and quantity into a radio frequency terminal. Upon completion of the receipt, the information is transmitted to the WMS and SKU labels are automatically printed at the receiver's portable printer.

Comparing the current to the proposed process, it is clear that radio frequency terminals, scanners, and printers will be required. However, there are additional process questions to investigate to further define the equipment requirements. For example, are the receiving operators using a fork truck to perform the receiving operation or will they be moving merchandise with pallet jacks? This will dictate whether the radio frequency terminal should be a hand-held unit with a short-range scanner or a vehicle-mount terminal with a long-range scanner. In this particular example, fork trucks are proposed, requiring vehicle-mount radio frequency terminals and long-range scanners to allow the operator to scan merchandise while seated on the truck.

After the equipment type for the operation is identified, the next step is to identify the quantity that must be purchased. In this example, a one-to-one ratio exists between equipment and receiving operators. The number of receiving operators is calculated through the design-year requirements. If there is planned volume growth, then a decision is made whether to purchase additional equipment as the volumes increase or to purchase all the equipment initially in one buy. Issues involved in this decision include available capital, confidence in volume growth, and quantity incentives offered by the vendors.

Another consideration in defining the equipment requirements is the number of printers that must be purchased. An analysis is completed to determine where to place the printers for maximum operator productivity. The proposed process flow dictates that the number of printers must match the number of radio frequency terminals, so six printers are added to the equipment cost estimate work sheet.

It can be seen from the prior example that much thought must go into the proposed process flow before the equipment requirements can be defined. This process analysis is used through the remaining functional areas of the warehouse to complete the equipment cost estimate.

In addition, if layout changes are part of the automation project, a proposed warehouse layout must be developed. New fork or clamp trucks, automated guided vehicles, pallet rack purchase or rearrangement, pick modules, carousels, and conveyor and sortation equipment are examples of important material handling elements that must be considered.

To assist with this step in the estimation process, it is recommended that equipment suppliers be contacted and asked to provide "ballpark" estimates of the equipment. Ballpark estimates are generally accurate to plus or minus 15 percent. Suppliers will usually provide this service at no cost, with the understanding that they will be allowed to submit a real bid (quotation) later in the project. Suppliers routinely estimate on the

FUNCTIONAL AREA	OFFICE	RECV.	Q/C	PUT-AWAY	PICK	CHECK	SHIP	OTHER	TOTAL QTY	COST/UNIT	TOTAL COST
Material handling equipment:											
Pallet jacks											
Counterbalance trucks											
Reach trucks											
ASRS											
AGV system											
Conveyor system											
Racking:											
Selective pallet rack											
Drive-in											
Push back											
Double deep											
Flow rack											
Shelving											
Picking system:											
Picking modules											
Pick-to-light											
Carousel											
Order selector											
Robotic											
Technology equipment:											
Computer display terminal											
Wedges											
Short-range scanners											
Long-range scanners											
Radio frequency system											
Voice recognition system											
Bar code printers											
Document printers											
Personal computers											
PC software											
Computer interfaces											
Hardware interfaces											
Total cost											

FIGURE 8-1 Equipment cost estimate work sheet.

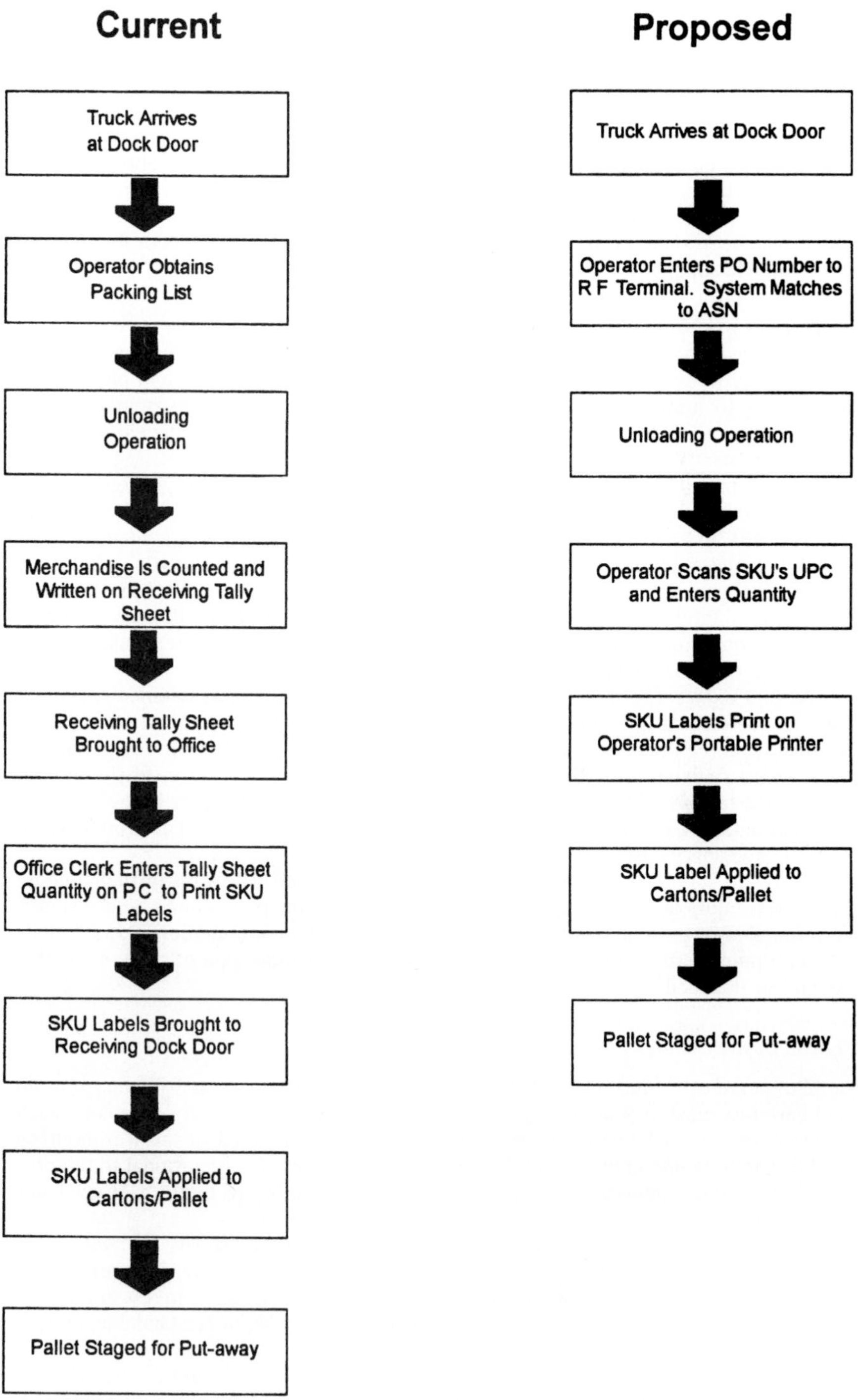

Current

```
Truck Arrives
at Dock Door
```
⬇
```
Operator Obtains
Packing List
```
⬇
```
Unloading
Operation
```
⬇
```
Merchandise Is Counted and
Written on Receiving Tally
Sheet
```
⬇
```
Receiving Tally Sheet
Brought to Office
```
⬇
```
Office Clerk Enters Tally Sheet
Quantity on P C  to Print SKU
Labels
```
⬇
```
SKU Labels Brought to
Receiving Dock Door
```
⬇
```
SKU Labels Applied to
Cartons/Pallet
```
⬇
```
Pallet Staged for Put-away
```

Proposed

```
Truck Arrives at Dock Door
```
⬇
```
Operator Enters PO Number to
R F Terminal.  System Matches
to ASN
```
⬇
```
Unloading Operation
```
⬇
```
Operator Scans SKU's UPC
and Enters Quantity
```
⬇
```
SKU Labels Print on
Operator's Portable Printer
```
⬇
```
SKU Label Applied to
Cartons/Pallet
```
⬇
```
Pallet Staged for Put-away
```

FIGURE 8-2   Current and proposed receiving processes.

high side when submitting a ballpark estimate. This is done for two reasons. One is to allow for the fact that the alternative plan is not fully developed, and changes and additions are likely. The second reason is that they hope to reduce the total cost during their quotation, when they have sufficient time to engineer the solution. Quotations can take several weeks to complete for complicated alternatives and should only be done after the preferred alternative has been selected and detailed specifications have been developed. The decision to perform a competitive bid with several suppliers or to "partner" with a single supplier is determined by company purchasing policy.

Software Cost Estimates.   A system specification for each alternative must be developed before software cost estimates can be determined. This specification is based on the same process flow charts used for determining equipment costs and on required interfaces to material handling equipment. Refer to Chap. 5, System Function Considerations, and Chap. 6, Requirements Definition, for help in developing these specifications.

The estimated cost of the software can be determined by in-house personnel if the proper resources are available. The other approach is to prepare a request for information, or RFI, to be sent out to prospective software suppliers. An RFI contains a high-level description of requirements and processes and should be sent to 10 to 12 suppliers. All that should be expected from these suppliers is information about the company and their solutions and a range of prices or average costs for their software. This provides the opportunity to learn more about the suppliers, and their responses should enable the potential vendor list to be narrowed down to five or less.

Next, a request for proposal (RFP) is prepared, which is a more detailed specification for the desired warehouse management system. The RFP is a request for a budget estimate that requires considerable work on the supplier's part. Once the estimates are received, a decision must be made as to which ones to use for the justification. If a preference for one of the suppliers has not yet developed, then typically the high and low estimates are eliminated, and the remaining ones are averaged to determine a reasonable budgetary cost.

Software suppliers should also include system-related hardware in their budget estimates. Equipment such as the computer hardware, printers, terminals, and scanners is a function of the warehouse management system under consideration. Use Fig. 8-1, the equipment cost estimate work sheet, as a guide to the type of equipment that should be included.

Labor Cost Estimates

Direct Labor.   Since many operations in the proposed processes will change or do not currently exist, it is necessary to estimate the workforce necessary to operate each alternative. Direct labor includes employees directly involved in the warehousing and distribution operation, where the number of employees is proportional to the production volume. Industrial engineers are generally required to perform direct-labor estimates.

In many cases, the labor savings due to automation are a primary motivation for undertaking the project. If management feels the labor estimates are not realistic and the potential reductions are not achievable, they will be reluctant to invest in the project. Therefore, it is essential that these estimates be accurate, understandable, believable, and verifiable.

Figure 8-3 presents the results of an analysis completed to determine the distribution of labor against the functional operations. This analysis highlights the opportunity areas for automation and maximum payback.

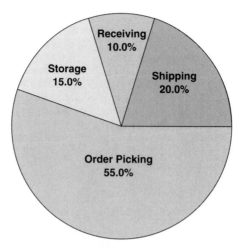

FIGURE 8-3   Warehouse costs by functional area.
(*Courtesy of The Logistics Institute at Georgia.*)

The following details a step-by-step procedure that can be used to calculate direct labor costs for the current process and the proposed alternatives. Included are example calculations for the fictitious ABC Company.

Establish the Current Process

Step A: Establish shipping volumes through the design year.   Determine the unit of measure to be used. This could be pallets, cartons, or items; e.g.; if products are shipped in a variety of ways, it will be necessary to convert everything to the lowest common denominator. For example, if full pallets of cartons are shipped as well as individual cartons, convert the full pallets to cartons. Collect data for the annual shipping total for the previous year. Multiply this total by the projected annual growth to establish the totals for each year through the design year. The current year becomes the base year. Next, calculate the average daily shipping volume by dividing by the number of work days in the year. Allow for warehouse shutdown for holidays and vacation as applicable.

Example:

Total 1994 shipping volume	6.9 million cartons
Total work days	254 days
Average 1994 daily ships	27,165 cartons per day

Projected daily shipments assuming 10 percent annual growth:

1995	29,882 cartons per day
1996	32,870 cartons per day
1997	36,157 cartons per day
1998	39,773 cartons per day
1999	43,750 cartons per day

Step B: Calculate the unit hour for each warehouse operation. A unit hour is nothing more than the length of time per unit of measure an operation takes to perform, expressed as a decimal hour. For example, if an operation takes 6 minutes to perform for each carton shipped, the unit hour is 0.10000 hour per carton, while an operation taking 0.5 minute has a unit hour of 0.00833 hour per carton.

Determine the hours spent on each operation in the warehouse for the previous year, or any other period for which data is handy. This data should be available from payroll records or claim sheets. Now divide the hours per operation by the shipping volume for the same period to determine the unit hour. Note that all hours for any operation should be divided by the shipping volume, not the volume for that operation. This ensures that all operations are expressed as weighted unit hours and can be added together to determine total time. Since the current process is the base or do-nothing case, it is normally assumed that the unit hours remain unchanged through the design year.

Example:

Total 1994 hours for receiving	13,800 hours
Total 1994 cartons received	6.9 million cartons
Unit hour for receiving	0.0020 hour per carton

Step C: Calculate the head count for each operation. *Head count* refers to the number of direct employees required to ship the volumes specified in step 1 during a standard work day. First, determine the total hours required by multiplying the daily shipping volume by the unit hour for each operation from step 2. Convert the total hours to a head count by dividing by hours in a standard work day, normally 8 hours. No adjustment for overtime is required, since the projection is trying to determine the number of "8-hour equivalent" employees required to ship the daily volumes. The head count determined by this method should be compared to the actual head counts. Remember to adjust the actual head count for overtime before comparing it to the calculations. For example, if there are six people working in receiving and they average 13 percent overtime, the adjusted headcount is 6.8 people (6 × 1.13). Baseline labor estimates should also be validated by the supervisors who have experience with the operations. Figure 8-4 shows an example of the receiving process head-count calculation.

Step D: Calculate the total annual labor cost. Once the direct-labor head count is estimated, the labor cost is calculated by multiplying the head count for each operation times the labor rate times the inflation rate. See Fig. 8-5 for a sample calculation. Finance or personnel departments should help determine the annual labor rate and inflation rate to use (generally 3 percent). The annual labor rate should include all out-of-pocket costs (base rate plus benefits such as sickness, accident, and vaca-

	1994	1995	1996	1997	1998	1999
Daily ship volume	27,165	29,882	32,870	36,157	39,773	43,750
Receiving unit hour	0.0020	0.0020	0.0020	0.0020	0.0020	0.0020
Hours per day	54.3	59.8	65.7	72.3	79.5	87.5
Head count	6.8	7.5	8.2	9.0	9.9	10.9

FIGURE 8-4 Receiving operation head-count calculation.

	1994	1995	1996	1997	1998	1999
Daily ship volume	27,165	29,882	32,870	36,157	39,773	43,750
Receiving unit hour	0.0020	0.0020	0.0020	0.0020	0.0020	0.0020
Hours per day	54.3	59.8	65.7	72.3	79.5	87.5
Head count	6.8	7.5	8.2	9.0	9.9	10.9
Annual rate ($ thousands)	21.0	21.6	22.3	22.9	23.6	24.3
Total cost ($ thousands)	143	162	183	207	235	266

FIGURE 8-5   Receiving operation annual labor cost calculation.

tion costs, medical, dental, and life insurance premiums, and retirement plans). Generally, the cost of benefits is in the range of 30 to 35 percent of the labor rate. Thus, if the labor rate is $10.00 per hour, the out-of-pocket cost to the company is $13.00 to $13.50 per hour.

Repeat the above steps for all operations to determine the total direct-labor cost of the current process, which is the base or do-nothing case. Finally, a sample summary sheet for all the operations for the ABC Company is shown in Fig. 8-6.

Proposed Alternatives.   Next, the direct labor for the proposed alternatives must be calculated.

Step A: Shipping volumes through the design year.   Use the same shipping volumes as the current process above.

Step B: Calculate the unit hour for each warehouse operation.   For operations that remain unchanged, use the unit hours developed for the current process. For operations that will be different from the base case, labor estimates must be done. It is critical that these estimates be realistic and accurate, since they are generally responsible for the major portion of the savings. If management is not comfortable with the labor estimates for the proposed process, they will be reluctant to approve the project.

Since these operations do not currently exist, the specialized skills of an industrial engineer are required to perform the analysis. Equipment suppliers have generalized rates for warehouse operations that can also be used as guidelines. However, for more accurate estimates, a detailed labor estimate should be completed. Labor estimating is outside the scope of this book, but additional information on developing labor estimates can be obtained from industrial engineering handbooks.

Any detailed labor estimate must include adjustments for lost time. Although people normally work 8 hours per day, they are not productive the whole time. Allowance must be made for personal, fatigue, and delay (PFD) time. This includes breaks, phone calls, department meetings, and so forth. Typically, an allowance of 15 percent is used. That means a person who puts in an 8-hour day is productive for 6.8 hours.

Another adjustment required is an allowance for sickness, accident, and vacation. This allowance can be determined by review of personnel records, but a good rule of thumb is 5 percent. Thus, if a workforce of 100 people is required, 105 employees must be hired to ensure 100 show up (on average).

ABC COMPANY
CURRENT PROCESS–UNIT HOURS/HEAD COUNT/COSTS

	1994	1995	1996	1997	1998	1999
CARTONS SHIPPED	(REFLECTS 10% ANNUAL GROWTH)					
ANNUAL (M)	6.90	7.59	8.35	9.18	10.10	11.11
DAILY TOTAL CARTONS	27165	29882	32870	36157	39773	43750
HOURS/CARTON						
RECEIVING	0.0020	0.0020	0.0020	0.0020	0.0020	0.0020
RETURNS	0.0011	0.0011	0.0011	0.0011	0.0011	0.0011
REC'VG CHECKER	0.0010	0.0010	0.0010	0.0010	0.0010	0.0010
PUT-AWAY	0.0045	0.0045	0.0045	0.0045	0.0045	0.0045
REPLENISH	0.0035	0.0035	0.0035	0.0035	0.0035	0.0035
PICKING	0.0128	0.0128	0.0128	0.0128	0.0128	0.0128
SHIPPING CHECKER	0.0010	0.0010	0.0010	0.0010	0.0010	0.0010
SHIPPING/LOADING	0.0073	0.0073	0.0073	0.0073	0.0073	0.0073
CYCLE COUNT	0.0006	0.0006	0.0006	0.0006	0.0006	0.0006
LOCATOR CLERK	0.0011	0.0011	0.0011	0.0011	0.0011	0.0011
LABEL PRINT	0.0008	0.0008	0.0008	0.0008	0.0008	0.0008
AVERAGE HOURS/CARTON	0.0357	0.0357	0.0357	0.0357	0.0357	0.0357
ADJUSTED HEADCOUNT						
RECEIVING	6.8	7.5	8.2	9.0	9.9	10.9
RETURNS	3.7	4.1	4.5	5.0	5.5	6.0
REC'VG CHECKER	3.4	3.7	4.1	4.5	5.0	5.5
PUT-AWAY	15.3	16.8	18.5	20.3	22.4	24.6
REPLENISH	11.9	13.1	14.4	15.8	17.4	19.1
PICKING	43.5	47.8	52.6	57.9	63.6	70.0
SHIPPING CHECKER	3.4	3.7	4.1	4.5	5.0	5.5
SHIPPING/LOADING	24.8	27.3	30.0	33.0	36.3	39.9
CYCLE COUNT	2.0	2.2	2.5	2.7	3.0	3.3
LOCATOR CLERK	3.7	4.1	4.5	5.0	5.5	6.0
LABEL PRINT	2.7	3.0	3.3	3.6	4.0	4.4
TOTAL ADJUSTED HEADCOUNT	121.2	133.3	146.7	161.4	177.5	195.2
CARTONS/EMPLOYEE/HOUR	28.0	28.0	28.0	28.0	28.0	28.0
ANNUAL LAB. COST (K$)	$21.0	$21.6	$22.3	$22.9	$23.6	$24.3
RECEIVING	$143	$162	$183	$207	$235	$266
RETURNS	$78	$89	$101	$114	$129	$146
REC'VG CHECKER	$71	$81	$92	$104	$118	$133
PUT-AWAY	$321	$364	$412	$467	$529	$599
REPLENISH	$250	$283	$320	$363	$411	$466
PICKING	$913	$1,034	$1,172	$1,328	$1,504	$1,704
SHIPPING CHECKER	$71	$81	$92	$104	$118	$133
SHIPPING/LOADING	$521	$590	$668	$757	$858	$972
CYCLE COUNT	$43	$48	$55	$62	$71	$80
LOCATOR CLERK	$78	$89	$101	$114	$129	$146
LABEL PRINT	$57	$65	$73	$83	$94	$107
TOTAL ANNUAL LABOR COST	$2,546	$2,884	$3,268	$3,703	$4,195	$4,753

FIGURE 8-6   Current process—unit hours, head count, and cost summary.

It is prudent to add contingency to these labor estimates, since these operations do not currently exist and there is some risk involved in trying to estimate them. Also, unavoidable and unplanned delays occur in all operations in a warehouse. Examples of unavoidable delays are lift-truck breakdowns or unexpected battery changes, trailers that are late for scheduled shipments, and conveyor-system problems. Additionally, process changes are likely to occur later in the detail design and implementation phases. Generally, a 15 percent contingency factor is acceptable and should be added to the unit hour estimate.

Step C: Calculate the head count for each operation. Follow the same procedure outlined in step C for the current process.

Step D: Calculate the total annual labor cost. Follow the same procedure outlined in step D for the current process.

Finally, a sample summary sheet for alternative 1 for the ABC Company example is shown in Fig. 8-7. Notice that the unit hours, head counts, and costs in 1994 for alternative 1 are the same as the current process in 1994. The example assumes that 1994 is the base year, and changes caused by implementing alternative 1 will not occur until 1995. The unit hour for the put-away operation remained the same for both the current process and alternative 1. In this instance, the use of RF technology did not reduce labor but did eliminate location errors. Also, the annual labor cost for two operations, shipping checker and locator clerk, becomes zero in 1995 since it is assumed they will be eliminated by the new process.

Indirect Labor. Calculating indirect-labor costs is not as straightforward as with direct-labor costs, but they can contribute significantly to the overall justification analysis. The cost of indirect labor, both during and after implementation of each alternative, must be estimated.

A preliminary implementation schedule can be created to assist in the estimation of any internal support required during implementation such as a full-time project manager, cadam operators, system architect and programmers, and purchasing and finance personnel. If new hires or outside consultants are to be used, their full cost should also be included. If the project is to be done entirely in-house, and the necessary people are already on board, the cost of their efforts during implementation should be calculated using a rate approved by the finance department. Since the objective is to determine the out-of-pocket cost differences between any alternative and the do-nothing base alternative, this rate reflects the fact that this indirect labor is already part of the company's overhead.

Indirect-labor costs to be incurred after implementation can be calculated by interviewing the managers of the support departments associated with the warehouse. An estimate is required for the head count needed to support the current process through the design year, with annual increases for projected growth. Labor rates from the finance department are multiplied by the head counts to determine indirect-labor costs for the current process. Shown below are some suggested support areas that may be affected by warehouse automation:

- Warehouse clerk
- Order entry
- Billing
- Accounts receivable
- Accounts payable

ABC COMPANY
ALTERNATIVE 1 UNIT HOURS/HEAD COUNT/COSTS

	1994	1995	1996	1997	1998	1999
CARTONS SHIPPED	(REFLECTS 10% ANNUAL GROWTH)					
ANNUAL (M)	6.90	7.59	8.35	9.18	10.10	11.11
DAILY TOTAL CARTONS	27165	29882	32870	36157	39773	43750
HOURS/CARTON						
RECEIVING	0.0020	0.0015	0.0015	0.0015	0.0015	0.0015
RETURNS	0.0011	0.0007	0.0007	0.0007	0.0007	0.0007
REC'VG CHECKER	0.0010	0.0003	0.0003	0.0003	0.0003	0.0003
PUT-AWAY	0.0045	0.0045	0.0045	0.0045	0.0045	0.0045
REPLENISH	0.0035	0.0027	0.0027	0.0027	0.0027	0.0027
PICKING	0.0128	0.0072	0.0072	0.0072	0.0072	0.0072
SHIPPING CHECKER	0.0010	0.0000	0.0000	0.0000	0.0000	0.0000
SHIPPING/LOADING	0.0073	0.0036	0.0036	0.0036	0.0036	0.0036
CYCLE COUNT	0.0006	0.0002	0.0002	0.0002	0.0002	0.0002
LOCATOR CLERK	0.0011	0.0000	0.0000	0.0000	0.0000	0.0000
LABEL PRINT	0.0008	0.0002	0.0002	0.0002	0.0002	0.0002
AVERAGE HOURS/CARTON	0.0357	0.0209	0.0209	0.0209	0.0209	0.0209
ADJUSTED HEADCOUNT						
RECEIVING	6.8	5.6	6.2	6.8	7.5	8.2
RETURNS	3.7	2.6	2.9	3.2	3.5	3.8
REC'VG CHECKER	3.4	1.1	1.2	1.4	1.5	1.6
PUT-AWAY	15.3	16.8	18.5	20.3	22.4	24.6
REPLENISH	11.9	10.1	11.1	12.2	13.4	14.8
PICKING	43.5	26.9	29.6	32.5	35.8	39.4
SHIPPING CHECKER	3.4	0.0	0.0	0.0	0.0	0.0
SHIPPING/LOADING	24.8	13.4	14.8	16.3	17.9	19.7
CYCLE COUNT	2.0	0.7	0.8	0.9	1.0	1.1
LOCATOR CLERK	3.7	0.0	0.0	0.0	0.0	0.0
LABEL PRINT	2.7	0.7	0.8	0.9	1.0	1.1
TOTAL ADJUSTED HEADCOUNT	121.2	78.1	85.9	94.5	103.9	114.3
CARTONS/EMPLOYEE/HOUR	28.0	47.8	47.8	47.8	47.8	47.8
ANNUAL LAB. COST (K$)	$21.0	$21.6	$22.3	$22.9	$23.6	$24.3
RECEIVING	$143	$121	$137	$156	$176	$200
RETURNS	$78	$57	$64	$73	$82	$93
REC'VG CHECKER	$71	$24	$27	$31	$35	$40
PUTAWAY	$321	$364	$412	$467	$529	$599
REPLENISH	$250	$218	$247	$280	$317	$359
PICKING	$913	$582	$659	$747	$846	$959
SHIPPING CHECKER	$71	$0	$0	$0	$0	$0
SHIPPING/LOADING	$521	$291	$330	$373	$423	$479
CYCLE COUNT	$43	$16	$18	$21	$24	$27
LOCATOR CLERK	$78	$0	$0	$0	$0	$0
LABEL PRINT	$57	$16	$18	$21	$24	$27
TOTAL ANNUAL LABOR COST	$2,546	$1,689	$1,913	$2,168	$2,456	$2,783

FIGURE 8-7 Alternative 1—unit hours, head count, and cost summary.

- Inventory control
- Sales analysis
- Purchasing
- General ledger
- Payroll
- Word processing
- Management information systems (MIS) department

See Fig. 8-8 for a sample summary sheet of the indirect-labor costs for the current process.

Next, a list of potential benefits resulting from the proposed automation alternatives is generated to identify areas for indirect-labor cost reduction. These potential benefits should be reviewed with the support departments to determine jointly the potential reductions and establish the head counts for each alternative plan through the design year. A reluctance to agree on the head-count reductions by the supporting

ABC COMPANY
CURRENT PROCESS—WAREHOUSE SUPPORT
INDIRECT SUPPORT

	1994	1995	1996	1997	1998	1999
HEAD COUNT						
WAREHOUSE CLERKS	6.0	6.0	7.0	7.0	8.0	8.0
ORDER ENTRY	2.0	2.0	2.0	3.0	3.0	3.0
BILLING	3.0	3.0	3.0	3.0	3.0	3.0
ACC'NTS RECEIVABLE/PAYABLE	4.0	4.0	5.0	5.0	6.0	6.0
INVENTORY CONTROL	2.0	2.0	2.0	3.0	3.0	3.0
SALES ANALYSIS	2.0	2.0	2.0	2.0	2.0	2.0
PURCHASING	1.0	1.0	1.0	2.0	2.0	2.0
PAYROLL	3.0	3.0	3.0	4.0	4.0	4.0
MIS	4.0	4.0	4.0	4.0	4.0	4.0
TOTAL HEAD COUNT	27.0	27.0	29.0	33.0	35.0	35.0
ANNUAL LAB. COST (K$)	$31.0	$31.9	$32.9	$33.9	$34.9	$35.9
WAREHOUSE CLERKS	$186	$192	$230	$237	$279	$287
ORDER ENTRY	$62	$64	$66	$102	$105	$108
BILLING	$93	$96	$99	$102	$105	$108
ACC'NTS RECEIVABLE/PAYABLE	$124	$128	$164	$169	$209	$216
INVENTORY CONTROL	$62	$64	$66	$102	$105	$108
SALES ANALYSIS	$62	$64	$66	$68	$70	$72
PURCHASING	$31	$32	$33	$68	$70	$72
PAYROLL	$93	$96	$99	$135	$140	$144
MIS	$124	$128	$132	$135	$140	$144
TOTAL INDIRECT COST	$837	$862	$954	$1,118	$1,221	$1,258

FIGURE 8-8 Current process—warehouse support.

departments should be expected since it is essentially a commitment to decrease their staffing. Often, decreases in head count due to automation are offset by increases for growth projections, such that head counts remain relatively unchanged even though shipping volumes increase.

The main focus is on determining the difference in cost between the current process and the proposed alternatives. Therefore, it is important to concentrate on those operations that are most likely to change if the proposed process is implemented. Examples of some areas where the warehouse management system can lead to indirect-labor cost reduction are described below. A WMS may require an increase in some areas, like MIS, which will offset some of these reductions.

Since a warehouse management system typically results in a significant reduction in paperwork, associated operations such as data entry, sorting, printing picking lists, report generation, filing, delivering, or copying can be reduced or eliminated. For example, most WMSs automatically print shipping documentation such as bills of lading, shipping instructions, manifests, and packing lists. In addition, the real-time information supplied by the WMS to all support departments eliminates the need for phone calls or faxes to and from the warehouse.

Once the indirect-labor differences between the current and proposed processes have been agreed upon, the labor rates from finance are multiplied by the head counts

ABC COMPANY
ALTERNATIVE 1–WAREHOUSE SUPPORT
INDIRECT SUPPORT

	1994	1995	1996	1997	1998	1999
HEAD COUNT						
WAREHOUSE CLERKS	6.0	2.0	3.0	3.0	4.0	4.0
ORDER ENTRY	2.0	0.0	0.0	0.0	0.0	0.0
BILLING	3.0	2.0	2.0	2.0	2.0	2.0
ACC'NTS RECEIVABLE/PAYABLE	4.0	3.0	4.0	4.0	5.0	5.0
INVENTORY CONTROL	2.0	0.0	0.0	0.0	0.0	0.0
SALES ANALYSIS	2.0	2.0	2.0	2.0	2.0	2.0
PURCHASING	1.0	1.0	1.0	2.0	2.0	2.0
PAYROLL	3.0	2.0	2.0	2.0	2.0	2.0
MIS	4.0	9.0	9.0	9.0	9.0	9.0
TOTAL HEAD COUNT	**27.0**	**21.0**	**23.0**	**24.0**	**26.0**	**26.0**
ANNUAL LAB. COST (K$)	$31.0	$31.9	$32.9	$33.9	$34.9	$35.9
WAREHOUSE CLERKS	$186	$64	$99	$102	$140	$144
ORDER ENTRY	$62	$0	$0	$0	$0	$0
BILLING	$93	$64	$66	$68	$70	$72
ACC'NTS RECEIVABLE/PAYABLE	$124	$96	$132	$135	$174	$180
INVENTORY CONTROL	$62	$0	$0	$0	$0	$0
SALES ANALYSIS	$62	$64	$66	$68	$70	$72
PURCHASING	$31	$32	$33	$68	$70	$72
PAYROLL	$93	$64	$66	$68	$70	$72
MIS	$124	$287	$296	$305	$314	$323
TOTAL INDIRECT COST	**$837**	**$671**	**$756**	**$813**	**$907**	**$934**

FIGURE 8-9   Alternative 1—warehouse support.

to determine indirect-labor costs for each of the alternatives. See Fig. 8-9 for a sample summary sheet of the indirect-labor costs for the ABC Company's alternative 1.

Other Costs.   Costs other than equipment, software, and labor can be combined in this category. Generally, these include facilities modifications, training, spare parts, and so forth. Some of these costs are one-time costs, and others are annual. For example, if a new three-level picking module is proposed, additional cooling and air handling may be needed to maintain acceptable working conditions on the upper levels. Also, roof structures may need reinforcing to support the weight of overhead conveyors. Power and air supplies may be required for proposed equipment. An environmentally controlled computer room with a raised floor might also be needed.

These miscellaneous costs can be significant for some alternatives and should not be overlooked. Repeatedly requesting additional capital for items that were missed during the justification process can cause project delays and raise management concerns. A sample summary sheet with a list of possible other costs is shown in Fig. 8-10.

Step 3: Calculate Savings

The benefits obtained from a successful automation implementation vary depending on the project scope and current operations. Figure 8-11 provides a reference list of savings frequently obtained with the implementation of warehouse automation.

Generally speaking, an outlay of investment capital at the beginning of an automation project yields cost savings after implementation has occurred. These savings are easily determined by subtracting the costs of the proposed alternative from those of the base or do-nothing alternative. Assuming the results are a cost reduction, they are the primary savings used to determine the return on investment. In many cases, good automation alternatives can be justified on tangible savings alone.

Because intangible savings involve more conjecture than tangible savings, they naturally contain more risk. Many potential savings require a commitment from management to help quantify expected gains. A thorough knowledge of the company's business, its competitors, and its customers is required. It is important to remember that small improvements can create large savings. For example, a 1 percent increase in market share can mean several million dollars in increased revenue.

Savings from an automation project can be separated into five main areas:

- Direct labor
- Indirect labor
- Inventory
- Space utilization
- Intangibles

To assist in the determination of the applicable savings, each area is described below. It is important to note that this discussion should serve as a general guideline for calculating savings and that each company's actual areas for savings will vary depending on the type of industry served and the level of automation being planned.

Direct Labor.   Direct-labor savings are calculated by subtracting the annual labor cost of an alternative from the annual labor cost of the current do-nothing process. Using the direct cost estimates from Figs. 8-6 and 8-7 for the ABC Company, the calculations in Fig. 8-12 apply. Repeat this step for each alternative versus the current process.

ITEM DESCRIPTION	INITIAL COST	ANNUAL COST
Heating		
Cooling		
Sprinklers		
Cabling		
Computer Room		
Power Supplies		
Air Supplies		
Battery Charging		
Spare Parts		
Spare Parts Storage		
Maintenance Labor		
Maintenance Area		
Totes		
Pallets		
Electrical Outlets		
Phones		
Roof Structure Modifications		
Fire Walls		
Fire Doors		
Offices		
Insurance		
Docks		
Lights		
Retraining		
Total Cost		

FIGURE 8-10 Other costs work sheet.

Indirect Labor. Indirect-labor savings obtained after implementation can be derived from all areas of the business that are affected by the automation project, including purchasing, inventory control, accounting, order processing, administration, and warehouse management.

Indirect-labor savings are calculated by subtracting the annual labor cost of an alternative from the annual labor cost of the current (do-nothing) process. Using the

Reduce	Improve
Excess inventory	Inventory accuracy
Misplaced or lost stock	Space utilization
Search and retrieve times	Throughput
Paperwork and forms	Order, lot, and expiration-date tracking
Human error	Order fill rates
Outside warehousing	Work planning and scheduling
Direct and indirect labor	Order cycle time
Overtime	Labor productivity
Supplemental labor	Equipment utilization
Backorders	Cross-dock
Taking physical inventory	Inventory turns
Freight costs	Stock rotation

FIGURE 8-11 Benefits of an automation project.

indirect cost estimates from Figs. 8-8 and 8-9 for the **ABC** Company, the calculations in Fig. 8-13 apply. Repeat this step for each alternative versus the current process.

Inventory. Increased inventory accuracy is a significant benefit resulting from warehouse automation. Inventory accuracy with a real-time warehouse management system can surpass the 99.9 percent level. Merchandise quantities received in the warehouse are immediately verified and a bar code label is applied, and from that point inventory is tracked 100 percent by the system. The tracking of all inventory movement by the bar code label virtually eliminates all entry errors and significantly reduces placement errors.

In addition, a warehouse management system ensures that warehouse procedures are completed as directed since the operator must follow the instructions presented and is not allowed to start a new task until the previous task has been validated. Operator errors associated with misplaced inventory are eliminated since both the inventory item and stocking location are scanned into the system and validated.

Since humans are involved, there is still the potential for inventory errors, such as an operator accidentally picking the incorrect quantity for an order. However, a systems real-time cycle counting function will catch these errors at the first opportunity. An inventory error identified during a cycle count is investigated by reviewing the past transactions on the system and determining where the problem occurred. The error can then be rectified with the customer, an adjustment is made, and the physical inventory again matches the system inventory quantity.

	1995	1996	1997	1998	1999
Current process total annual labor cost ($ thousands)	2884	3268	3703	4195	4753
Alternative 1 total annual labor cost ($ thousands)	1689	1913	2168	2456	2783
Annual savings ($ thousands)	1195	1355	1535	1739	1970

FIGURE 8-12 Direct-labor savings calculation.

	1995	1996	1997	1998	1999
Current process total indirect-labor cost ($ thousands)	862	954	1118	1221	1258
Alternative 1 total indirect-labor cost ($ thousands)	671	756	813	907	934
Annual savings ($ thousands)	192	197	305	314	323

FIGURE 8-13   Indirect-labor savings calculation.

The benefits of this achievable and maintainable inventory accuracy have a dramatic impact on the warehouse as well as the total business. Increases in inventory accuracy result in the following benefits:

1. *Increased productivity.*   Warehouse operators, supervisors, and receiving and shipping office personnel reduce time spent trying to resolve the problems caused by inventory problems. Personnel checking locations, reviewing paperwork, scouting the warehouse, and checking with other operators for missing inventory can be reduced.

2. *Reduction in order shipment shortages.*   If the merchandise cannot be located due to inaccurate inventory records, the customer order is changed to reflect the shortage.

3. *Reduction in backorders.*   When there are shortages due to inventory misplacement or quantity errors, the merchandise quantity that was needed is backordered. Lost sales can also be attributed to inaccurate inventory, since there are certain customers who will not accept a backorder.

4. *Decreased expediting.*   Some customers cannot wait to have backordered items shipped with their next shipment. Therefore, the unshipped portion of the order is expedited in the next day or two. This results in time and money spent tracking, processing, and shipping orders overnight.

5. *Decreased inventory.*   Companies without an automated WMS increase the safety stock levels held in the warehouse in an effort to minimize problems associated with inventory errors such as shipping shortages, backorders, and expediting. They are essentially planning for inefficiencies. However, with improved accuracy, safety inventory levels can be minimized, resulting in lower carrying costs and space requirements. For example, if an average inventory is $10 million, a 5 percent reduction yields a $500,000 one-time cash savings, in addition to annual space, labor, and damage savings. A WMS also reduces the lead time between when the inventory on hand reaches the reorder point and the reorder is received by the supplier of the item.

6. *Minimized physical inventory disruptions.*   Taking physical inventory is a necessary but disruptive procedure with batch or manual systems. Because the system inventory updates lag the physical inventory movement within a batch environment, a physical inventory requires work-flow stoppage in an area or sometimes the entire facility. Downtime is measured in days while inventory is counted, verified, and updated in the system.

A real-time warehouse management system enables the concept of rotating inventory counts, i.e., cycle counting. In a cycle count system, the physical inventory process is an integral part of the ongoing operations since real-time system updates allow concurrent counting and processing. The DC personnel are directed to verify the contents in selected inventory locations. The cycle count frequency is adjusted within the system to suit business needs. It is typically determined by the activity rate, unit costs, and the cost associated with an incorrect inventory count. For example, there is a greater benefit in checking stereo equipment than that in checking a carton of socks. Benefits resulting from real-time system cycle counting include:

1. Elimination of downtime to complete physical inventories. Cycle counting can pay for itself by eliminating the traditional annual or semiannual physical inventory.

2. Enhanced tracking of high-dollar or high-risk inventory by identifying these as high-frequency cycle count items.

3. Reduced downtime to resolve problems due to early identification of inaccurate inventory records (for example, cycle counts automatically generated if the system record goes negative).

4. Fewer "short ships" because order fulfillment is created from a highly accurate inventory file.

5. Productivity improvements because personnel spend their time distributing the merchandise instead of searching for it.

Space Utilization. Warehouse expansion can be delayed or avoided entirely by implementing an automated warehouse management system. Space savings are achieved through increased inventory accuracy and system-directed put-away, rewarehousing, and cross-docking operations.

If inventory records are inaccurate, the system does not have an accurate view of available storage space for incoming inventory. Empty locations are bypassed as put-away candidates by a system that inaccurately reflects inventory. In addition, as discussed previously, inventory accuracy leads to safety stock reductions, and thus less storage space.

System-directed put-away activity enables higher storage densities. The attributes that are described in designing the put-away module (Chap. 5) assist in maximizing storage utilization and optimization. Automated systems are capable of storing individual products with different lot numbers, expiration dates, and inventory dates (FIFO) in the same location and still maintain the product attributes separately. This significantly increases storage volume utilization over manual warehouse operations that must use separate locations to maintain separate product attributes. In addition, real-time put-away systems maximize the location usage rate since there is visibility to available storage locations as soon as a product is picked, versus systems that must wait for batch system updates.

Rewarehousing, or *profiling,* is a term used for the warehouse reorganization logic used in automated systems that identifies partially utilized locations that can be efficiently combined. The WMS directs the operator to reorganize the inventory in order to create new empty locations for incoming merchandise.

A WMS with real-time visibility to both incoming merchandise and customer orders promotes a cross-docking operation where merchandise is received and staged but never put in a storage location. Orders are completed by picking or moving the product directly from the staging area for shipment. Cross-docking reduces storage requirements and operator and equipment time associated with double handling.

Space utilization costs are significant and should be included in the return on investment (ROI). After determining the amount of floor space that can be saved, the following methods are used to calculate the cost savings:

Building cost. The estimated floor-space area saving is multiplied by a construction and fit-up (furniture and fixtures) rate. The resultant one-time floor-space saving is used when calculating the justification.

Occupancy cost.   Floor-space savings are multiplied by the costs of occupying warehouse space, such as heating, air conditioning, gas and electric, insurance, maintenance, and so forth. A good estimate can be obtained from the finance department by using the current cost for occupying the warehouse and calculating a cost per square foot.

Cost of money.   If a warehouse expansion or new building is delayed for a certain number of years, the averted after-tax mortgage payment is claimed as potential savings during the years the building costs were delayed.

Lease avoidance.   If expansion plans include leasing space versus building, the lease cost per square foot and the in/out handling charges are reduced or eliminated.

Intangibles.   Depending on company policy, conservative estimates can be calculated for intangible savings and the return on investment can be calculated with and without these savings. Even if an investment can be justified with tangible savings alone, the intangible benefits should be listed to increase management's confidence and reduce perceived risk in other costs or savings. Although it is more difficult to calculate intangible benefits accurately, it would be wrong to ignore them and assume their value is zero.

Customer service.   Failure to consider service issues can result in loss of market share and reduced revenues. If the competition is delivering orders faster, with fewer quantities, invoice errors, and backorders, it follows that they will garner more business. Customers will delay payment of invoices until all shipping problems have been rectified. It is probably incorrect to assume that continuing the current process unchanged will maintain the current market share, selling prices, or product costs. In fact, the opposite is likely to be more accurate.

The company's distribution policy is an important influence on customer service. Industry trends reviewed in Chap. 1 show that customers are requesting more frequent shipments and smaller quantities, along with unique requirements (i.e. special labeling, kitting, and split-case picking).

Warehouse and distribution automation allows these new requests to be met more efficiently. Although difficult to calculate, factors to consider are increased revenue through new value-added services and by maintaining and increasing the customer base. In addition, customer and sales-force surveys can provide further insight into these calculations. If necessary, these benefits can be estimated and included in the ROI. Some important customer service factors that are affected by warehouse automation include:

- *Order cycle time.*   How long is an order in the system from the time received until it is shipped?
- *Inventory availability.*   Is the DC able to fill all the line items on the order?
- *Delivery reliability.*   Does the order arrive on time or as requested by the customer?
- *Order accuracy.*   Does the customer receive what they order?
- *Condition of merchandise.*   Does the order arrive undamaged, with the appropriate labels and markings?
- *Order-status information.*   Can the customer obtain the latest, up-to-the-minute status of their order?

Miscellaneous.   Also worth mentioning are various benefits that may or may not yield quantifiable savings (hard or soft), but should be considered. Material handling automation often improves employee safety and workstation ergonomics and can reduce absenteeism. Employee morale is often improved by warehouse automation. There is a sense of pride from having a state-of-the-art warehouse process. The different jobs created by the new process are perceived as better jobs with more responsibility. Management control and planning improve as a result of reports and data available from the WMS.

Step 4: Compare the Alternatives

Most companies use present worth (or net present value) and internal rate of return (or ROI). Many companies calculate the payback period, but less than 10 percent base their decisions entirely on it. These methods are discussed in more detail later in this chapter.

The method and assumptions used to calculate the return on investment must be agreed to by the finance department and upper management. There are various financial methods that exist, each with programs and work sheets available to assist in the calculation. The following are common methods used in the industry.

Payback Period. The payback period is simply the length of time required to recover an investment (the number of months or years it takes for savings to equal the costs). For example, suppose a $100,000 investment yields a savings of $10,000 per month. The payback period for this example is $100,000/$10,000 = 10 months.

Measuring the success of an investment using payback can be misleading because any savings that occur after the payback period and the time value of money are both ignored. However, because it is easy to understand and calculate, it is handy for high-level decisions and disqualification of alternatives at the beginning of a project.

The Time Value of Money. The concept of the time value of money is basic to understanding and calculating the return on an investment. Obviously it is better to have $100,000 today than $100,000 one year from today. From an investment point of view, if today's $100,000 is invested at an interest rate (say 10 percent), then after a year passes the investment is worth about $110,000. This illustrates the time value of money; to have a return an investment must pay back the investment *plus* interest. This is just as true for capital investments as it is for putting money in the bank; an investment should provide an adequate return or it should not be made.

Net Present Value and Internal Rate of Return. The concept of net present value (NPV) is important for justifying an automated system because it overcomes the shortcomings of "payback" by considering the time value of money and by allowing today's investment to be compared to savings that occur at a later date—over a period of time. Net present value is a simple concept; it identifies what tomorrow's money is worth today. In the preceding paragraph $100,000 was invested at 10 percent, which was worth $110,000 after one year. The NPV of the $110,000 one year in the future, at 10 percent interest, is $100,000. Simply put, next year's $110,000 is worth only $100,000 today.

The internal rate of return (IRR) is a way to calculate an interest rate. If an investment is made that yields an uneven or variable return (an example might be the labor savings during a period of increasing sales), IRR can be used to calculate the interest rate that would be equivalent to putting the money in the bank.

Both the IRR and the NPV use the same basic formula and inputs for calculation (the actual formula used to compute the IRR and NPV may be found in standard finance or engineering economy books). Computer spreadsheets contain functions for IRR and ROI. Each requires estimating the project life, annual net cash flow, and initial investment amount. The last input, or variable, required to evaluate an investment using IRR or NPV is the discount rate. The discount rate is the interest rate that is used in the NPV and IRR analysis and is the rate used to reduce future savings or costs to their equivalent present value.

Hurdle Rate. The *hurdle rate* is a company-determined discount rate that reflects the minimum acceptable rate of return on a company investment. It is sometimes referred to as the cost of capital. Hurdle rates typically range from 9 to 25 percent. Historically, the cost of capital over the last 40 years has averaged 13 percent, based on the mean annual returns of the stock market. The IRR calculation determines the rate at which future cash flows must be discounted to give a present value exactly equal to the value of the investment. If the IRR is greater than the hurdle rate, then the investment proposal should be accepted. If two projects are being compared against each other, then the one with the higher IRR is a better choice. Financial officers use the hurdle rate as a benchmark to eliminate unattractive investments. Some companies use a lower hurdle rate for warehousing projects because warehousing projects may have a lower risk than other projects competing for available capital dollars. For example, the risk of introducing a new product may be much higher than the risk associated with modernizing a warehouse.

NPV may also be used to evaluate a proposal. The NPV technique does not solve for a particular rate. Rather, the hurdle rate is used to discount future flows to their present value. These flows are then totaled and compared to the required investment amount. If the NPV of the future flows (savings) is greater than the present value of the investment, then the project is acceptable. If two projects are being compared, the one with the larger NPV is considered to be more attractive.

Depreciation. Depreciation is the cost of tangible assets, less prospective salvage value, that is written off annually on the accounting books. There are several methods for calculating this write-off, including straight line, declining balance, sum of the years digits, and sinking-fund depreciation. These methods all give different answers, and some are more aggressive (taxwise) than others. The most conservative method is straight line. Finance departments generally have a preferred method of depreciation and should provide input on this decision.

Cost-Savings Summary. The results from the previous steps are now used as input to obtain the return on investment. A cost-savings summary sheet is created for the current process, as well as each alternative, which documents the assumptions and lists the cost and savings estimates for each year through the design year. Lastly, a cost-savings summary sheet showing the difference between each alternative and the current process is created. Shown in Figs. 8-14 to 8-16 are examples of these sheets for the ABC Company.

These figures are then entered into the return on investment (ROI) spreadsheet to obtain the financial results. Note that the intangible savings are not included at this time. Figure 8-17 shows a completed spreadsheet.

These results of the cost and savings cash flow are calculated and displayed in the lower left-hand corner of this spreadsheet. The results are then entered on the benefit and justification summary sheet shown in Fig. 8-18. This process is used for all the alternatives that were selected for analysis.

ABC COMPANY
CURRENT PROCESS
COST/SAVINGS SUMMARY (K$)

	1995	1996	1997	1998	1999	2000	2001	2002
CAPITAL EQUIPMENT COST								
FORK TRUCKS	75	85	96	109	124	140	159	180
RACKS								
CONVEYOR								
WMS EQUIPMENT								
OTHER								
TOTAL CAPITAL	75	85	96	109	124	140	159	180
EXPENSE								
CONVEYOR MAINTENANCE								
CONVEYOR SPARE PARTS								
CONSULTANT STUDY								
WMS SOFTWARE/SERVICES								
TOTAL EXPENSE	0	0	0	0	0	0	0	0
ANNUAL LABOR COST								
DIRECT	2884	3268	3703	4195	4753	4896	5042	5194
INDIRECT	862	954	1118	1221	1258	1296	1335	1375
TOTAL LABOR	3746	4222	4821	5416	6011	6192	6377	6569
SAVINGS								
INVENTORY								
SPACE								
INTANGIBLES								
TOTAL SAVINGS	0	0	0	0	0	0	0	0

FIGURE 8-14   Cost-savings summary—current process.

Step 5: Perform Supplemental Analysis

The method described above is based on the assumption that all of the cost and savings cash flows for the project are known and certain. However, in most cases the actual amount and timing of these cash flows are estimated, and uncertainties exist. There are many factors that contribute to the uncertainties in the amount and timing of the estimates such as delivery, construction or installation delays, equipment changes due to unexpected bottlenecks, risk of integrating new equipment technologies, labor rates, and inflationary or recessionary pressures.

Usually, there are different levels of uncertainties and risk that are contained within the project. Therefore, sensitivity analyses are used to help identify the effects of uncertainties on the outcome of an alternative. "What if" questions are generated, and the economic analysis is recalculated. "What if" scenarios typically evaluate sensitivity to volume growth changes from what was initially estimated as well as sensitivity to fluctuations in interest rates and labor savings.

Another variable that can be evaluated is the intangible savings and their effect on the ROI. If a project has a marginal payback, net present value, or rate of return,

ABC COMPANY
ALTERNATIVE 1
COST/SAVINGS SUMMARY (K$)

	1995	1996	1997	1998	1999	2000	2001	2002
CAPITAL EQUIPMENT COST								
FORK TRUCKS			96	109	124	140	159	180
RACKS	1030							
CONVEYOR	2100							
WMS EQUIPMENT	420		40		42		45	
OTHER	500							
TOTAL CAPITAL	4050	0	136	109	166	140	204	180
EXPENSE								
CONVEYOR MAINTENANCE		50	52	54	56	58	60	62
CONVEYOR SPARE PARTS	80	40	41	42	43	44	45	46
CONSULTANT STUDY	90							
WMS SOFTWARE/SERVICES	650							
TOTAL EXPENSE	820	90	93	96	99	102	105	108
ANNUAL LABOR COST								
DIRECT	1689	1913	2168	2456	2783	2866	2952	3041
INDIRECT	671	756	813	907	934	962	991	1021
TOTAL LABOR	2360	2669	2981	3363	3717	3828	3943	4062
SAVINGS								
INVENTORY	500	52	54	56	58	60	62	64
SPACE (LEASE AVOIDANCE)		14	28	42	56	70	70	70
INTANGIBLES	200	227	257	291	330	374	424	480
TOTAL SAVINGS	700	293	339	389	444	504	556	614

FIGURE 8-15   Cost-savings summary—alternative 1.

adding intangible benefits a little at a time will reveal when the project becomes acceptable. Since intangible benefits are difficult to quantify accurately, this method identifies the incremental amount that will make the financial analysis acceptable. If the intangible savings used in this step are a small portion of the potential savings, management will feel that the risk is minimal.

Upon completion of analysis of each of the alternatives, a sensitivity analysis summary sheet is created. An example for alternative 1 is shown in Fig. 8-19.

Step 6: Specify the Preferred Alternative

The last step in conducting a benefit and justification analysis involves the selection of the alternative plan the team will recommend. If the above-noted steps were followed, each alternative now has a cost-savings summary, an ROI spreadsheet, a benefit and justification analysis summary, and a sensitivity analysis summary. This provides the summary level analysis and supporting detail required for executive review.

ABC COMPANY
DELTA ANALYSIS
ALTERNATIVE 1 VS. CURRENT PROCESS
COST/SAVINGS SUMMARY (K$)

NOTE: OUTFLOWS ARE NEGATIVE, INFLOWS POSITIVE

	1995	1996	1997	1998	1999	2000	2001	2002
CAPITAL EQUIPMENT DELTA								
FORK TRUCKS	75	85	0	0	-0	0	-0	-0
RACKS	-1030	0	0	0	0	0	0	0
CONVEYOR	-2100	0	0	0	0	0	0	0
WMS EQUIPMENT	-420	0	-40	0	-42	0	-45	0
OTHER	-500	0	0	0	0	0	0	0
TOTAL CAPITAL	-3975	85	-40	0	-42	0	-45	-0
EXPENSE DELTA								
CONVEYOR MAINTENANCE	0	-50	-52	-54	-56	-58	-60	-62
CONVEYOR SPARE PARTS	-80	-40	-41	-42	-43	-44	-45	-46
CONSULTANT STUDY	-90	0	0	0	0	0	0	0
WMS SOFTWARE/SERVICES	-650	0	0	0	0	0	0	0
TOTAL EXPENSE	-820	-90	-93	-96	-99	-102	-105	-108
ANNUAL LABOR DELTA								
DIRECT	1195	1355	1535	1739	1970	2030	2090	2153
INDIRECT	191	198	305	314	324	334	344	354
TOTAL LABOR	1386	1553	1840	2053	2294	2364	2434	2507
SAVINGS DELTA								
INVENTORY	500	52	54	56	58	60	62	64
SPACE	0	14	28	42	56	70	70	70
INTANGIBLES	200	227	257	291	330	374	424	480
TOTAL SAVINGS	700	293	339	389	444	504	556	614

FIGURE 8-16 Cost-savings summary—difference analysis.

In the example for the ABC Company, alternative 1 is the best choice in economic terms. It has the highest NPV ($2,359,000) and IRR (43.3 percent) and the shortest payback (3.0 years). The payback period elapsed time starts when the project begins. However, the savings do not begin accruing until after implementation is complete. On major projects, implementation can last a year or more. Assuming a 1-year implementation for the ABC Company example, payback is really 2 years from the start-up of the new process.

Other factors can affect the final decision a company may make. Certainly, the capital budget is a primary consideration. The value of customer service improvements is another. Perceived risk by management can also enter into the final decision. All of these factors should have been considered and included, as far as possible, in the benefit and justification analysis process.

Figure 8-20 shows 10 principles, which, if followed, help ensure that all the fundamental requirements have been met by the analysis.

INPUT FACTORS :	ECONOMIC LIFE (YEARS) 7	TAX RATE 0.35
	GENERAL INFLATION 0.00	HURDLE RATE 0.20
	MONTH NO. EXPENSE BEGIN 2	SCRAP VALUE 0

RETURN ON INVESTMENT	INVESTMENTS (OUTFLOWS)	1995	1996	1997	1998	1999	2000	2001	2002
PROJECT NAME... ABC COMPANY	FORK TRUCKS	-75.0	-85.0						
ALTERNATIVE 1 VS. CURRENT	RACKS	1030.0							
WITHOUT INTANGIBLES	CONVEYOR	2100.0							
	WMS EQUIP/OTHER	920.0		40.0		42.0		45.0	
ANALYST.....									
DATE	SUB-TOT	3975.0	-85.0	40.0	0.0	42.0	0.0	45.0	0.0

TAXABLE SAVINGS OR REVENUE	INFLAT'N. ADJUST	1995	1996	1997	1998	1999	2000	2001	2002
				(Savings or Revenue items must have positive values.)					
INPUT VALUES REFLECTING	DIRECT LABOR	0.00	1195.0	1355.0	1535.0	1739.0	1970.0	2030.0	2090.0	2153.0
SAVINGS OR REVENUE	INDIRECT LABOR		191.0	198.0	305.0	314.0	324.0	334.0	344.0	354.0
IMPROVEMENTS.	INVENTORY		500.0	52.0	54.0	56.0	58.0	60.0	62.0	64.0
(Assumed to be Positive)	SPACE			14.0	28.0	42.0	56.0	70.0	70.0	70.0
	INTANGIBLES		0.0	0.0	0.0	0.0	0.0	0.0	0.0	0.0

		1995	1996	1997	1998	1999	2000	2001	2002
	DIRECT LABOR	1195.0	1355.0	1535.0	1739.0	1970.0	2030.0	2090.0	2153.0
EXTENTIONS OF THE ABOVE	INDIRECT LABOR	191.0	198.0	305.0	314.0	324.0	334.0	344.0	354.0
VALUES INCLUDING THE	INVENTORY	500.0	52.0	54.0	56.0	58.0	60.0	62.0	64.0
INFLATION IDENTIFIED	SPACE		14.0	28.0	42.0	56.0	70.0	70.0	70.0
	INTANGIBLES								
	SUBTOT.	1886.0	1619.0	1922.0	2151.0	2408.0	2494.0	2566.0	2641.0

		1995	1996	1997	1998	1999	2000	2001	2002
	CONVEYOR MAIN	12.5	50.0	52.0	53.0	55.0	56.0	58.0	60.0
ESTIMATED EXPENSE (INVESTMENT)	CONV. SPARE PART	80.0	40.0	41.0	42.0	44.0	45.0	46.0	48.0
OR OTHER ITEMS FOR WHICH ANY	CONSULTANT STUD	90.0							
INFLATION FACTOR HAS ALREADY	WMS SOFTWARE/SER	650.0							
BEEN INCLUDED.									
(Assumed to be negative)									

DEPRECIATION ITEMS	DEPRECIATION	568.0	568.0	573.7	573.7	579.7	579.7	586.1	18.1
(Assumed to be negative)									

RESULTS	SUB-TOTALS	-1400.5	-658.0	-666.7	-668.7	-678.7	-680.7	-690.1	-126.1
PAY BACK......... 3.0 YEARS	TOTAL TAXABLE	485.5	961.0	1255.3	1482.3	1729.3	1813.3	1875.9	2514.9
IRR .... 43.3 %	INCOME TAX EXPENS	169.9	336.4	439.4	518.8	605.3	634.7	656.6	880.2
IRR START EST. 60%	AFTER TAX CASH FLO	883.6	1192.7	1389.6	1537.2	1703.7	1758.3	1805.4	1652.8
NET PRES. VAL.. 2359.0	TOTAL CASH FLO	-3091.4	1277.7	1349.6	1537.2	1661.7	1758.3	1760.4	1652.8
	DISCOUNT CASH FLO	-3091.4	1064.7	937.3	889.6	801.4	706.6	589.6	461.3

FIGURE 8-17 ROI spreadsheet.

Alternative 1 vs. Current Process		Alternative 2 vs. Current Process	
Capital	$4,102,000	Capital	$4,502,000
Payback	3.0 years	Payback	3.1 years
IRR	43.3%	IRR	38.1%
NPV	$2,359,000	NPV	$2,044,000
First-year expense	$1,401,000	First-year expense	$1,500,000
Total savings	$17,687,000	Total savings	$18,087,000
Total after tax	$11,923,000	Total after tax	$12,118,000
Alternative 3 vs. Current Process		Alternative 4 vs. Current Process	
Capital	$2,652,000	Capital	$1,702,000
Payback	4.1 years	Payback	5.4 years
IRR	28.9%	IRR	9.3%
NPV	$505,000	NPV	−$429,000
First-year expense	$1,250,000	First-year expense	$693,000
Total savings	$7,456,000	Total savings	$4,350,000
Total after tax	$5,370,000	Total after tax	$2,245,000

FIGURE 8-18   ABC Company benefit justification and analysis summary.

Case I—25% Increase in Capital Cost		Case II—25% Decrease in Labor Savings	
Capital	$5,128,000	Capital	$4,102,000
Payback	3.8 years	Payback	4.2 years
IRR	26%	IRR	24%
NPV	$792,000	NPV	$385,000
Case III—25% Increase in Capital Cost, 25% Decrease in Labor Savings		Case IV—Case III with Intangible Savings	
Capital	$5,128,000	Capital	$5,128,000
Payback	4.8 years	Payback	3.8 years
IRR	15%	IRR	26%
NPV	−$625,000	NPV	$788,000

FIGURE 8-19   ABC Company sensitivity analysis summary for comparison of alternative 1 and the current process.

1. Challenge and justify company philosophies and assumptions:

   • Rethink and question every assumption.
   • Review the logic underlying the categorical elimination of certain warehousing alternatives.
   • Analyze the cost and benefit of all predispositions and "favored" approaches.

2. Analyze every warehousing alternative:

   • Diligently compare every feasible warehouse option, applying the same level of scrutiny and detail to each option.
   • Goal: find the *very best* option.

3. Include all costs:

   • Use the same level of cost-determination rigor for every alternative.
   • Full costing of each alternative assures an apples-to-apples comparison and a better decision.

4. Make every effort to quantify the "service effects" of warehousing alternatives:

   • Develop methods to relate service enhancements to changes in revenue.
   • When appropriate, include incremental revenue from service enhancements in the rate-of-return calculations.
   • Challenge unreasonable service-revenue assumptions.

5. Recognize "soft" dollar and "hard" dollar impacts:

   • Investigate all opportunities for soft savings—the effect of changes in the warehousing network extend beyond precisely measured changes in storage, handling, and transportation costs.
   • Include revenues that can be directly and logically associated with enhanced warehouse capacity.
   • Challenge the inclusion of excessive or unjustified revenue increments.

FIGURE 8-20   Ten fundamental principles for effective warehousing financial analysis.[1]

REFERENCE

1. "The Financial Evaluation of Warehousing Options: An Examination and Appraisal of Contemporary Practices," Warehousing Education and Research Council, Oak Brook, IL, 1988.

BIBLIOGRAPHY

Cassimatis, P., *A Concise Introduction to Engineering Economics,* Chapman & Hall, London, 1988.

Kearney, A. T., Inc. Staff, *Measuring and Improving Productivity in Physical Distribution,* Council of Logistics Management, Oak Brook, IL, 1984.

Thuesen, H. G., Fabrycky, W. L., and Thuesen, G. J., *Engineering Economics,* 5th ed., Prentice-Hall, Englewood Cliffs, N.J., 1977.

White, J. A., Agee, M. H., and Case, K. E., *Principles of Engineering Economic Analysis,* Wiley, New York, 1977.

6. Compare warehouse ownership options to third-party options on the basis of return on investment:

- Assess the full costs of each option.
- Relate dollar savings (with ownership) to the level of assets committed to the facility.
- Calculate the rate of return.

7. Develop decision rules for comparing warehousing options that have unequal levels of service:

- Is a warehouse option with "X" percent higher fill rates equal to "Y" percentage points of return on investment?
- Treat differences in service levels in a more objective manner.

8. Do not confuse bookkeeping and accounting rules with cash-flow effects:

- Accounting methods established for external consumption or tax purposes are not always appropriate for capital budgeting analysis; e.g., a capitalized lease is recorded on the balance sheet, but on a cash-flow basis, the lease payment is an annual expense.
- Focus on incremental cash flows.
- Exclude sunk costs.

9. Use post-audits to evaluate asset management performance:

- Use independent auditors from other functional areas to evaluate the realized rate of return.
- Assess the warehouse investment performance against forecasted performance.
- Use the resulting information to improve future estimates and decisions.

10. Periodically review current warehousing approaches:

- Question and challenge existing warehouse systems on a regular basis.
- Review rates of return and operating performance to ensure conditions are unchanged and previously targeted objectives are being achieved.

FIGURE 8-20   (*Continued*)

CHAPTER 9

DESIGN PHASES

INTRODUCTION

After the business goals have been established, the warehouse automation strategy is defined, and the project team is assembled, the next phase in implementing the strategy is the design specification development. Design specifications for a warehouse automation project can be divided into two categories: system and material handling.

Prior to entering this design phase, the decision of whether to purchase an existing package or develop a WMS from scratch has been made. Purchasing a WMS package usually entails minor to average modifications and frequently requires new functions that are unique to the business environment. Development of a WMS from scratch has obvious design requirements. Thus, in both scenarios, some level of design specification is required though the package environment, assuming quality documentation, provides a more solid starting point.

Regardless of which scenario a company selects for implementing a WMS, system and material handling design specifications are needed. The system specification focuses on the WMS software. In a structured implementation environment, the first phase incorporates an external design and analysis to ensure that the programming group, or WMS supplier, has a firm understanding of system requirements. The second phase revolves around the internal (or detailed) design and analysis aspects. New system development techniques are available, such as object-oriented design and analysis, which alter how these phases are conducted and improve the long-term system maintainability.

The material handling (MH) design specification provides additional design background and sample tables to support the WMS Requirements Report MH analysis presented in Chap. 6. The detailed design process refines the proposed MH equipment or hardware and ensures that it is properly implemented for the warehouse's environment. Interfaces for warehouse control subsystems (WCSs) that may be required are finalized during this process. The MH detailed design phase is conducted simultaneous to the system design to ensure that both halves are integrated to optimize warehouse operations.

OVERVIEW

This chapter presents commonly used methodologies for developing design specifications. These methodologies also provide a mechanism to assist in evaluating the quality of potential WMS package suppliers. The components of a typical design specification are reviewed and examples presented. A structured (or functional/data) design approach for developing a system specification is compared to an object-oriented design approach. Finally, these design approaches are compared and an alternative (hybrid) approach is presented.

The MH design process analyzes MH equipment requirements in two categories: mobile and fixed position. Mobile equipment (fork lifts, AGVs, and so forth) is usually purchased to fit within the warehouse configuration and requires little or no design modification. Fixed-position MH equipment is subcategorized as being static or dynamic. Static MH equipment (such as pallet racking, shelving, and so forth) is directly controlled by the WMS. Dynamic fixed-position MH equipment (such as conveyors, ASRSs, pick-to-light systems, and so forth) requires the most design time due to the added complexity of having a WCS to contain equipment-specific logic and to interface with the WMS.

EXTERNAL DESIGN PHASE

The purpose of the external design phase is to formulate, in detail, how the WMS functions will support the current and future business requirements. The prerequisite for beginning this phase is that the proposed process flow is finalized. The proposed process flow, discussed in Chap. 6, provides the starting point for the external design. The external design refines, expands, and documents the proposed process flows into discrete WMS functions. Figure 9-1 shows the WMS functions required to support the

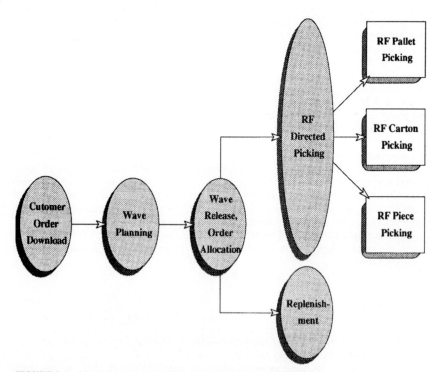

FIGURE 9-1   Block diagram of WMS functions required to support customer order picking.

process for picking a customer order. This section reviews methodologies commonly used to develop an external design, presents an example of an external design, and discusses planning considerations for developing an external design.

External Design Approaches

Two common approaches to developing an external design are structured (or functional/data) and object oriented. The traditional structured design approach utilizes a top-down methodology to document requirements. System rules and exceptions are specified as part of a function's behavior, and references are made to system files and data needed for support and definition purposes. The object-oriented design approach treats objects (or data and behavior) as the basis for system functionality. Rules and exceptions are incorporated into the system and defined by the data elements that they affect.

The structured analysis and design approach stresses functional decomposition and organizing the system's functionality around functional procedures. The object-oriented analysis and design approach organizes the system's functionality in terms of conceptual and/or real-world objects that exist within the system's domain. A brief description and comparison of these two design methodologies is presented, along with a hybrid methodology on how these methodologies can coexist. Further detailed information on the definition, history, and programming impact of the above-noted design methodologies can be obtained in the references listed at the end of this chapter.

Structured Design Methodology

The structured design methodology is similar to the traditional "waterfall" model of system development; see Fig. 9-2. The classic waterfall model begins by creating a requirements definition for the system. Once the system requirements are known, they are analyzed to determine the system functions that are required. These functions are then taken to the design phase, which further defines the rules and exceptions, and the data needed to support these requirements. After design, the implementation phase defines the architecture and develops the application code. This process continues through the integration, testing, and maintenance phases where each phase completes and "flows downstream" to the next phase.

The structured external design methodology follows a series of steps similar to the waterfall project philosophy and starts the functional definition in external, or user, terminology. This begins with high-level process flows, which define the system's functional requirements and are then decomposed into system functions. Subfunctions are extracted from higher-level functions when the design team recognizes a common set of reusable subfunctions. The system functions and subfunctions are then defined in terms of behavior rules and exceptions. Finally, the file and data structures are identified for the purpose of further defining the rules and the information required to support system functions.

The structured design approach provides the ability to plan the design phase with discrete milestones associated with each function's design completion. A firm design plan helps keep the design process moving forward and trackable. The structured design approach also provides a traditional method of problem solving and problem decomposition that allows easy acceptance by the design team. The design team's acceptance and the implementation (or programming) team's experience with structured techniques provide the shortest implementation schedules in today's environment.

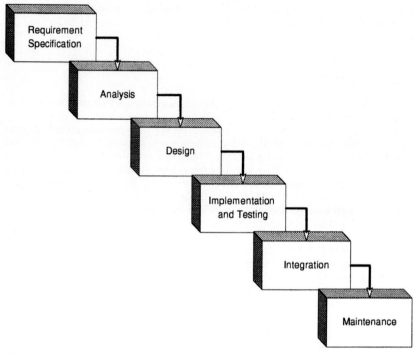

FIGURE 9-2   Waterfall model for a traditional project approach.

The structured design approach, however, has the inherent problem of "flowing upstream" when viewed in waterfall terms. This is where the basic waterfall assumptions begin to deteriorate. When the design team reaches the data level in a function's definition, changes may be uncovered that require restructuring of the function's design or the addition of a new function. This problem of flowing upstream becomes more apparent when cross-functional changes are required, such as customer order picking requiring a change to receiving, for example, due to the requirement for an additional inventory attribute. Many times these cross-functional changes, if recognized, are absorbed into the downstream functions to avoid delaying the design process. Usually, changes recognized during the design process can be managed if the overall waterfall philosophy to the project is followed. Downstream design changes have the greatest project impact if the implementation phase begins on upstream functions prior to design completion for the entire system.

Object-Oriented Methodology

The object-oriented methodology evolved from the need to manage large integrated systems. This methodology extends beyond the bounds of external design and addresses the analysis and design for the business domain being addressed. This methodology addresses the need to increase code reuse and to decrease the complexity of system maintenance. The object-oriented approach focuses on data decomposition and eliminates arbitrary data access in an effort to resolve the fundamental problems that are associated with the structured methodology.

The object-oriented approach requires a paradigm shift from traditional software development concepts. The concept of object orientation is to promote system stability by associating system behavior rules and exceptions with the data, which is considered the system's most stable element. This concept promotes the reuse of behavior rules by defining services that an object provides. Where the structured approach begins by considering the desired system behavior and migrates to data definition, the object-oriented approach begins by considering the system's behavior as integrated with data.

Several theories and methodologies exist on the process of developing an object-oriented design. A commonly used first step in the object-oriented approach is object modeling. *Object modeling* is the identification of objects and their relationships. Many objects, generally "real world," are easily identified as nouns within the process requirement statements. Examples of real-world objects include a fork truck, PO, location, SKU, customer order, and so on. Other objects, usually "conceptual," require more investigation and are identified throughout the design process. Examples of conceptual (or abstract) objects include a receipt transaction or customer shipment record that may not physically exist but is required to support the system.

The relationships between objects are used to define how these objects interact to meet the system requirements; see Fig. 9-3. Object-oriented design brings with it a new set of terms and definitions. The following is an abbreviated list that describes some common terms associated with the object-oriented technology:

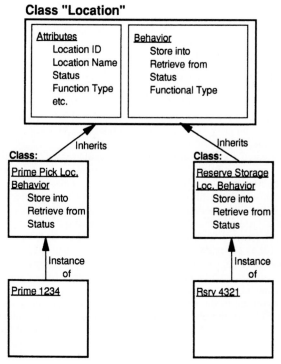

FIGURE 9-3   High-level view of class objects and object relationships.

- *Object (or instance).* A real-world, or conceptual, entity that has a unique identification, state, and behavior. An example of an object would be a stocking location, customer, customer order, item (or SKU), and so forth.
- *Class.* An abstraction of a set of objects that have common behavior or structures. An example of a class is a location that has common behavior for different location types (i.e., drive-in, push-back, selective, and so forth).
- *Method (or service).* A set of logical operations that are provided by an object. Add, change, and delete are services common to most objects. Another example of a method is status control, such as locked or held, which relates to inventory and location objects.
- *Instance.* An object belonging to a specific class. A replenishment location is an instance of the location class.
- *Inheritance.* A class relationship that allows a class to adopt the characteristics of a superclass (or parent). This concept allows for reuse and commonality and expresses the principles of generalization and specialization. An example of inheritance is a location superclass, which is a generalization that allows a class (or child) to inherit its characteristics. A replenishment location class is a specialization of a location and will inherit established characteristics, such as a location address and zone, common to all locations.
- *Encapsulation.* A principle of information hiding that deals with separating the external aspects of a subject (entity) from the implementation (internal) details of that object. The external aspects are visible to the outside world, while the internal aspects are hidden. The external aspects of the "locator" object provide the method of finding a location for inventory, when requested from put-away activity, for example, and returns the location found. The internal aspects of how the location was found are not known to the calling object.

Objects and methods identified during the modeling phase are continuously refined and carried through the analysis and design phases, and can be found in the programming (or construction) and implementation phases. This regimented bottom-up process reduces the massive testing phases associated with the structured methodology because an object's integrity is continuously validated from analysis and design and through construction.

The object-oriented approach utilizes data encapsulation to ensure that an object's data can only be accessed by the object's own functions. This improves the system's integrity by reducing the sources of how data is manipulated. The object-oriented approach promotes code reuse through the use of object classes. Objects within a class inherit the class' behavior. Services offered by a class can be easily accessed where needed through the use of messages. The extensive analysis and design time associated with the object-oriented methodology results in simpler and faster construction and implementation for an experienced object-oriented programming team.

The object-oriented approach requires a different mindset toward system development. Difficulties often arise in adapting from the concept of basic top-down problem solving to the concept of starting with the basic objects that constitute the problem domain and working up to the operations (functions) that will interface with the end user. The task of identifying objects is very involved and requires increased analysis and design over traditional approaches. This can lead to frustration in the project cycle when results associated with the implementation cycle are not realized until close to the end of a project (reference the external design plan). Additional project overhead must be accounted for in the project plan when an organization is switching to the object-oriented approach.

Hybrid Methodology

The object-oriented design approach bypasses the identification of functions associated with the structured design approach. This is done to reduce the interpretation of requirements and the potential loss (or overlooking) of information. Adopting a pure object-oriented approach requires commitment from the entire project team starting at the project's inception.

However, implementation of a WMS requires the involvement of a large and diverse project team. Acceptance of concepts and terminology associated with the object-oriented approach by the entire project team can be time consuming and difficult. The top-down functional identification and decomposition associated with the structured approach is often more easily understood and accepted.

The concept behind a hybrid methodology is to realize the code reuse and maintainability benefits associated with object-oriented design techniques during the project's implementation (programming) phase, while realizing the early acceptance and participation by all team members associated with structured design techniques. This is done by limiting the number of team members that must adopt the object-oriented methodology to the programming design and construction staff (or vendor). The system requirements are refined and behavior rules are identified using the structured method to develop an external design. Figure 9-4 depicts the hybrid design methodology.

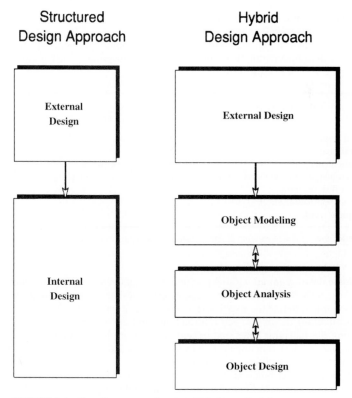

FIGURE 9-4   Transition process for a hybrid design methodology.

The external design is developed by describing what a system must do by identifying the functions that constitute the system. After the external design is complete, it is passed to the construction phase through the use of an internal design where the object-oriented design techniques are used. The programming team continues the use of object-oriented techniques to construct the operations that will meet the functional requirements identified in the external design. The user interfaces, or screens, associated with a structured external design are redefined as a result of object-oriented analysis and design. For this reason, the external design is created such that the project design team avoids defining screen flows and screen formatting so the programming team has flexibility for implementing objects that will ultimately define the user interfaces.

EXTERNAL DESIGN EXAMPLE

The purpose of the external design is to refine the requirements from the proposed flow process charts and to document the behavior rules required to meet the system's objectives. The external design format depends on the design methodology being used. This section highlights the various design methodologies' informational requirements and gives a sample format for an external design.

The format for an external design can vary significantly depending on the implementation team. It is incumbent on the project leader to identify the external design format and to ensure that the design team adequately describes each function according to the format's specification.

The table shown in Fig. 9-5 illustrates format elements commonly specified within an external design. As computer-aided software engineering (CASE) approaches are

External design format element	Structured	Object-oriented	Hybrid
Introduction and purpose	X	X	X
Operational narrative	X	X	X
Requirements outline	X	X	X
Operational process flow diagram	X	Optional	X
Screen flow diagram	X		X
Screen specification sheet	X		X
Sample screens	X		

FIGURE 9-5   Table of external design format elements comparing design methodologies.

utilized, the detail required within an external design decreases. Object-oriented techniques frequently do not recognize a true external design since they focus on the data and behavior that define the user interfaces and operations (or functions). CASE technologies allow for flexible screen reformatting, and the external design is focused on identifying the behavior rules and exceptions. Since the structured external design approach decomposes functions down to the support data required, a more extensive and detailed design specification is required.

A RF pick by order function is used to illustrate an external design format. The following specifications (see Figs. 9-6 and 9-7) highlight the object-oriented and hybrid design methodologies' departure from this structured external design format.

RF Order Picking

Introduction

The RF picking module is a real-time method for transferring product from storage locations to the shipping area. It directs the user, via RF terminals, to move the product from the storage locations to pick staging areas or shipping lanes, consolidated picking drop zone areas, or packing stations. A picker can perform all picking for an order or selectively pick portions of an order by unit of measure (full pallet, carton, or piece), or pick only portions of an order within a warehouse pick zone (or for a particular equipment type).

Operational Narrative

A customer order is released by the order processing clerk, who will allocate inventory and create picking queue instructions for RF picking. The picking queue will identify the location and movable unit required by an order along with the picking zone and equipment type specified by the location.

A warehouse associate is assigned to the picking operation by the shipping supervisor. The picker will log on to the WMS, which will present the picker with the RF picking menu. The picker identifies the warehouse zone, the equipment type being used, and whether the full pallet, carton, or piece picks should be presented. The WMS determines the highest priority and the oldest order that has not been assigned for picking that meets the requested pick zone and equipment type.

After the WMS determines the pick queue entry to work, the assigned picks are restricted from use by another picker. The WMS directs the picker to the first location to pick by the location's address label sequence. The WMS also displays the SKU, quantity, and movable unit to pick. The picker travels to the directed location and scans the directed movable unit. If the directed movable unit is not available, the picker can place the location on cycle count, or scan an alternate movable unit. If the location is put on cycle count, the WMS attempts to re-allocate picks for the location and direct the picker to the next pick location. If an alternate movable unit is scanned, the WMS ensures the movable unit meets the customer's order requirements and swaps the movable units. An error message is displayed if the movable units cannot be swapped or the unit is not in the directed location.

After the WMS confirms the movable unit being picked, the picker is requested to confirm the SKU and enter the quantity picked. After the quantity is entered, the picker is requested to scan a movable unit that the customer order is being picked onto if the pick is for a carton or piece quantity. If the location is picked empty, the picker is asked to confirm that the location is empty and performs a cycle count on the location.

After the pick is complete, the picker is directed to the next location for the customer order. If no picks remain, the picker is directed to drop the picked movable unit in ship staging or packing (if only picking a portion of the customer order). The picker can request to drop the picked movable unit if it becomes full. All directed and nondirected picks are logged into a picking productivity file for performance reporting.

FIGURE 9-6   RF order picking sample external design format, introduction, and operational narrative.

Operational Process Flow Diagram

The operational process flow diagram is used to illustrate the system and person interactions that are documented in the operational narrative. Figure 9-8 illustrates an abbreviated process flow chart for how a warehouse picker is directed by the WMS to pick and stage an order for shipping. Note that both the hybrid and structured design methodologies contain process documentation up to and including this type of dia-

---

Requirements Outline

RF picking is summarized in the following list of functions:

1. Pick full pallets, full cartons, or pieces.
2. Display special picking instructions with the first pick of order.
3. Submit a location to the cycle count queue and automatically reallocate inventory for pick.
4. Generate cycle count transactions and/or discrepancies for locations picked empty.
5. Deliver units to shipping lanes, consolidated picking drop zone areas, packing stations docks, or drop zones.
6. Change the lane assignment for an order if the current lane is not usable.
7. Call "after pick" interface programs if required for the location from which inventory was picked. Call an "after pick" interface if required for the carrier upon dropping the picked inventory into a staging location.
8. Restrict the eligible picks by selecting a particular order or wave to work.
9. Accommodate zone picking where the pickers remain in an area and the shipping unit is passed from one zone to the next.
10. Capture serial numbers, lot numbers, and/or expiration dates for picked items if necessary.
11. Capture productivity tracking information.

FIGURE 9-7   RF order picking sample external design format and requirements outline.

gramming. The diagramming technique is optional but recommended for the object-oriented design methodology.

The operational process flow diagram shows the interaction between the warehouse operator (M: denotes manual) and the WMS (S: denotes system). This diagramming technique represents the conceptual functional flow and gives the program design team a starting point for defining system interactions. Figure 9-8 shows the standard symbols used in an operational process flow diagram.

Screen Flow Diagram

The screen flow diagram is used to illustrate the WMS screens that are presented to a user in completing a task. Use of screen flow diagrams, rather than actual screen formats, has the advantage of not requiring specific hardware selection prior to designing functions since the selected hardware, especially for RF terminals, will restrict the amount of information that can be displayed on a single screen. Figure 9-9 illustrates the functions and screens used to support picking and staging a customer order for shipment. Note that screen flow diagrams are used to yield a "one-page snapshot" for a function's design. This diagramming technique is used to replace screen layouts to allow the program design team flexibility in creating screens.

The screen flow diagram is used primarily when a CASE programming tool is used. This diagramming technique is used for hybrid and structured designs. The box numbering refers to a unique "screen specification" sheet that outlines the screen for-

RF Order Picking Process

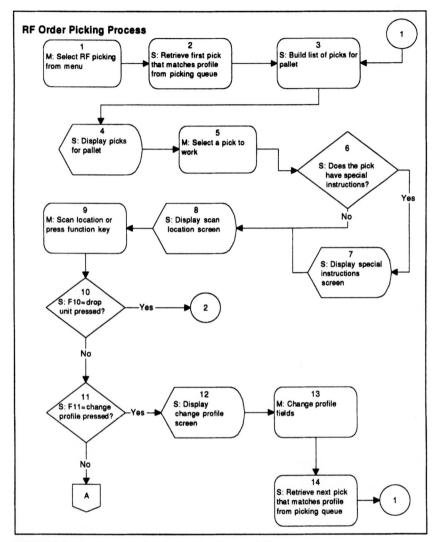

FIGURE 9-8 Sample process flow diagram for RF order picking.

mat and related behavior rules. The screen flow diagram represents the conceptual functional flow and gives the program design team a starting point, but does not represent a required design.

Screen Specification Sheets

The screen specification sheets are used in conjunction with the screen flow diagrams to define the support data and behavior rules associated with a screen. Figures 9-10

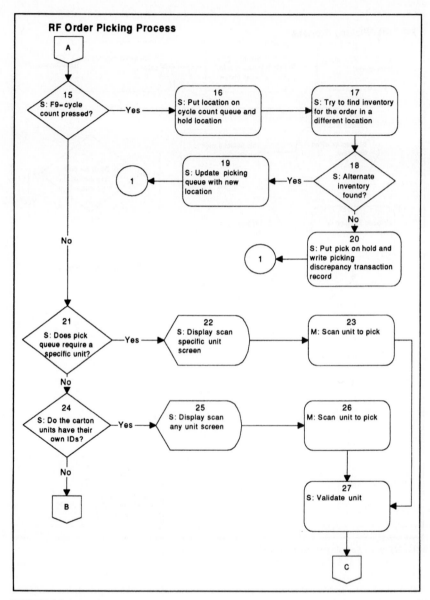

RF Order Picking Process

A

15
S: F9= cycle count pressed?

— Yes →

16
S: Put location on cycle count queue and hold location

→

17
S: Try to find inventory for the order in a different location

19
S: Update picking queue with new location

← Yes —

18
S: Alternate inventory found?

1

No

20
S: Put pick on hold and write picking discrepancy transaction record

1

No

21
S: Does pick queue require a specific unit?

— Yes →

22
S: Display scan specific unit screen

→

23
M: Scan unit to pick

No

24
S: Do the carton units have their own IDs?

— Yes →

25
S: Display scan any unit screen

→

26
M: Scan unit to pick

No

B

27
S: Validate unit

C

FIGURE 9-8 (*Continued*)

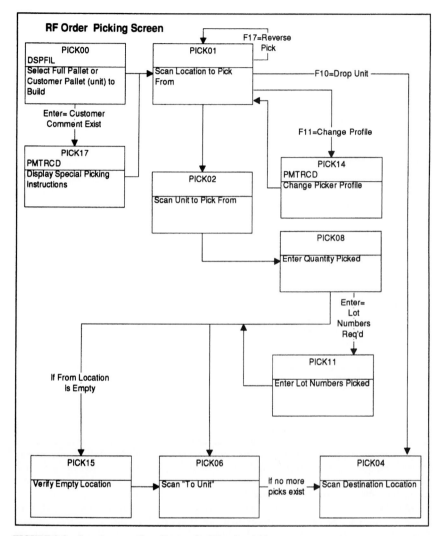

FIGURE 9-9 Sample screen flow diagram for RF order picking.

and 9-11 show screen specification sheets for two of the screen designators shown in Fig. 9-9. Behavior rules and exceptions are with the function key, option key, or data element as appropriate. General screen specifications are found under "Special Function Rules and Procedures" as required.

Sample Screens

Screen formats are only used if the structured approach is required. Object-oriented and hybrid methodologies do not use sample screens due to the probability of function

---

I. RF Picking – Select Pick for Pallet   **Screen Designator: PICK00**
This program displays a list of carton picks (sorted by cube) for a pallet ID. From this list, the user selects a pick to work.

II. Files.Fields
Subfile Fields:
        Pick Task.Quantity to Pick (display)
        Pick Task.Location (display, aisle bay only)
        Pick Task.Pallet ID (hidden, subfile restricted by pallet ID)
        Pick Task.Sequence Number (hidden, subfile sorted by sequence number)
        Pick Task.Cube (hidden, subfile sorted by cube within a sequence number)
        Item.Item Description (display)

III. Function Keys
F02=View picks in location sequence
F03=Exit
      - If a pallet has been started when F3 is pressed, PICK04 is displayed for delivery of the pallet before exiting the
      function.
Enter=Confirm
      - If this pick has special instructions, go to PICK17. Otherwise, go to PICK01.

IV. Selection Options
1=Select

V. Special Function Rules and Procedures
Initially, this program will retrieve the highest priority, oldest pick record from the Pick Task and will assign all picks for the
same pallet ID to the picker. This "pallet ID" corresponds to the internally generated pallet ID assigned during the pallet
building portion of wave processing. This screen displays a subfile of the pallet's picks. The user selects a pick to work and
control passes to PICK01.

The screen will display the quantity to pick, the location, and then the item description. On an RF screen, the description
field will overflow into the undisplayed portion of the screen. When the same screen is displayed on a truck mount, the entire
size description will show.

FIGURE 9-10   Sample screen specification sheet for screen function PICK00.

redefinition and redesign as the analysis and construction phases take place. When the external design format requires screen layout design, caution must be exercised not to overdesign the sequence of fields and their headings to avoid stalling the design process. Frequently, the design team's focus on defining behavior characteristics and the functions required to support the warehouse processes is lost while team members debate the optimal name for a column heading for a subfile display.

Specifying the screen format within the external design document requires advanced knowledge of screen sizes that are supported by the hardware vendors. Typical CRT screens typically support a 24×80 character format. RF screen sizes not only vary greatly between vendors, but also depend on the font that is chosen during terminal setup.

The example RF screens shown in Figs. 9-12 and 9-13 do not display options and function keys. The external design guides used for these sample screens specify that option and function keys for all RF functions can be viewed by selecting the F1 key, which displays help text for the screen. This is done to maximize the available display area for hand-held RF terminals, which are much smaller that the CRT screens. This example uses a 8×20 character format.

Figure 9-12 is a RF display format that prompts the picker to select an item to pick for a pallet that was assigned during order processing. The picker selects the item to pick by keying in a "1" and pressing the "Enter" key. This display allows the picker to select the best item to pick for building a "good" carton stack. Figure 9-13 is a RF dis-

---

I. RF Picking – Scan Location to Pick From   **Screen Designator: PICK01**

This function retrieves the first pick instruction that matches the picker's profile and displays the picking instruction to the picker, prompting for the location or reference ID to be entered only if the location does not track individual units. By scanning the location, or entering the reference ID, the picker verifies that he is at the correct location.

II. Files.Fields

Screen Fields:
Pick Task.From Location (display)
User Field.Location (input)
User Field.Location Reference (input)

III. Function Keys
F03=Exit
F09=Cycle Count
        - Put the picking location on the cycle count queue and hold the location. Try to find inventory for the pick in another location. If alternate inventory can be found, then change the Pick Task record and redisplay the pick. The alternate inventory must conform to the same lot number restrictions as the original pick. If alternate inventory cannot be found, then put the pick on hold, write a picking discrepancy transaction record, and display the next pick on the Pick Task.
F10=Deliver (only valid for carton or piece picking, go to PICK04)
F11=Change Profile (go to PICK14)
F17=Reverse Pick.
        - Reverse the order (ascending to descending bays or vice-versa) of picks within the current aisle.
Enter=Confirm (go to PICK02)

IV. Selection Options
(none)

V. Special Function Rules and Procedures
For pallet and piece picking, this function retrieves the first pick instruction that matches the picker's profile. The pick records will be sequenced by priority. For carton picking, the pick instruction for the size and quantity selected on PICK00 is displayed.

If special instructions exist for the pick, invoke PICK17.

The scanned location must match the Pick Task location.

        Special Case for Primary Pick Locations: If the picking location is a prime, then this program will check the quantity on hand in the prime to ensure that the entire pick quantity can be satisfied. If there is not enough in the prime and a replenishment move has been generated, this program will direct the picker to the replenishment's from location for the picking of the merchandise. A message will be issued to the picker if the replenishment task is currently being performed. The Pick Task record may have to be updated to reflect this change, and the replenishment record will be unavailable to other users until the pick is made. (If there is not a replenishment move, no changes will be made and the picker will be directed to the prime location.) The picker will pick as much as he can, and put the location on cycle count. The system will find inventory elsewhere in the warehouse for the remaining quantity to pick.

        Productivity Recording: Records are written to the productivity tracking file when the operator picks up a unit, delivers/drops a unit, and/or deviates from the normal picking process (i.e., skip a pick, put a location on cycle count).

FIGURE 9-11   Sample screen specification sheet for screen function PICK01.

play that directs the picker to a location and prompts the picker to scan (or enter) the location for validation. The picker can scan the location's bar code label or enter the location's reference identification (refer to Chap. 5 for definitions). This ensures the picker is at the correct pick location.

EXTERNAL DESIGN PLAN

The external design specification process typically begins at the point that inventory is introduced into the warehouse (i.e., receiving), and progresses through inventory leav-

```
Select Pick for Plt              .
PLT PLT1000
?
_   Loc AA1B2L3S3
    Qty 5 SKU AP362
_   Loc AA1B214S1
    Qty 1 SKU CK901
                                  +


    .
```

FIGURE 9-12   RF order picking sample screen format for PICK00.

```
Pick Queue From Loc         .

    AA1B2L3S3


Scan Loc: _____

or

Enter Ref ID: ____



    .
```

FIGURE 9-13   RF order picking sample screen format for PICK01.

ing the warehouse (i.e., order fulfillment and shipping). It is imperative that the external design team appoint a project lead who utilizes a schedule to ensure that the design team does not spend excess time within a particular warehouse function or process. The tendency of a design team is to expend an inordinate amount of energy and time designing receiving and put-away functions. This can cause the team to overlook picking and shipping requirements due to being rushed to complete the design on time or due to overall fatigue and diminishing focus.

The duration for developing a WMS's external design is dependent on the selected time methodology. A structured external design approach typically takes 20 to 30 percent of the overall project's time before internal design and programming can begin. An object-oriented analysis and design methodology (which includes the structured external and internal design phases) can take as much as 65 to 75 percent of the overall project's time. While the design phase for the object-oriented approach is longer than the structured approach, the construction (or programming) phase is proportionally shorter due to the direct use of information documented during the analysis and design phase and the improved function commonality.

The long design and analysis time period associated with the object-oriented methodology has yielded an alternative approach to provide quicker results. The alternative approach, sometimes referred to as a *spiral* construction, allows smaller groups of programs to be designed, built, and put into use prior to the whole system's completion. As more and more of these program groups are completed, they are coupled together to form the system.

It is important to keep in mind the primary reasons for choosing an object-oriented approach over the traditional structured approach. Object-oriented design and analysis are focused on long-term reduced maintenance and flexibility. If two experienced implementation teams, one structured and one object-oriented, start design and programming from scratch, the structured team will likely complete the system first. However, the object-oriented system's improved ability to fix problems and maintain the system may outweigh its initial shortcomings, especially if the system is planned for a 3- to 7-year life span.

The external design phase is initiated by organizing and orientating the project team. Orientation is used to inform all team members of the design process and how the external design specification will be documented and used. The orientation is especially important if new members, such as a system supplier or additional warehouse personnel, have been added to the project. The orientation process provides a current operations walkthrough, and a current process review. The duration of the orientation process typically ranges between 1 to 2 weeks.

After the design team completes orientation, they should jointly identify and review the functions required to support the requirements. This acts as an outline for the design process. Using the duration guidelines identified for the functions in Chap. 5, an external design development schedule is put in place at this time. An example of external design tasks is shown in Fig. 9-14.

An external design issues list and change list are started at the same time that the schedule is being developed. The issues list is used to identify work items that require further information and research or that must be implemented in another function. A person and target resolution date should be assigned to each issue added. When an issue is resolved, the answer, or solution, is recorded for future reference. The change list is used to track items that are out of the current design scope. Change-list items should be reviewed after the external design is complete to determine if other systems need to be changed or the current scope expanded.

Once team orientation and the external schedule are complete, the design team is ready to begin the process of creating an external design. This process begins with the creation of inventory (receiving) and stocking, and progresses through shipping. The time required to document a function's design depends on whether an existing package is being used or the function is being designed from scratch. In either case, standard operating procedure (SOP) changes are identified during each function's design so that the SOP can be documented and put in place while the WMS is being implemented.

After the initial pass of all functions has been completed, the external design first draft document is finalized and all open issues are resolved (either answered or deferred to the implementation phase). The design team is now ready to begin the design of host interface programs and conduct a design review. Design requirements for interfaces are well defined at this point by the functional design process. Timings for the interfaces to the host need to be resolved and the interface data stream finalized. The review process gives the design team the opportunity to identify commonality between functions, eliminate redundant processes, incorporate cross-functional require-

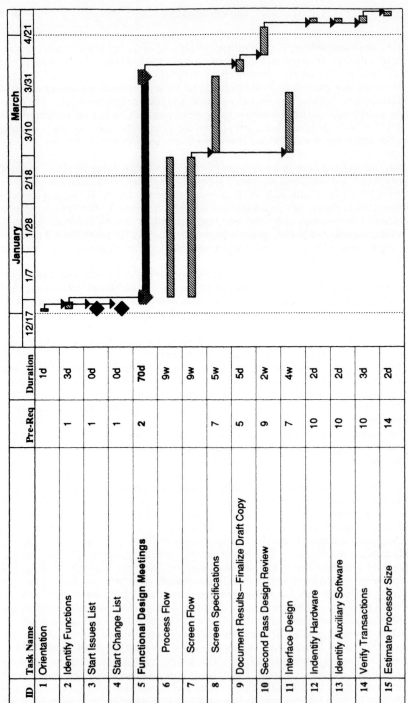

The table portion of the Gantt chart:

ID	Task Name	Pre-Req	Duration
1	Orientation		1d
2	Identify Functions	1	3d
3	Start Issues List	1	0d
4	Start Change List	1	0d
5	Functional Design Meetings	2	70d
6	Process Flow		9w
7	Screen Flow		9w
8	Screen Specifications	7	5w
9	Document Results—Finalize Draft Copy	5	5d
10	Second Pass Design Review	9	2w
11	Interface Design	7	4w
12	Indentify Hardware	10	2d
13	Identify Auxiliary Software	10	2d
14	Verify Transactions	10	3d
15	Estimate Processor Size	14	2d

FIGURE 9-14 Gantt chart external design plan for a structured approach.

9-18

ments, or design any new functions that may have been overlooked. The review and interface design duration typically takes 20 to 40 percent of the original design time.

The final step in the external design phase is to identify any hardware and additional software needed to support the WMS. Specifications for peripheral hardware equipment such as RF terminals, CRTs, scanners, and line and bar code printers is documented. Any subsystem software required, such as a freighting or routing system, is also specified. Finally, the transaction data for each function is reviewed to allow the computer system to be sized (memory, DASD, processor size, controllers, and so forth).

INTERNAL DESIGN PHASE

The scope of the internal (or detailed) design phase, as with the external design, depends on the design methodology being used. The internal design phase objective is to identify the data-base requirements and architecture, translate the external design into program and logic flow, and incorporate programming and screen design standards.

The internal design phase, in a traditional sense, supports the structured design methodology. During internal design, the support files and fields identified in the external design are extrapolated and designed into a data base. The internal design translates the external design requirements and provides the road map for programming by defining program structure through the use of programming standards.

However, the object-oriented design methodology does not require this level of definition. The object-oriented design approach incorporates the data-base design within the object definition processes. A translation of an object-oriented design into an internal design and construction simply builds from the existing detailed information.

The hybrid design approach utilizes the internal design phase to transition the external design functional definitions into an object-oriented design. The internal design is used to conduct the object-oriented analysis and design to create objects, classes, and so forth. The internal design team uses the external design created from the structured design approach rather than beginning with a process requirements statement. The external design gives the internal design team a more detailed and refined design direction than a process requirements statement, along with assurance to the design team that the system's requirements are being achieved by the objects being defined by the construction team.

MATERIAL HANDLING DESIGN PHASE

The purpose of the MH design phase is to specify, in detail, how MH equipment types are to be used within the warehouse or distribution center. Many design elements used in specifying the MH equipment are included within the WMS Requirements Report material handling analysis presented in Chap. 6. This section provides samples of MH specifications that are commonly used in designing a MH system.

Material handling equipment is classified as being either mobile or fixed position. Mobile equipment is typically purchased from a vendor with little or no design modifications (and includes fork lifts, AGVs, and so forth). However, the functional capabilities of mobile equipment play a pivotal role in the warehouse layout design and positioning of fixed equipment (such as aisle width and racking height). Fixed-position equipment requires more detailed design than mobile equipment since other equipment and resources interact with it.

The remainder of this chapter reviews the fixed MH equipment subcategories of static and dynamic types. Static equipment is considered warehouse control system (WCS) independent (directly controlled by the WMS), while dynamic equipment is WCS dependent to allow interface with the WMS and/or warehouse operator. A WCS is used to provide dedicated control logic to manipulate the static (and some mobile) MH equipment. However, a WCS must be integrated, or interfaced, to the WMS to retrieve data that directs material movement to ensure accurate and timely delivery of product to the next warehouse area so that the equipment does not become an island of automation.

Design category A described below represents MH equipment that is WCS independent (or static). These types of equipment are manipulated by the WMS to optimize their use. For instance, the WMS is designed with an equipment type table that specifies the type of equipment required to work within a location or to handle a specific SKU. The WMS determines the type of equipment required when a work task is generated. Location type, product processing type, environment, weight, capacity, and so forth are additional attributes within the WMS that provide control over how locations are used within the warehouse.

Design category B described below represents MH equipment that is WCS dependent (or dynamic). The WCS logic to interface with this MH equipment type can reside on the same processor (computer) as the WMS. However, it is common to have a distributed processor that is dedicated to controlling a piece of equipment. A distributed WCS processor ensures that the MH equipment has the system response time required to optimize its use and is not burdened with other warehouse activities and workload. For instance, a conveyor sortation system can accept input from a scanner that identifies the SKU or customer order for which a carton has been picked. Based on this information, the WCS communicates with a programmable logic controller (PLC) that causes the sortation system to divert the picked carton to the correct shipping spur. The timing of this communication requires split-second response time and can involve complex logic specifically designed for the environment in which the MH equipment is being used.

Material Handling Design Category A

Designing for WCS-independent MH equipment requires a variety of information. Chapter 4, Material Handling Considerations, discusses options that exist for a variety of MH equipment and how the different equipment types are applied within the warehouse. This section presents sample formats for information used to specify equipment that falls within this category.

Building Layout.   A layout drawing conveys and defines all material handling equipment operating boundaries included in the warehouse. The layout drawing, illustrated in Fig. 9-15, specifies the location of columns, racking configuration, lift-truck constraints, and receiving or shipping dock orientation. This is the main communication vehicle to equipment vendors for defining and explaining equipment specifications.

Concept Drawings.   Concept drawings are used to show a building layout cross-sectional schematic where multiple MH storage and transport equipment coexists. Concept drawings further define equipment requirements and dimensions. An example pick module configuration is shown in Figs. 6-24 and 6-25. This three-level mezzanine pick module configuration shows the uses of pallet flow rack for high-volume carton picking on the top level, carton flow rack for high-volume broken case (piece) picking, and the use of pick tunnels which include power and gravity conveyors. The

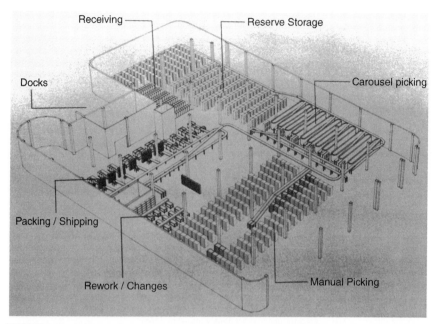

Receiving ——
—— Reserve Storage
Docks
—— Carousel picking
Packing / Shipping
Rework / Changes
—— Manual Picking

FIGURE 9-15   A 3D building layout design illustrates the entire warehouse and material handling design.

racking dimensions show the planned position and warehouse constraints along with product storage requirements.

Product Metrics.   This is a series of tables that describe the product dimension and weight boundaries that must be handled by the MH equipment. For example, summary tables are used to show the range of pallets and cartons that the MH equipment must be able to support. Figure 9-16 shows an example of the maximum pallet weight requirement for the warehouse racking and fork lifts. Figure 9-17 shows an example of carton dimensions that exist within the product line. It is important to include the maximum and minimum boundaries to ensure that the MH vendor's equipment can meet the range of product specifications. For example, the carton weights of 4 and 30 pounds must be conveyed equally in a carton flow rack. Each carton weight must convey by gravity forward from a standstill yet not build up too much momentum that might cause product damage or become a safety hazard.

Carton weight, lb	Pallet weight, lb
4	520
5	477
10	900
16	640
18	720
22	1100
30	990

Note: Measurements are maximum.

FIGURE 9-16   Carton and pallet requirements for the design of racking, conveyors, and fork lifts.

Design Criteria.   This is a series of descriptive paragraphs, and/or tables, that describe how the MH equipment is utilized. A design criteria statement is written for

Length	Width	Height	% Total
23.38	19.50	17.88	20.30
18.50	11.25	9.50	0.10
18.50	11.25	7.50	0.10
18.25	13.25	7.00	0.20
18.25	11.25	6.50	0.10
13.25	9.25	9.00	0.20
13.25	9.25	8.00	0.70
13.25	9.25	7.00	0.90
13.25	9.25	6.00	0.70
13.25	9.25	5.00	0.70
13.25	9.25	4.00	1.60
11.25	9.25	10.50	0.60
11.25	9.25	9.50	1.60
11.25	9.25	8.50	7.20
11.25	9.25	7.50	8.70
11.25	9.25	6.50	15.00
11.25	9.25	5.50	14.50
11.25	9.25	4.50	17.80
9.50	6.00	5.50	0.10
9.50	6.00	4.50	8.70

FIGURE 9-17   Carton dimension distribution used for material handling equipment design.

each functional warehouse area that utilizes specific MH equipment. For instance, the following statement describes the above-noted high-volume carton picking area:

Two modules of gravity pallet racks flow racks will be installed. The racks are 144 in deep and store 40- by 48-in pallets three deep in each location. The flow rack is angled to automatically convey the pallets forward upon the entry or removal of a pallet. These two modules are 200 ft long by 30 ft 3 in deep by 28 ft high. These individual carton pick modules are replenished from the rear by pallet lift trucks.

Material Handling Design Category B

Designing for WCS-dependent MH equipment requires the same informational detail specified for category-A MH equipment with the addition of WCS software control, process descriptions, and throughput requirements. Software control includes design options for the WMS interface to MH equipment through the use of bar code labels and data files. Next, process descriptions are used to indicate the required warehouse operator interfacing that may be required. Finally, sample volume throughput formats are provided to review the data requirements for specifying this equipment category.

WCS Software Alternatives.   The WCS software provides the interface communication between the WMS and a PLC. Interfacing requirements depend on the type of MH equipment being specified. For instance, interfacing to a carton sortation system is accomplished by using a bar code that is scanned to identify the carton's contents and where it should be diverted. The bar code is either an existing product label that identifies the SKU or a label applied during picking that identifies the product or customer order.

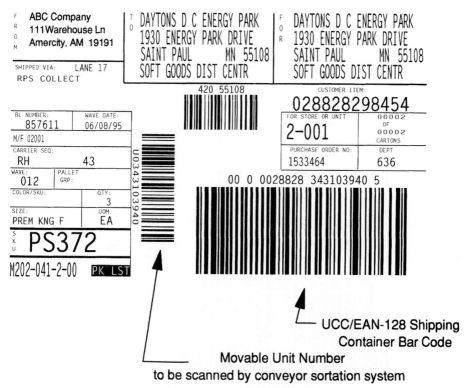

FIGURE 9-18 Pick/ship label design utilizing a "dumb" bar code label where the movable unit number is used to reference a WCS sortation file.

The bar code format for a pick label can be either "dumb" or "smart." A dumb bar code simply provides carton (or movable unit) identification that is then matched to information downloaded to the WCS that gives the customer order and sortation requirements. Figure 9-18 shows an example of a dumb bar code label where the movable unit is scanned by the conveyor sortation system. The movable unit number is matched to the downloaded information whose format is shown in Fig. 9-19. The

Data fields	Number of characters	Description
Record type	1 A	Record type—add, change, or delete
Wave number	2 A	Wave number assigned in the order processing system
License plate number	10 A	Unique movable unit number associated with the carton
SKU number	7 A	SKU number of the item
Lane	10 A	Name of the shipping lane to divert to
Carrier code	4 A	Assigned code for a carrier name
Divert location	10 A	Used to identify actual divert location

FIGURE 9-19 Typical WCS conveyor sortation interface file format.

information downloaded to the WCS indicates the divert spur where the order is to be loaded (or staged).

A smart bar code provides encoded information that is interpreted by the WCS to determine the action to perform. A sample of a smart bar code is shown in Fig. 9-20. The bar code indicates the wave and order sequence numbers for which the carton was

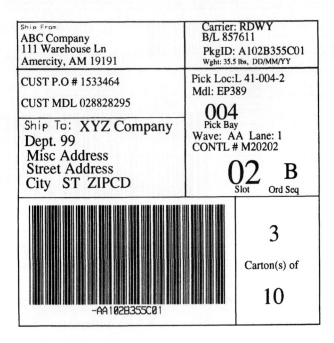

Movable Unit "Smart" Text
Wave AA
Divert Spur 1
Pallet Build Slot 02
Slot Sequence B
Carton Weight 35.5 lb
Carton ID Sequence C01

FIGURE 9-20   Pick or ship label design utilizing a "smart" bar code label.

picked, the carton weight, the assigned divert spur, and a carton sequence number. The wave and order sequence numbers uniquely identify the customer order number, which is matched to the information downloaded by the WMS to determine if the carton can be diverted. Since the information downloaded to the WCS is summarized to the order level (i.e., all cartons for an order are treated the same), then less information needs to be exchanged between the WMS and WCS. In addition, WCS look-up time is decreased over the dumb bar code scenario.

The use of a dumb bar code has the advantage over a "smart" bar code if existing carton labeling is used for sortation scanning. It may be undesirable to apply a pick label due to product appearance when it arrives to the customer. For instance, if the bar coded SKU is available on all products, the WCS can determine the order(s) that requires the SKU and perform the proper divert according to the "customer order to SKU" information provided by the WMS.

The smart bar code has the added advantage over a dumb bar code of allowing pick verification through the use of an in-line scale to weigh the carton as it is scanned. If the carton weight does not match the weight indicated within the bar code, then the carton is rejected for manual handling and investigation. The carton sequence number makes the bar code label unique to the carton for use as a movable unit label for later operations (such as building a pallet or loading a trailer). Note that this example includes the divert spur within the bar code. If the warehouse operations require the ability to change the divert spur after a pick label has been printed, then the divert spur information should be included in the information downloaded from the WMS.

The information contained within the WCS interface record not only depends on the type of product labeling used, but also varies according to the type of operation being performed by the WCS. For instance, if the WCS is controlling a picking operation (such as pick-from-carousel or pick-to-light systems), the information downloaded from the WMS identifies the customer order, SKU, and the quantity to pick. The information captured by the WCS and sent back to the WMS after the pick is completed indicates the movable unit built, the picker who built it, and the quantity picked.

WCS Process Description.   Process descriptions are used when the WCS interacts with warehouse operators to perform an operation (such as picking) or to specify the procedural logic required within the WCS. As illustrated below, a process description contains a written format and a flow diagram.

Pick/ship labels are printed in location sequence by the WMS and distributed to pickers according to their assign zone. The pick label indicates the order being picked and the associated movable unit. A picker signs onto the pick-to-light (PTL) system by typing his or her user ID and password. The picker then scans the movable unit number on the pick/ship label into the PTL system. The PTL must ensure that only one order is being actively picked within a zone, and multiple pickers cannot sign onto the same pick zone.

After a pick label is scanned into the PTL system, all picks for the zone are illuminated. Each pick quantity is displayed, and the picker is responsible for picking the correct quantity from the illuminated location. Picks are confirmed by pressing the button at the location. After all picks are complete for a zone and no picks remain in subsequent zones for the customer order (or tote), the PTL displays a "done" message to the operator and sends the completed pick information to the WMS. The operator pushes the completed carton onto the exit conveyor. If picks are required in a subsequent zone(s) for the customer order, then the PTL displays a "pass to #" message where the # indicates the next pick zone.

An abbreviated example of an exception processing description for a PTL system is as follows:

If the quantity to pick does not exist in the illuminated location, the operator has the ability to request a replenishment or to adjust the quantity picked. A replenishment is requested by using a terminal connected to the WMS. After a replenishment is completed, the operator picks the required quantity. If a replenishment cannot be made to the location, the operator adjusts the quantity picked at the location and indicates that the pick is complete. The PTL updates the picked quantities to the WMS. The WMS will manage backorders for the customer order if an order line cannot be picked completely.

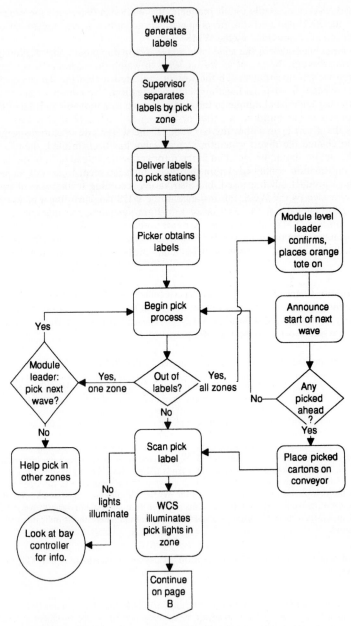

FIGURE 9-21 Sample (abbreviated) process flow diagram for a WCS pick-to-light system dialog.

The flow diagram portion of a process description pictorially represents the process described by the written format by using flow chart symbols and conventions. Both normal and exception operational requirements are reflected in the flow chart format. An abbreviated example of an operational flow chart from a PTL system is shown in Fig. 9-21.

Throughput Requirements. Throughput requirements are developed for all WCS-dependent MH equipment types. Throughput requirements provide the data to the MH supplier to define detailed specifications and to ensure no misunderstandings or assumptions are made regarding the MH performance criteria. This data is also used in acceptance testing upon completion of equipment installation. A minimum of three types of information is usually required to ensure that performance is achieved by the MH design.

First, system sizing data specifies the average and peak volumes by the number of orders, lines, picks, and shipping containers (or totes). It is important to remember to calculate material specifications based on the design year, which is usually 3 to 5 years beyond the present year. This ensures that the MH equipment will perform adequately and not require short-term changes to increase performance. A system sizing data sample table is shown in Fig. 9-22.

Description	Average	Peak
Orders per day	3,308	6,287
Orders per hour	236	449
Lines per day	32,507	48,244
Lines per shift	16,253	24,122
Lines per hour	2,322	3,446
Picks per day	103,789	145,426
Picks per shift	51,894	72,713
Picks per hour	7,414	10,388
Containers per day	7,596	9,288
Containers per shift	3,798	4,644
Containers per hour	543	633

FIGURE 9-22　System sizing table.

Next, customer order configuration data specifies the average order size. A sample order configuration table is shown in Fig. 9-23. This information is used in conjunction with the system sizing data to determine the average weight per shipping container and the number of picks per shipping container.

Containers per order	2.5
Order weight	30.5 lb
Items per container	10.3 lines
Order size	16.9 pieces
SKUs in slots	6,800
Number of slots	10,500

Note: SKUs may have more than one slot if required.

FIGURE 9-23　Order configuration table.

Finally, SKU picking velocity data shows the SKU profile according to customer order forecasting (or history). A sample SKU picking velocity table is shown in Fig. 9-24. This information is used to indicate if specialized picking modules should be created according to customer buying habits. For example, if a significant volume of shipments is concentrated to a few SKUs, then a high-volume carton picking area serviced by a carton conveyor and PTL system located close to the shipping docks may be warranted.

SKUs	Number of picks	% SKUs	% Picks
185	25,621	2.7	24.7
425	19,490	6.3	18.8
809	17,381	11.9	16.7
1,847	23,705	27.2	22.8
3,534	17,592	51.9	17.0
Total 6,800	103,789	100.0	100.0

FIGURE 9-24   SKU picking velocity table.

SUMMARY

Design for an automated warehouse must consider WMS and MH detail design to achieve an integrated system where information moves with material. A documented design process ensures that proper consideration is given to all business areas and that plans are properly documented. The implementation section, discussed in the following chapters, ensures that all of the activities under the planning and design section are properly implemented to achieve a successful automated warehouse installation.

BIBLIOGRAPHY

Jacobson, I., Chisterson, M., Jonsson, P., and Övergaard, G., *Object-Oriented Software Engineering,* Addison Wesley, Reading, MA, 1992.

Moad, Jeff, *Object Methods Tame Reengineering Madness,* Datamation, Newton, MA, 1995, p. 43.

Rumbaugh, J., Blaha, M., Premerlani, W., Eddy, F., and Lorensen, W., *Object-Oriented Modeling and Design,* Prentice-Hall, Englewood Cliffs, NJ, 1991.

P · A · R · T · 3

IMPLEMENTATION

CHAPTER 10
PROJECT MANAGEMENT

INTRODUCTION

Managing a project is a very different environment from managing day-to-day operations of a functional department. Projects are characterized by having a finite lifetime and a defined set of goals. Much of the project must be planned far in advance, even though the details of this work to be done are not known when the work is first planned. Many of the project team members do not report to the project manager but rather to different functional managers and are "on loan" to the project. As a result, the project manager, without any direct employees, is frequently looked upon as a non-manager by the company. In spite of this, the project manager must direct and evaluate everyone's work while keeping them motivated and boosting their morale. In addition, the project manager must perform many of the other normal management functions.

Because of all these factors, project management requires a special set of skills and training. This chapter identifies the techniques used to manage a warehouse automation project. There is, however, no substitute for experience. Any company beginning a large automation project should make every effort to find an experienced project manager to lead the effort.

OVERVIEW

This chapter discusses the techniques used to manage a warehouse automation project. The content of the project manager's job is reviewed, and the tasks needed to initiate and plan the project are described along with an example of a plan for an automation project. Each step of the implementation phase is described in detail along with examples of how to plan these steps.

THE PROJECT MANAGER'S JOB

In a warehouse automation project, the project manager is responsible for seeing that the project is well organized and runs properly so that it is completed on time and within budget and meets the project goals. In order to do this, the project manager uses five primary techniques:

1. *Planning.* Developing a project schedule, or plan, that shows what work must be done and by whom and when the work must be started and completed.

2. *Tracking.* Following through to be certain all tasks are being completed as planned.

3. *Reporting.* Keeping the appropriate parties informed of problems and overall project status.

4. *Leading.* Building, directing, and motivating a project team.

5. *Controlling.* Taking actions to keep the project on schedule and within budget.

Planning

Planning is a process of identifying all of the activities that are required to complete the project and then breaking the project down into a series of tasks. The time and resources needed for each task are identified, as well as the sequence in which the tasks must be done. The person or group responsible for each task is then identified and assigned. Through a continuous planning process, the project plan grows and changes as the project proceeds and more details are defined about future activities. This project plan becomes the primary document that the project manager uses to perform his or her responsibilities.

Part of planning is the accommodation and management of changes to the project's scope as planned and budgeted. In any project, events will occur that require deviations from the original plan. For example, senior management may request a change to the project's scope or goals. A new requirement may be discovered during system design that will require revisiting the new process design steps. Team members may not be available when planned due to changes in commitments. Anything that causes the project to deviate from the plan should be expected to have an effect on the schedule and the budget. The project manager must have processes and procedures in place to allow changes to be introduced into the project plan in a controlled manner. Under a formal change control process, the effects of a proposed change, including schedule, budget, and functional ramifications, are evaluated before the change is made. Then, company management is able to make an informed decision on whether to implement the change or not. This process also gives all affected parties advance notice of the change, thereby reducing confusion and further delays.

Tracking

Tracking is comparing the actual progress of each task against what is expected in the project plan. Thus, a comprehensive project plan is a vital prerequisite to tracking. On a regular basis, the project manager determines if tasks are on schedule. Have tasks started as planned? Have tasks been completed as planned? Are all tasks currently underway on time according to the project plan? Is each task being performed using the planned amount of resources? Using this information, the project manager determines if the project is falling behind, is ahead of schedule, or is on time.

Reporting

On a regular basis, the project manager reports on the project's status to everyone involved in or affected by the project. At times, reporting is limited to those members involved in a particular phase. At other times, summary reports are sent to all project team members including the executive sponsor and senior management. Reporting encompasses various reporting styles and includes project review meeting minutes and status meeting minutes.

Leading

This technique comprises all of the traditional aspects of management leadership. The project manager is the leader of the project team. He or she determines who should be a part of the team and enlists their support and participation. The project manager assigns project tasks to various team members and ensures that all team members have meaningful work while not being overloaded. Team building, ensuring that the project team works as a real team and that every member feels that his or her contribution is valuable, is an important facet of team leadership. The project manager is responsible for making sure that the project team understands the project's goals and their responsibilities and is well motivated and has high morale. As a part of this, the project manager must keep the team well informed about the tasks to be completed and their importance to the project's success.

Controlling

Controlling a project involves taking action on the information gathering during tracking and making sure that the project is on schedule, is within budget, and is achieving the stated goals. It may be necessary to assign additional, or different, people to a task that is falling behind schedule. Additional team members may be needed or overtime may be required. Priorities of various tasks frequently need to be rearranged. While sometimes challenging, all of these variables must be juggled while still keeping the project team motivated and working effectively.

Additional Resources

Numerous books devoted entirely to the subject of project management have been written to assist the project manager. Several of these are listed at the end of this chapter. These books should be referenced for more in-depth knowledge of general project management techniques and tools such as Gantt charts, project evaluation and review technique (PERT) charts, the critical path method (CPM), and so on. The remainder of this chapter is dedicated to topics in project management as they specifically relate to warehouse and distribution automation projects.

PROJECT START-UP ACTIVITIES

Once the company's senior management has established the core of the project team and appointed the project manager, much of the responsibility for moving the project forward passes to the project manager. The project manager now must take a number of steps to establish the proper foundations for the project without losing project momentum. The following topics are discussed below with an aim of assisting the project manager.

Review the high-level project scope and goals with management.

Establish the project high-level plan.

Develop the project organization:

- Organize the project team.
- Develop the staffing plan.
- Conduct the team orientation meeting.

Establish project processes:

- Documentation standards.
- Change control.
- Project issues and problem lists.
- Set up project files.
- Internal reporting and review.
- External reporting and review.

Set up project war room.

Conduct project kickoff meeting.

Although somewhat tedious and complicated, the above tasks should not be put off because they establish a framework of standards and procedures to ensure consistency, efficiency, and timeliness of the project.

Review the High-Level Project Scope and Goals with Management

When the automation project is approved by the company's management, the project has a general scope and a set of high-level goals. In most cases, some of the goals are explicitly stated. Many times, however, other goals are indirectly implied or assumed by management. The project manager must be sure that all of the project's goals are well understood and clearly documented from the onset of the project. In addition to making the project manager's job easier, this will allow for a smooth transition if there is a later change in the project manager or in executive sponsorship.

Establish the Project High-Level Plan

The project high-level plan is a document that summarizes the information currently known about the automation project. This plan provides the project manager with a vehicle for disseminating key project information to both the functional manager who will be involved in the project and the direct project team members. The high-level plan then becomes the starting point for generating the project plan, which is discussed later in this chapter. An example of a high-level plan is shown in Fig. 10-1. At a minimum, the plan should contain the following:

- A description of the project.
- The project's goals and scope.
- A project schedule overview. Although the details of the schedule have not been planned at this point, the basic steps of the overall project, as described in Chap. 2, are known. The plan should describe the process then known and include estimated starting dates and durations for each phase and step. The overview should note important project milestones as well as key dates when the project must integrate with other company projects or important business events. A simple bar chart is usually sufficient for this type of schedule.

XYZ Corporation
Warehouse Automation Project
High-Level Plan

Project Description

In today's business environment, our company has become increasingly aware of the importance of efficient warehouse operations. With greater customer demands for products and services on a more timely basis at lower costs, business trends indicate that our warehouses have become the focal point of our distribution network.

Based on customer needs, our company's recent growth, and new industry requirements, the apparent solution is a warehouse automation project designed to improve inventory control, manage warehouse space, and increase employee productivity. In addition to improving internal warehouse operations, we must ensure that the latest data is available for operations at our headquarters and within the warehouse.

Project Scope and Goals

Success in warehousing today is measured in reduced cost and improved quality and customer service. New business is attained through the establishment of oneself as a leader in warehouse automation. In order for our company to continue evolving and managing our distribution operations, we must take steps to implement warehouse automation. This new comprehensive and flexible system is to automate our current manual operations, while it centrally controls the diverse technologies and equipment in our warehouse.

The computer systems created by this project will enable warehouse management to view, plan, and control the operations of the facilities with real-time information. These systems will also bridge between various other business systems such as production scheduling, purchasing, logistics planning, and order management systems, allowing us to provide the proper customer order response time critical in today's competitive market.

The warehouse automation project will encompass the following functional areas within the warehouse:

- Receiving and dock management
- Put-away operations
- Location management
- Cycle count
- Order processing
- Picking
- Replenishment
- Packing
- Shipping

Key benefits to be realized from the warehouse automation project include:

1. Improving customer service through the reduction of order cycle times and better shipping accuracy
2. Improving productivity by reducing product handling, optimizing travel time of storage and retrieval equipment, optimizing warehouse tasks, and eliminating errors
3. Increasing inventory accuracy through real-time activity confirmation
4. Maximizing facility utilization through location maintenance and rewarehousing

Schedule Overview

The warehouse automation project will be accomplished in three distinct phases: preparation, definition, and implementation. The *preparation* phase will establish a set of detailed project goals and a team will be formed to define and implement the business changes needed to accomplish these goals. During the *definition* phase, documents will be written that describe in detail

FIGURE 10-1   An example high-level plan.

how the warehouse will operate with automation. There are two major steps in this *definition* phase. First, define the new business processes for the facility. Second, specify the material handling hardware and system software required to support the new processes. During the *implementation* phase, the specified hardware and software will be procured and installed. Then, the new business processes will be brought into operation. The proposed timetable is as follows:

- *Phase 1.* Preparation January 1995–June 1995.
- *Phase 2.* Definition June 1995–November 1995.
- *Phase 3.* Implementation November 1995–June 1996.

Project Constraints

- Warehouse operations must not be disturbed during the rush season October 1995 to January 1996.
- Distribution operations must be automated and back into full operation prior to the start of the 1996 rush season.
- The start-up of the automated warehouse must be coordinated with the new order entry system project scheduled to be completed in June of 1996.
- The project must be able to show a 2-year payback period before implementation will be approved.

Participants

The following organizations will be supplying full-time project team members:

- Distribution
- MIS
- Industrial engineering

The following organizations will be supplying part-time project team members:

- Finance
- Order processing
- Accounts payable
- Invoicing
- Customer service

FIGURE 10-1 *(Continued)*

- Key schedule and budget constraints. For example, sometimes an automation project's completion must coincide with the completion of a new building. If so, then it is vital that all project participants be aware of this information.
- A list of the functional organizations that will be involved in the project and a description of their planned participation.

Develop the Project Organization

The project manager performs this activity at the beginning of each project phase. Although many of the team's members continue from phase to phase, the nature of the work changes as the project moves from planning to definition to implementation. Therefore, the project manager revisits the project team's organization at the start of each phase.

Review Personnel. Frequently, team members have already been assigned to the project team. If so, the project manager determines if these assignments are still valid, and verifies when the people will actually be available. Normally, team members are completing other assignments and will not all be available on the planned project start date. The project manager also ascertains what events have the potential to affect these planned availability dates. For example, a later completion of another project may delay the availability of some members.

The project manager must review the strengths and weaknesses as well as the experience of the assigned team members. The factors considered include breadth of knowledge of the facility operations, awareness of automation technologies, experience in distribution process design, knowledge of company logistics operations, and so forth.

Organize the Project Team. The project manager analyzes the work that must be done, particularly for the upcoming project phase, and determines the project team positions required, the skills required for each position, and when each position must be filled. The project manager also determines how to organize the team into functioning subteams. Of course, the skills and weaknesses of the existing team members are taken into account when determining positions and subteams.

The project manager then compares the positions required against the personnel already assigned to the project and determines who fits best into each position. When necessary, the project manager decides if any team members need extra training and arranges for this training. For any unfilled positions, the project manager must work with functional management to get people with the necessary skills assigned to the project.

Develop the Staffing Plan. Using the personnel requirements prepared above, the project manager draws up a written staffing plan for the current phase of the project. This document shows, at a glance, when team members are expected to be working on the project. Figure 10-2 shows an example of a staffing plan. For any team positions that are new to the company, for example, a business analyst, the project manager should draw up job descriptions to go with the staffing plan. These job descriptions help the project manager communicate the skills needed when seeking candidates to fill these positions. Figures 10-3 to 10-6 are samples of job descriptions that would accompany a staffing plan.

A written staffing plan helps the project manager easily communicate the project team's needs to functional and senior management. At the beginning of the preparation phase, the details of the succeeding two phases are not yet known. Thus, the staffing plan prepared at the start of the project only shows details for the first phase. A revised staffing plan is then prepared prior to the start of the other two phases.

Conduct the Team Orientation Meeting. Although the team orientation meeting logically falls under developing the organization of the project, it should not be held until the project manager has completed the activities of establishing the project processes. With these processes established, the project manager has the what, who, and how of the project in hand and can then effectively impart this information to the project team at the first formal team meeting. The objectives of the team orientation meeting are to introduce the team members and discuss each of their roles; give an overview of the project; and review what is to be accomplished, the basic working environment, and procedures to be used. At this meeting, the project manager should present the following to the team:

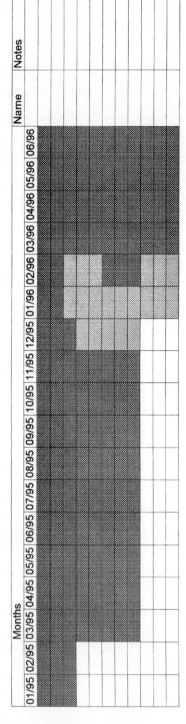

FIGURE 10-2 An example staffing plan.

Project Manager

Job Description

The Project Manager plans, tracks, and coordinates the project and directs the activities of the project team members. The Project Manager's duties include the following:

- Writing project plans and schedules

- Determining project staffing requirements and writing the project staffing plan

- Tracking project status

- Establishing project procedures and working environment

- Administering the project change control procedure

- Leading project team meetings

- Presenting project status to senior management

FIGURE 10-3   Job description for a project manager.

- Project background and purpose
- Goals and scope
- Schedule overview
- Team organization, staffing plan, and initial tasks
- Project working methodologies, processes, and standards

Technical Team Leader

Job Description

The Technical Team Leader is responsible for the overall technical integrity of the project. The Technical Team Leader establishes, with the help of the project team, the overall architecture of the warehouse automation system. The Technical Team Leader is responsible for the following:

- Evaluation of project change requests

- Selection of system operating platforms and languages

- Coordination of interface designs

- Validity, completeness, feasibility and technical integrity of project design documents

- Completeness and thoroughness of project design review meetings

---

FIGURE 10-4   Job description for a technical team leader.

Business Analyst

Job Description

A Business Analyst works under the direction of the Project Manager. The Business Analyst performs the following tasks:

• Reviews current warehouse business processes and determines deficiencies.

• Establishes process designs for assigned business processes within the warehouse automation project.

• Assists System Analysts in determining the computer system requirements to support the process designs.

• Provides project team expertise on current facility processes and procedures.

• Helps install new systems and train warehouse personnel in their operation.

• Writes operating procedures for new business processes.

FIGURE 10-5   Job description for a business analyst.

Establish Project Processes

Because of a number of various factors, it is very important for the project manager to establish a written set of procedures and working methods for the team from the start of the project. Team members come from various departments, each with their own operating styles. Many of the members are from functional departments that are primarily concerned with day-to-day operations. A warehouse automation project has a working environment very different from operations, and many team members have a

---

System Analyst

Job Description

A System Analyst works under the direction of the Project Manager. The System Analyst performs the following tasks:

• Assists Business Analysts in establishing business process designs.

• Determines the computer system requirements to support business process designs.

• Directs Programmer/Analysts and Programmers in the design of modules and programs to satisfy system requirements.

• Provides project team expertise on computer systems.

• Helps install new systems and train warehouse personnel in their operation.

• Writes user manuals and technical reference manuals for new systems.

FIGURE 10-6   Job description for a system analyst.

difficult time making the transition. In order to get the team functioning with minimal lost effort, the project manager should establish the basic operating framework for the team and explain the operating methods to all members.

Documentation Standards.   The project team will be generating many documents during the course of the project. So that these documents will be easier to write, can more easily be shared, and will have a common look, standards for project documents should be set. Standards include the following:

- Word processing program to be used
- Basic outlines for all routine documents such as meeting notes, status reports, and designs
- Glossary of special terms used in the distribution center or warehouse and related areas of the company

As the project proceeds, the pool of basic document outlines and the project glossary continually grow as new documents are created and as new terms are encountered.

Change Control.   Change control is one of the most critical aspects of a properly run project. As the project moves through its various phases, each step builds upon the work that was done in the preceding steps. It cannot be expected, however, that prior decisions are never revisited. As more information is learned, or as the business environment around the project changes, it will be found that some choices were not correct or that some aspect of operations was overlooked. As the inevitable need for changes arises, a procedure must be in place to accommodate changes smoothly into the project with as little disruption as possible. Thus, the project manager must issue a change control procedure prior to the project kickoff. Figure 10-7 shows a sample change control procedure document.

Changes always have some effect on both the project's schedule and budget. The change control procedure should include provisions for evaluating these effects prior to the implementation of the change. This allows management to make an informed decision on whether the benefits of a proposed change outweigh its costs in time and material.

The change control procedure also should have provisions to ensure that everyone involved in the project is informed about and is aware of any proposed change and its implications before the change takes effect. This will allow all of the team members to be in the best position to implement changes as they occur and minimize confusion.

Project Issue and Problem Lists.   During the course of the project many issues and problems will arise that cannot be quickly resolved. *Issues* are things that affect the project process. For example, information may be needed relating to a project in some other area of the company that will not be available when needed by the warehouse automation project. *Problems* are things that do not conform to the design of the automated facility as designed by the project team, for example, a conveyor that does not divert packages correctly. For some items, the project manager's judgement is required to determine if an item is an issue or a problem.

Because problems and issues come up very frequently during a project, they have a tendency to be forgotten if a list is not kept to track them. Each item on the list is assigned to a team member with a due date for resolution. (It is common to keep separate lists of problems and issues since these are generally assigned to different groups of team members.) All unresolved items on the problem and issues lists are reviewed at each weekly team meeting, and their current status is discussed by the team member assigned.

Overview

The intent behind the Change Tracking Report (CTR) process is to provide visibility and focus on project change activity to ensure that deviations from original project scope are authorized and approved as well as to ensure that cost and schedule implications associated with change are agreed upon prior to undertaking the work. The main benefit of this focus is to keep a high degree of confidence relative to cost and schedule commitments.

The CTR process is to be used for tracking and controlling any change to the agreed upon scope of work.

1. Change in deliverables

2. Change in requirements

3. Change in design

4. Change that would affect schedule of implementation

At project inception, the project implementation team should review the CTR process and document any planned deviations from the stated process.

At project inception, the project implementation team decides who will be performing the following key roles.

Change Requestor - Any member of the implementation team.

Project Technical Lead - Person who has technical responsibility for the project in question. This person is responsible for ensuring that the evaluation of the CTR is conducted, ensuring that approved CTRs are implemented, entering the CTR into the CTR system and tracking the progression of the CTR from inception to completion.

Project Manager - Person who has schedule responsibility for the project and who can commit company funds. Typically this will be the project manager.

Steps in the CTR Process

1. **The CTR is submitted for approval.**

   The *Change Requestor* fills out a Change Tracking Report form that describes the change and the impact if not implemented, signs the form and forwards it to the person performing the role of *Project Technical Lead.* It is important to include as much supporting detail as possible with the CTR to minimize follow-on requests for additional information.

FIGURE 10-7   Sample change control procedure.

Set Up Project Files.   Any distribution center or warehouse automation project will generate dozens of documents, such as design specifications, status reports, meeting minutes, change tracking reports, schedules, and so forth. A central repository should be created where the official version of each document can be found. This ensures that every team member can get correct information quickly without interrupting other

2. **The cost and schedule impact of the CTR evaluation effort is assessed.**

The *Project Technical Lead* signs the initiation section of the CTR form to indicate receipt of the CTR, enters it into the system that tracks CTRs with a status of open (OPN), assesses the schedule impact of the evaluation, determines resources needed for the evaluation and forwards the CTR form to the *Project Manager.*

3. **The Project Manager approves the schedule and resource cost impact of the evaluation.**

The *Project Manager* can either accept, reject or defer the evaluation. This is denoted by checking the appropriate action block on the CTR form, signing the CTR form and forwarding the CTR form to the *Project Technical Lead.* Checking the accept block in the evaluation section of the CTR form authorizes the evaluation work and indicates acceptance of the cost and the schedule impacts of the CTR evaluation. Work on the CTR form is returned to the *Project Technical Lead.*

If the CTR is rejected for evaluation, the CTR process is complete and the CTR is updated to be in a status of INV.

If the CTR is deferred for evaluation, the *Change Requestor* should resubmit the CTR when he or she would like it to go through the process again. In the interim, the CTR is updated to be in a status of INQ. CTR status updating is the responsibility of the *Project Technical Lead.*

If the CTR is approved for evaluation, the CTR process proceeds.

4. **The impact of the CTR implementation effort is assessed.**

The *Project Technical Lead* coordinates the evaluation of the CTR, completes the implementation section of the CTR form, updates the status of the CTR to RCV and forwards the CTR form to the *Project Manager.*

5. **The Project Manager authorizes the implementation of the CTR.**

The *Project Manager* reviews the impact of the implementation and the description of the change being implemented with the original *Change Requestor*, and approves, rejects or defers the CTR. Checking the accept block in the implementation section of the CTR form and signing the implementation section of the CTR form authorizes the project team to begin the implementation effort and indicates acceptance of the price associated with the CTR implementation. The completed CTR form is forwarded to the *Project Technical Lead* who updates the status of the CTR as appropriate to correspond to the disposition of the CTR: accepted, deferred or rejected. Work on the approved CTR implementation will not begin until the signed CTR form is returned to the *Project Technical Lead.*

Once a CTR has been accepted for implementation, the *Project Manager* incorporates the schedule and resource effects of the change into the official Project Plan.

FIGURE 10-7   *(Continued)*

members to look for a document. By keeping the central repository up to date, this will minimize instances of the team working with outdated versions of designs.

The conventional method of keeping a project central file has been a set of file cabinets kept in a central, secure location. Under this method, a log is kept of the documents in the file and their revision dates. A sign-out sheet shows who has removed any documents from the file. Each team member removes the master copy from the file

Change Tracking Report Form

The CTR form is the vehicle for communicating the content of the change request. It records the request and the authorizations for evaluation and implementation. Copies of the form are available from the *Project Technical Lead*.

The Change Tracking Report form is divided into three sections:

1. **Initiation Section**

   The Initiation Section describes the change that is being requested, the reason for the change and the impact if not implemented. Signatures in the Initiation Section denote:

   Change Requestor - Identifies the person who originated the CTR.

   Project Technical Lead - Signifies that this person has received the change request.

2. **Evaluation Section**

   The Evaluation Section describes the schedule impact and the price associated with the evaluation of the CTR. It also provides the vehicle for authorizing the evaluation to occur. Signatures in the Evaluation Section:

   Project Technical Lead - Denotes agreement with the impact associated with the evaluation.

   Project Manager - Denotes acceptance, rejection or postponement of the CTR evaluation. If the CTR is accepted it denotes the acceptance of the schedule and cost impact of the evaluation and authorizes the evaluation.

3. **Implementation Section**

   The Implementation Section describes the schedule impact and cost associated with the CTR implementation. It also provides the vehicle for authorizing the implementation. Signatures in the Implementation Section:

   Project Technical Lead - Denotes agreement with the impact associated with the implementation.

   Change Requestor - Denotes approval of the description of the change being implemented.

   Project Manager - Denotes acceptance, rejection or postponement of the CTR implementation. If the CTR is accepted it denotes the acceptance of the schedule and cost impact of the implementation and authorizes implementation of the CTR.

FIGURE 10-7   *(Continued)*

only long enough to make a working copy and then immediately returns the master copy to the file.

With the widespread use of personnel computers, the traditional central file of paper documents is usually replaced by an electronic repository. Typically, a project disk is established on a local area network (LAN) server. All project documents are then stored in soft-copy form on this disk. This allows all project participants rapid access to any needed project information. By linking the project LAN into corporate

CTR Severity Codes

Severity codes indicate the severity of the change request and are assigned by the *Change Requestor*. Severities are used by the *Project Manager* to prioritize the CTR evaluation work and implementation work if more than one CTR is being scheduled for evaluation or implementation concurrently.

1. **Severity 1**

   Major concern. Application is functional; however, it is very inefficient without this change, e.g., an overlooked database field will prevent invoicing for the shipped merchandise.

2. **Severity 2**

   Medium concern. Application has taken a minor impact and most normal functions are operational, e.g., a need for maintaining Item Master was overlooked. Put-away looks for certain item attributes for put-away location criteria. SQL can be used to modify the Item Master; however, it may be very time-consuming based on the number of attributes to modify/update and the total number of items.

3. **Severity 3**

   Low concern. Application is not impacted and the change is of more of an informational (usage) nature, e.g., a screen sequence change.

4. **Severity 4**

   Cosmetic change. Application is fine, but preference exists for something to be changed, e.g., right-justify an output filed on a screen display rather than left-justify.

CTR Status Values

During the weekly project team meetings reports of CTR status will be presented. The following is a list of potential CTR status values and meanings.

1. **RCV - Received**

   The CTR has been received by the *Project Technical Lead*. This status is recorded by the *Project Technical Lead.*

FIGURE 10-7 *(Continued)*

wide area networks (WANs) or other networks, project information can then be obtained by authorized personnel anywhere in the company. Owing to the increased team productivity attributed to "electronic" central files, this has became the preferred working methodology for medium to large projects. Making the necessary investments in hardware, software, and training has been found to produce rapid payoffs in terms of project costs.

2. **ASN** - Assigned

   The CTR has been assigned for implementation. This status is recorded by the *Project Technical Lead.*

3. **INQ** - Inquiry

   The CTR has been deferred at the request of the *Change Requestor.* This status is recorded by the *Project Technical Lead.*

4. **INV** - Invalid

   The CTR has been cancelled at the request of the *Change Requestor.* This status is recorded by the person performing the role of *Project Technical Lead.*

5. **FXD** - Fixed

   Design and code are complete. This status is recorded by the programmer assigned to the CTR.

6. **TST** - Test

   Software project lead verification is complete. This status is recorded by the software project leader.

7. **RDY** - Ready

   Test group verification is complete. This status is recorded by the test group.

8. **SND** - Send

9. **CLO** - Closed

   Change Requestor has accepted the implementation of the CTR. This status is recorded by the *Project Technical Lead.*

CTR Classification Codes

CTR classification codes are assigned by the *Project Technical Lead.* Classification codes are used in postproject analysis exercises.

1. NR - New requirement

2. CR - Change in existing requirement

3. CD - Change in existing design

4. IR - Inconsistency in requirement

5. ID - Inconsistency in design

FIGURE 10-7 *(Continued)*

CHANGE TRACKING REPORT (CTR)

********************************INITIATION********************************

Project: _____ CTR#: _____

Requestor name: _____ Severity: _____

Affected function: _____

Description of proposed change (add attachments, if necessary):

_____

_____

_____

Reason for change: _____

_____

Impact if not implemented: _____

Change Requestor: _____ Date: _____

Project Manager: _____ Date: _____

Project Technical Lead: _____ Date: _____

*************************************EVALUATION************************************

Schedule impact to evaluate, if any (add attachments, if necessary): _____

_____

Resources to evaluate: _____

Accept for evaluation ( )   Reject ( )   Defer ( )

Project Technical Lead: _____ Date: _____

Project Manager: _____ Date: _____

**********************************IMPLEMENTATION*******************************

Description of change to be implemented (add attachments, if necessary):

_____

_____

_____

Impact of change to be implemented (add attachments, if necessary):

_____

_____

_____

Change classification: _____

Projected implementation date: _____

Resources to implement: _____

Accept for evaluation ( )   Reject ( )   Defer ( )

Project Technical Lead: _____ Date: _____

Project Manager: _____ Date: _____

Change Requestor: _____ Date: _____

Note: The above estimate will be withdrawn if not accepted by: /   /

FIGURE 10-7   *(Continued)*

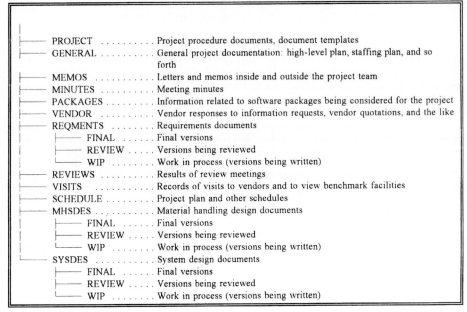

```
├─── PROJECT  . . . . . . . . . Project procedure documents, document templates
├─── GENERAL . . . . . . . . . General project documentation: high-level plan, staffing plan, and so
│                              forth
├─── MEMOS  . . . . . . . . . . Letters and memos inside and outside the project team
├─── MINUTES  . . . . . . . . . Meeting minutes
├─── PACKAGES . . . . . . . . Information related to software packages being considered for the project
├─── VENDOR  . . . . . . . . . Vendor responses to information requests, vendor quotations, and the like
├─── REQMENTS  . . . . . . . Requirements documents
│      ├─── FINAL  . . . . . . Final versions
│      ├─── REVIEW . . . . . Versions being reviewed
│      └─── WIP  . . . . . . . Work in process (versions being written)
├─── REVIEWS  . . . . . . . . Results of review meetings
├─── VISITS  . . . . . . . . . . Records of visits to vendors and to view benchmark facilities
├─── SCHEDULE . . . . . . . . Project plan and other schedules
├─── MHSDES . . . . . . . . . Material handling design documents
│      ├─── FINAL  . . . . . . Final versions
│      ├─── REVIEW . . . . . Versions being reviewed
│      └─── WIP  . . . . . . . Work in process (versions being written)
└─── SYSDES  . . . . . . . . . System design documents
       ├─── FINAL  . . . . . . Final versions
       ├─── REVIEW . . . . . Versions being reviewed
       └─── WIP  . . . . . . . Work in process (versions being written)
```

FIGURE 10-8   Sample project file directory structure.

Before the project begins, the project manager establishes the directory structure of the project disk and issues standards to the team indicating where documents are to be placed. Approved and finalized documents are stored separately from documents that are being written or are under review. Figure 10-8 shows a sample directory structure that can be used as a starting point. The project manager decides who will have authority to place documents into the "final version" directories. This authority should be limited to only a few people, usually the project manager and the technical team leaders. This prevents unexpected changes to final versions and ensures that the information in the repository is correct and up to date. The project manager also issues a procedure for revising documents in the repository. This procedure includes a method for "checking out" a document to be revised that prevents two people from attempting to revise the same document simultaneously.

If an electronic central file is used for the project, the project manager also establishes a procedure for storing any paper documents that are received by the project. The preferred method is to scan and store all paper documents on the project disk. Current image storage software and optical character recognition (OCR) software have made this a practical proposition even for small companies. Alternatively, paper documents can be kept in a central file cabinet. However, if this option is used, a log of the documents with an abstract of their contents should be kept on the project disk.

The project manager also should consider the use of a groupware software tool for project communications. Groupware is a communications-facilitating software program that organizes discussions on various topics and keeps them in a central repository. Most groupware programs include tools to allow team members to locate information quickly on any aspect of the project. These features distinguish groupware from electronic mail (e-mail). Groupware allows team members in widely separated locations to collaborate on project activities and keep track of information that has been exchanged. Many project teams have found groupware to be an invaluable tool.

Internal Reporting and Review.   Well-informed team members are more involved in the project and are able to focus their efforts on the work that is important to completing the project. Therefore, procedures for regular reporting and communications within the project team are essential. A team meeting is scheduled for the same day, time, and place every week for the duration of the project. At this meeting, the project manager presents the overall project status, discusses any project schedule changes, summarizes any external review meetings, passes on any corporate management comments to the team, and discusses with the team any external events that have affected or may affect the project.

Following the project manager's remarks, each team member presents his or her status report. Each member's status report is brief and follows a standard format:

- Activities scheduled to be accomplished during the last week and their status
- Any other activities that were accomplished during the last week
- Action plan and expected completion date for late activities
- Activities scheduled for the next week
- Any known changes to the schedules for work items assigned to the team member

Each report is informal and prepared charts are discouraged as these do not provide any added value to the project. Excessive time is not spent solving problems or discussing why activities are late. Usually, most items are not a matter of concern for the majority of the team, and lengthy discussions only serve to waste the team's time. Late activities and their effect on the project are handled by the project manager outside of the weekly team meeting. If a late activity is caused by or results in an issue or problem, an item should be added to the appropriate tracking list.

Following the member presentations, the issues and problem lists are reviewed. As each item is called off, the team member responsible gives a brief report on its status. If the issue or problem has not been resolved, the member reports the current expected resolution date.

The final portion of the weekly team meeting is the roundtable. The project manager calls on each team member in turn for comments. When his or her turn comes, the team member can raise any issues or concerns or discuss how he or she feels about the project and its progress. Valid issues or problems are added to the appropriate list and assigned to a team member for resolution. Concerns are discussed among the team members and either resolved on the spot or added to the issues list. The project manager is responsible for managing discussions and keeping them constructive, not general griping sessions. Properly conducted, the roundtable is a valuable project management and team building tool.

External Reporting.   Company management must be kept up to date on the status of the project. The project manager makes arrangements with company management to apprise them of the project's status on a regular basis. To do this, the project manager employs two methods of reporting. On a weekly basis, the project manager issues a project status memo to those managers most directly involved in the project. The results of the weekly team meeting are used to help prepare this report. The memo summarizes the following:

- Schedule status
- Budget status

- Resource requirements changes
- Major issues affecting schedule or budget
- Major changes affecting schedule or budget
- Vendor problems

These weekly status memos are followed up by a monthly, more formal, status report. This monthly report has a wider distribution than the weekly memos and goes to all company managers who may have an interest in the project. In writing this report, the project manager never assumes that the reader will be aware of the real project status or the existence of a concern or understands its ramifications. The project manager documents all deviations and explains their consequences in detail. Of course, the project manager never surprises company management with bad news and always gives affected managers prior warning before bad news is documented and distributed.

In addition to the status memos and reports, the project manager also schedules monthly status meetings with the project's executive sponsor and other company management to keep them informed of the project's progress. Some of the topics presented are as follows:

- Schedule status (are the major milestones still valid?)
- Budget status (how are expenditures tracking against the plan?)
- Major activities scheduled within the next month
- Summaries of the changes and issue and problem lists
- Concerns or issues that require executive sponsor awareness or action
- Action plans to resolve issues or problems

Although the project manager always invites all appropriate members of company management to each meeting, the actual attendees of the monthly status meeting vary, depending upon the perceived "health" of the project. If the project is going smoothly, perhaps only the executive sponsor, the project manager, and the technical team leader attend. If the project is having problems or during critical periods, the project manager should expect several members of the company's management to attend the meetings and prepare accordingly.

Set Up a Project War Room

Many meetings are held during the course of a warehouse automation project. In addition to the weekly team review meeting, there are design reviews, issue resolution meetings, planning sessions, and so on. Various company conference rooms could be scheduled for each meeting; however, having a room reserved for the project is a valuable asset for promoting good team communications. A dedicated room allows team members to conduct impromptu meetings whenever the need arises and allows the team to maintain progress charts, status charts, a large-scale project plan, and various project displays where they are available to all and ready for meetings. With these considerations in mind, the project manager sets up what is known as the project war room. This is a room large enough to hold the entire project team, with chairs for all, a large worktable, abundant blackboard space, corkboards for posting charts, a speaker phone for telephone conferences, and provisions for projecting and displaying data including computer workstations with large displays. In order to foster a team esprit de corps, the project manager posts the project team name prominently outside the door to the room. All members of the team are encouraged to use the war room for project meetings. As the project progresses, the project team and company management come to associate the war room with the project, and discussions of the project will gravitate to this room. This phenomenon helps to foster good communications and builds a good working team.

Conduct the Project Kickoff Meeting

The final project start-up activity is to conduct the project kickoff meeting. This meeting should occur after the team orientation meeting, but prior to any actual substantive project work. This meeting becomes the official project starting point and signals all parts of the company that the project has begun. Managers from all of the key operational areas of the company are invited to this meeting. Because many of the attendees have very little knowledge of the project, one of the goals of this meeting is to provide them with sufficient information so that they can determine how the project will affect their organizations. At this meeting, the project manager does a formal presentation of the following:

- Project background and purpose
- Goals and scope
- How the project will affect the company
- Schedule overview
- Introductions of team members
- Team organization and staffing plan

GENERATE A PROJECT PLAN

The project manager now draws up a detailed project plan, beginning with the schedule overview in the high-level plan. The project plan is the key document around which most of the project manager's efforts in tracking and controlling the project will revolve. Because of this, at the beginning of each phase, the project manager concentrates most of his or her efforts on preparing the project plan. Since the process of creating the project plan is very complex, the project manager assigns portions of the plan to various team members. Of course, the plan is not prepared in a vacuum, and the project manager involves the functional managers who will be supporting parts of the project. In so doing, the project manager, the team members, and the functional managers are all forced to think through what will be required to complete the project. The steps to creating a project plan are to:

- Develop work breakdown structure
- Assign durations
- Determine interrelationships among tasks
- Assign resources

Each of these steps is discussed in turn, below, followed by an example of a warehouse automation project plan.

Develop a Work Breakdown Structure

The first step of writing the project plan is to break down each phase of the project into a series of high-level tasks. Each of these tasks must then be broken down into a series of subtasks and so forth. This process continues until each subtask is small enough to allow realistic labor and cost estimates to be made and to permit the project manager to detect schedule slippages before they become serious. A good project management goal is to be able to detect a schedule slippage six time periods before it occurs. In other words, a one-week schedule slip should be detected 6 weeks before it occurs, and a 1-month skip should be detected 6 months before it occurs. Although this is a difficult goal to meet, project activities never suddenly become a week late 1 week before they are due. Tasks frequently become late in little increments over time. The project manager can uncover these small slips if the project plan is sufficiently detailed. By detecting problems well in advance, the project manager is able to take corrective actions before they grow and seriously affect the overall project schedule. A common rule of thumb is that no task in the project plan should be more than 40 hours in planned duration. This level of detail allows detecting slippages when tasks are 1 day or less late. This also ensures that no task will go for more than two team meetings if it is not behind schedule.

This series of tasks and subtasks is known as the project work breakdown structure (WBS). In order to be able to identify each item, many of which will have similar names, each is assigned a unique identification number, the WBS number. A common technique for assigning WBS numbers is to use an outline style of numbering. In the system the third major step of phase 1 has the WBS number 1.3. The second major activity within this step is then 1.3.2. This activity's third subtask is then 1.3.2.3, and so forth.

Of course, the details of phases 2 and 3 (see Chap. 2) cannot be determined at the start of phase 1 since the definition of the work in these phases depends upon the definitions and designs created in their predecessor phases. However, most if not all of the work to be done in phase 1 is known at the start of the project, and phase 1 can be planned in detail in the first version of the project plan. Although the details of phases 2 and 3 are not known at the start, many of the major steps are known and have been explained in this book. Thus, the major tasks of phases 2 and 3 are planned at a high level in the initial project plan. As more information is defined, the details of these phases are refined in the project plan and a phase is completely planned shortly before it begins. A project plan is not a static document, and each task definition becomes more and more precise as the task's start date draws nearer. A task that is 1 month in duration and is planned to occur 6 months from now may be an acceptable level of detail, particularly if the task will be further defined by design work that will take place in the intervening time. On the other hand, a 1-month task that will start next week is simply too large to permit adequate tracking.

As the tasks are determined, they are entered into a project management program. Many good project management software packages are available for personal computers. It is not the intent of this book to teach the use of any particular package, but all of the commonly available packages support the techniques discussed in this chapter. These programs are of great help to the project manager. They take care of many of the tedious aspects of creating a project plan. For example, once all of the tasks and their durations, interrelationships, and resource requirements have been entered, the program will automatically draw Gantt charts and PERT charts, determine critical paths, and level resources. Once again, if these terms are not familiar, the reader is referred to the references at the end of the chapter.

Assign Durations

Once the project tasks have been defined, the project manager determines how long each task will take to complete. This is known as the *task's duration*. For many tasks, determining the duration is relatively simple. For example, the time required to procure bar code labels can be determined by asking various label vendors for their standard lead times. Of course, it is well known that standard lead times can always be improved by paying a premium or by selecting a vendor with especially short lead times, albeit at a higher price. These techniques are reserved, however, for situations in which a schedule slippage must be prevented. For the purposes of the project plan, the project manager only uses standard lead times. In this way, improvements can be scheduled when necessary and when the extra cost is justified by the time saved.

For other tasks, the functional department responsible for completing the tasks provides the project manager with a time estimate. The project manager then reviews the estimate with the functional manager to ensure that the estimated duration allows sufficient time to complete the task under normal working conditions but is not overly conservative.

For other tasks, the duration must be estimated by the project team. A typical example is writing the software to support a put-away operation. A similar module has never been written before, and the time required is unknown. Figure 10-9 shows the subtasks of this activity. Each of the subtasks in this example is fairly typical of those required for the generation of a software module. Many of the software modules in a project will have the same, or similar, subtasks. In fact, most of the tasks involved in any warehouse automation project fall into various categories, each constituent of which will follow a common set of subtasks. The project manager takes advantage of this commonality by creating templates for many tasks that are used to break the project into detailed subtasks.

Although all of the tasks in a category have similar subtasks, this does not mean that these similar subtasks will have the same durations or require the same resources. Different tasks may be similar in structure, but the work to be done will vary widely due to different degrees of difficulty and complexity. For example, installing carton shelving requires many of the same steps as installing pallet racking. However, pallet racking installation is much more difficult and complex and requires very different equipment.

Returning to the example in Fig. 10-9, a duration must be determined for each subtask. The project manager could review each subtask with the department that will be writing the module. When a project contains dozens of such modules requiring estimation, this process is very time consuming and subject to too much variability. A simpler, more consistent approach is through the use of complexity size matrixes. These allow the project manager to easily determine the duration for common subtasks. Figure 10-10 shows such a matrix for the subtask of code design. The project manager has previously met with the involved functional departments, and agreements have been reached on the durations in the matrix based upon their experience. Similar matrixes have been prepared for each subtask typically involved in software module construction.

To use this matrix, the task under consideration is reviewed to determine if its complexity is simple, average, or complex. The definitions of these levels were written down and agreed upon when the matrix was created. The complexity definitions corresponding to the example software matrix are shown in Fig. 10-11.

Similarly, the task under consideration is reviewed to determine if its size is little, medium, or big. These size definitions have also been documented and agreed upon. Figure 10-12 shows the size definitions for the example software matrix.

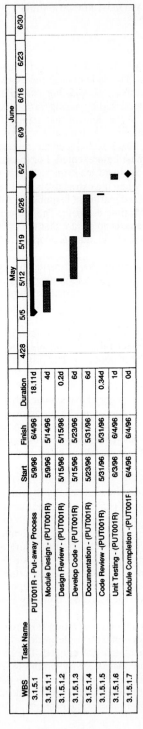

WBS	Task Name	Start	Finish	Duration	4/28	May 5/5	May 5/12	May 5/19	May 5/26	6/2	June 6/9	June 6/16	6/23	6/30
3.1.5.1	PUT001R - Put-away Process	5/9/96	6/4/96	18.11d										
3.1.5.1.1	Module Design - (PUT001R)	5/9/96	5/14/96	4d										
3.1.5.1.2	Design Review - (PUT001R)	5/15/96	5/15/96	0.2d										
3.1.5.1.3	Develop Code - (PUT001R)	5/15/96	5/23/96	6d										
3.1.5.1.4	Documentation - (PUT001R)	5/23/96	5/31/96	6d										
3.1.5.1.5	Code Review -(PUT001R)	5/31/96	5/31/96	0.34d										
3.1.5.1.6	Unit Testing - (PUT001R)	6/3/96	6/4/96	1d										
3.1.5.1.7	Module Completion -(PUT001F	6/4/96	6/4/96	0d										

FIGURE 10-9 The subtasks of a software program development.

10-26

Module design task for software system construction

	Estimated days			
	Function complexity	Simple	Average	Complex
Function size	Little	2	4	5
	Medium	3	6	8
	Big	6	10	14

FIGURE 10-10   An example complexity/size duration matrix.

Function complexity	Function complexity definition
Simple	Stand-alone operation, subordinate to another warehouse function with no dependent functions. A data entry point.
Average	Subordinate function with two or three related or dependent functions. Defines a process such as trailer unload or detailed receiving.
Complex	Focal point function; controls a set of warehouse processes with three or more dependent (subordinate) functions. Examples would include receiving log maintenance, order processing, and work with shipments.

FIGURE 10-11   Example complexity definitions.

Function size	Function size definition
Little	One to six screens or functions used within the process
Medium	In between
Big	More than 11 screens used within the process

FIGURE 10-12   Example size definitions.

The project team examines the example put-away process and determines whether the software required is small in size and is of medium complexity. Thus, to determine the planned duration of the code design subtask for the put-away software, the project manager refers to the matrix and determines that code design will take 3 days. The project manager uses similar matrices to determine the planned duration for each subtask of the put-away software.

This technique of using a complexity/size matrix can be applied to the estimation of many different types of tasks. Initially determining the proper duration values for the matrix and obtaining agreement on the correct values from the associated functional management can be a lengthy process. Once this is complete, however, then using the matrix to assign durations to activities proceeds quite quickly.

If a task's duration cannot be easily estimated, the project manager reexamines the content of the task to determine if the task can be broken into smaller subtasks. Difficulty in estimating a task's duration is an indicator that the task is either too large or too poorly defined to properly track and control.

Determine Relationships among Tasks

In most cases, a project activity cannot start until some other activity is completed. For example, the computer software to support a process cannot be designed until the process design is complete. In other cases, the start of an activity must wait until some other activity is partially completed. For example, bar code labeling of pallet racking cannot start until the first aisle of the racking has been installed. The project manager and the team determine these relationships among the tasks in the project and enter them into the project plan. This is generally a fairly straightforward job, although rather tedious.

Assign Resources

The next step in creating a project plan is to determine the resources required for each task. Resources include anything that is available in limited quantities to the project. Typically, the resources included in a project plan are people and equipment. For each task, the project manager determines how many people and the skills required to complete the task. This is entered into the project plan. If any special equipment is required, for example, an order picker to install rack labels, this is also included in the plan. In addition, the project manager determines what resources are actually available for the project and during what time periods. People assigned to the project are never assumed to be 100 percent available to work on project activities. Administrative work, vacations, sick time, meetings, training, and so forth always consume time that is not listed in the plan. The project manager never assigns more than 75 percent of a person's time to project plan activities. The project manager also never plans on personnel working overtime in the original plan. Overtime is reserved for situations when a late activity must be brought back on schedule. Planning on overtime removes one of the project manager's key tools for keeping the project on schedule.

The project management tool used to record the plan will indicate when the resources used exceed those available. The project manager then either modifies the plan to resolve the overutilization or tries to obtain additional resources.

Example Project Plan

A high-level view of a project plan for a typical warehouse automation project is shown in Fig. 10-13. This plan shows phases 1 and 3 and their major steps, along with some key milestones. Preceding chapters have covered the details of the activities that take place during phases 1 and 2. Therefore, the discussion in this chapter will be primarily concerned with phase 3, the implementation phase.

As Fig. 10-13 shows, the major steps of phase 3 are the following:

- Software system customization or construction
- MHS and technology hardware procurement
- Installation
- Testing
- Training
- Implementation

Most project management tools allow the project manager to alter the view of a project plan by collapsing a task, which hides its subtasks from view. Figure 10-13 shows

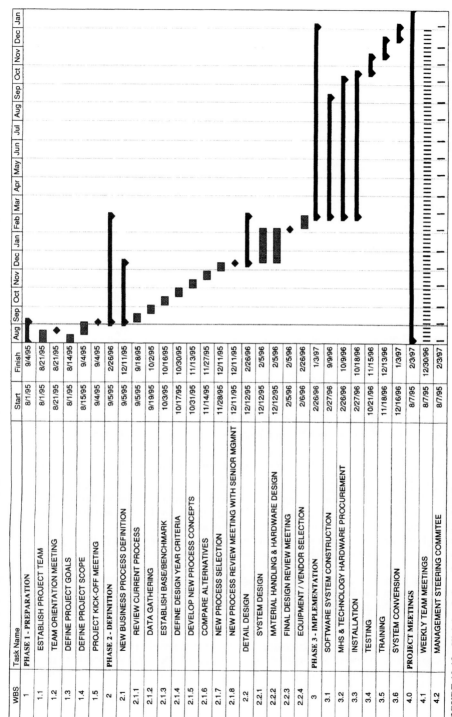

WBS	Task Name	Start	Finish
1	PHASE 1 - PREPARATION	8/1/95	9/4/95
1.1	ESTABLISH PROJECT TEAM	8/1/95	8/21/95
1.2	TEAM ORIENTATION MEETING	8/21/95	8/21/95
1.3	DEFINE PROJECT GOALS	8/1/95	8/14/95
1.4	DEFINE PROJECT SCOPE	8/15/95	9/4/95
1.5	PROJECT KICK-OFF MEETING	9/4/95	9/4/95
2	PHASE 2 - DEFINITION	9/5/95	2/26/96
2.1	NEW BUSINESS PROCESS DEFINITION	9/5/95	12/11/95
2.1.1	REVIEW CURRENT PROCESS	9/5/95	9/18/95
2.1.2	DATA GATHERING	9/19/95	10/2/95
2.1.3	ESTABLISH BASE/BENCHMARK	10/3/95	10/16/95
2.1.4	DEFINE DESIGN YEAR CRITERIA	10/17/95	10/30/95
2.1.5	DEVELOP NEW PROCESS CONCEPTS	10/31/95	11/13/95
2.1.6	COMPARE ALTERNATIVES	11/14/95	11/27/95
2.1.7	NEW PROCESS SELECTION	11/28/95	12/11/95
2.1.8	NEW PROCESS REVIEW MEETING WITH SENIOR MGMNT	12/11/95	12/11/95
2.2	DETAIL DESIGN	12/12/95	2/26/96
2.2.1	SYSTEM DESIGN	12/12/95	2/5/96
2.2.2	MATERIAL HANDLING & HARDWARE DESIGN	12/12/95	2/5/96
2.2.3	FINAL DESIGN REVIEW MEETING	2/5/96	2/5/96
2.2.4	EQUIPMENT / VENDOR SELECTION	2/6/96	2/26/96
3	PHASE 3 - IMPLEMENTATION	2/26/96	1/3/97
3.1	SOFTWARE SYSTEM CONSTRUCTION	2/27/96	9/9/96
3.2	MHS & TECHNOLOGY HARDWARE PROCUREMENT	2/26/96	10/9/96
3.3	INSTALLATION	2/27/96	10/18/96
3.4	TESTING	10/21/96	11/15/96
3.5	TRAINING	11/18/96	12/13/96
3.6	SYSTEM CONVERSION	12/16/96	1/3/97
4.0	PROJECT MEETINGS	8/7/95	2/3/97
4.1	WEEKLY TEAM MEETINGS	8/7/95	12/30/96
4.2	MANAGEMENT STEERING COMMITEE	8/7/95	2/3/97

FIGURE 10-13 High-level view of a typical project plan.

a view where all of the major tasks have been collapsed and their subtasks are not visible. This allows the entire project to be viewed in a condensed form. If all of the subtasks were visible, the project plan would cover dozens of pages. The subtasks of an activity can be easily viewed by expanding only that activity.

Software System Construction. Expanding the first major step of phase 3, software system construction, shows the main tasks within this step. This is shown in Fig. 10-14. It can be seen that this step has been subdivided into a set of modules or functional groups of software. Each module represents the software to support a major business process or a logical group of processes. The project manager groups the software in this way so that when the software in a module is complete, the module can be tested in a manner that simulates the associated process. This type of testing is known as functional testing.

The subtasks within each of these modules are very similar. So, to avoid repetition, only one of these modules will be expanded to examine in greater detail and this will serve as a model for the other modules. Figure 10-15 shows the subtasks within the cycle count module. This module consists of the development of several individual programs followed by a functional test of the entire module. The other modules all have a similar subtask structure.

All of the programs within the cycle count module also have similar subtasks within them. An expansion of one of these programs, cycle count on demand, is shown in Fig. 10-16. Note that the subtasks are the same as those that appeared above in the example put-away program. These subtasks are typical of the software development for warehouse automation projects. The first step in developing the program is an internal design. During this internal design, the program designer does the following:

- Defines the file structures and fields used by the program.
- Verifies any external interfaces. These were initially defined in phase 2, but are verified and the details defined during this phase.
- Defines the program in terms of a high-level functional description of what the program does.
- Creates the flow diagrams for the program that detail the step-by-step logic the program is to follow.

The next subtask is an internal design review. A meeting is scheduled in which several of the program designers review the completed internal design and verify that the program design will accomplish its goals and objectives.

Following the design review, the program is developed by a programmer who writes the actual computer language statements using the internal design as a blueprint. When this activity is complete, there is a code review. This is another meeting where several of the project programmers read the computer program to ensure that the program code matches the internal design, that proper programming techniques are used, that the program follows the project's programming standards, and that the program functions as intended.

After the code review, the program is unit tested. During this task, the program is tested using special test data to ensure that the program works properly. Following this step, there is a rework task to allow the programmer time to fix any problems uncovered in unit testing. Once this is complete, the program is ready for the functional test of the entire module.

The remaining program development tasks within the cycle count module have a similar subtask structure. Returning to the cycle count module, Fig. 10-15, the only other task within this module is the module's functional test. Figure 10-17 shows an

WBS	Task Name	Start	Finish	Feb	Mar	Apr	May	Jun	Jul	Aug	Sep	Oct	Nov	Dec	Jan	Feb	Mar	Apr	May	Jun
3	**PHASE 3 - IMPLEMENTATION**	2/26/96	1/3/97																	
3.1	SOFTWARE SYSTEM CONSTRUCTION	2/27/96	9/9/96																	
3.1.1	Location Management	2/28/96	4/16/96																	
3.1.2	Receiving	2/28/96	5/14/96																	
3.1.3	Replenishment	2/28/96	5/20/96																	
3.1.4	Cycle Count	3/27/96	7/3/96																	
3.1.5	Putaway	4/24/96	6/12/96																	
3.1.6	Order Processing	4/24/96	7/3/96																	
3.1.8	Picking	4/24/96	8/5/96																	
3.1.7	Shipping	5/22/96	9/9/96																	
3.1.9	Interfaces	2/28/96	9/2/96																	

FIGURE 10-14 Software system construction.

10-31

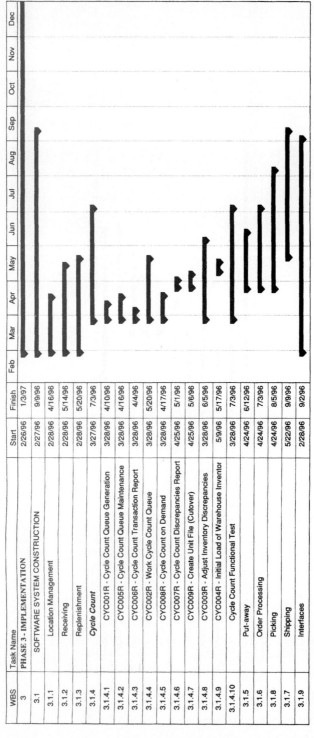

WBS	Task Name	Start	Finish	Feb	Mar	Apr	May	Jun	Jul	Aug	Sep	Oct	Nov	Dec
3	PHASE 3 - IMPLEMENTATION	2/26/96	1/3/97											
3.1	SOFTWARE SYSTEM CONSTRUCTION	2/27/96	9/9/96											
3.1.1	Location Management	2/28/96	4/16/96											
3.1.2	Receiving	2/28/96	5/14/96											
3.1.3	Replenishment	2/28/96	5/20/96											
3.1.4	*Cycle Count*	3/27/96	7/3/96											
3.1.4.1	CYC001R - Cycle Count Queue Generation	3/28/96	4/10/96											
3.1.4.2	CYC005R - Cycle Count Queue Maintenance	3/28/96	4/16/96											
3.1.4.3	CYC006R - Cycle Count Transaction Report	3/28/96	4/4/96											
3.1.4.4	CYC002R - Work Cycle Count Queue	3/28/96	5/20/96											
3.1.4.5	CYC008R - Cycle Count on Demand	3/28/96	4/17/96											
3.1.4.6	CYC007R - Cycle Count Discrepancies Report	4/25/96	5/1/96											
3.1.4.7	CYC009R - Create Unit File (Cutover)	4/25/96	5/6/96											
3.1.4.8	CYC003R - Adjust Inventory Discrepancies	3/28/96	6/5/96											
3.1.4.9	CYC004R - Initial Load of Warehouse Inventor	5/9/96	5/17/96											
3.1.4.10	Cycle Count Functional Test	3/28/96	7/3/96											
3.1.5	Put-away	4/24/96	6/12/96											
3.1.6	Order Processing	4/24/96	7/3/96											
3.1.8	Picking	4/24/96	8/5/96											
3.1.7	Shipping	5/22/96	9/9/96											
3.1.9	Interfaces	2/28/96	9/2/96											

FIGURE 10-15 Cycle count module.

WBS	Task Name	Start	Finish
3	PHASE 3 - IMPLEMENTATION	2/26/96	1/3/97
3.1	SOFTWARE SYSTEM CONSTRUCTION	2/27/96	9/9/96
3.1.1	Location Management	2/28/96	4/16/96
3.1.2	Receiving	2/28/96	5/14/96
3.1.3	Replenishment	2/28/96	5/20/96
3.1.4	Cycle Count	3/27/96	7/3/96
3.1.4.1	CYC001R - Cycle Count Count Queue Generation	3/28/96	4/10/96
3.1.4.2	CYC005R - Cycle Count Queue Maintenance	3/28/96	4/16/96
3.1.4.3	CYC006R - Cycle Count Transaction Report	3/28/96	4/4/96
3.1.4.4	CYC002R - Work Cycle Count Queue	3/28/96	5/20/96
3.1.4.5	CYC008R - Cycle Count on Demand	3/28/96	4/17/96
1.6.2.4.8.1	Module Design - (CYC008R)	3/28/96	3/28/96
1.6.2.4.8.2	Design Review - (CYC008R)	3/28/96	3/28/96
1.6.2.4.8.3	Develop Code - (CYC008R)	3/28/96	4/4/96
1.6.2.4.8.6	Code Review -(CYC008R)	4/4/96	4/5/96
1.6.2.4.8.4	Documentation - (CYC008R)	3/28/96	3/28/96
1.6.2.4.8.5	Unit Testing - (CYC008R)	4/5/96	4/12/96
3.1.6.6.7	Module Rework - (CYC008R)	4/12/96	4/16/96
1.6.2.4.8.8	Module Completion -(CYC008R)	4/16/96	4/17/96
3.1.4.6	CYC007R - Cycle Count Discrepancies Report	4/25/96	5/1/96
3.1.4.7	CYC009R - Create Unit File (Cutover)	4/25/96	5/6/96
3.1.4.8	CYC003R - Adjust Inventory Discrepancies	3/28/96	6/5/96
3.1.4.9	CYC004R - Initial Load of Warehouse Inventor	5/9/96	5/17/96
3.1.4.10	Cycle Count Functional Test	3/28/96	7/3/96
3.1.5	Put-away	4/24/96	6/12/96
3.1.6	Order Processing	4/24/96	7/3/96
3.1.8	Picking	4/24/96	8/5/96
3.1.7	Shipping	5/22/96	9/9/96
3.1.9	Interfaces	2/28/96	9/2/96

FIGURE 10-16 Subtasks within cycle count on demand.

WBS	Task Name	Start	Finish
3	PHASE 3 - IMPLEMENTATION	2/26/96	1/3/97
3.1	SOFTWARE SYSTEM CONSTRUCTION	2/27/96	9/9/96
3.1.1	Location Management	2/28/96	4/16/96
3.1.2	Receiving	2/28/96	5/14/96
3.1.3	Replenishment	2/28/96	5/20/96
3.1.4	Cycle Count	3/27/96	7/3/96
3.1.4.1	CYC001R - Cycle Count Queue Generation	3/28/96	4/10/96
3.1.4.2	CYC005R - Cycle Count Queue Maintenance	3/28/96	4/16/96
3.1.4.3	CYC006R - Cycle Count Transaction Report	3/28/96	4/4/96
3.1.4.4	CYC002R - Work Cycle Count Queue	3/28/96	5/20/96
3.1.4.5	CYC008R - Cycle Count on Demand	3/28/96	4/17/96
3.1.4.6	CYC007R - Cycle Count Discrepancies Report	4/25/96	5/1/96
3.1.4.7	CYC009R - Create Unit File (Cutover)	4/25/96	5/6/96
3.1.4.8	CYC003R - Adjust Inventory Discrepancies	3/28/96	6/5/96
3.1.4.9	CYC004R - Initial Load of Warehouse Inventor	5/9/96	5/17/96
3.1.4.10	Cycle Count Functional Test	3/28/96	7/3/96
3.1.4.10.1	Test Case Writing	3/28/96	4/10/96
3.1.4.10.1.1	Write Cases 1 to 7	3/28/96	4/3/96
3.1.4.10.1.2	Write Cases 8 to 15	4/4/96	4/10/96
3.1.4.10.2	Testing	6/5/96	7/3/96
3.1.4.10.2.1	Test Cases 1 to 5	6/5/96	6/12/96
3.1.4.10.2.2	Test Cases 6 to 10	6/12/96	6/19/96
3.1.4.10.2.3	Test Cases 11 to 15	6/19/96	6/26/96
3.1.4.10.2.4	Code Rework & Regression Test	6/26/96	7/3/96
3.1.5	Put-away	4/24/96	6/12/96
3.1.6	Order Processing	4/24/96	7/3/96
3.1.8	Picking	4/24/96	8/5/96
3.1.7	Shipping	5/22/96	9/9/96
3.1.9	Interfaces	2/28/96	9/2/96

FIGURE 10-17 Cycle count functional test plan.

expansion of the subtasks of the cycle count module's functional test. This is a fairly simple task breakdown. While the individual programs of the module are being developed, the tester is writing the test cases for the module. Note that the test case writing task is neither dependent on nor refers to any of the programs' internal designs. The tester uses only the external designs developed in phase 2. The intent of functional testing is to verify that the module will support the associated business processes and not to verify merely that it follows the internal designs. A program may be free of errors and still fail to support its intended business process properly. Functional testing is intended to uncover this type of flaw early in the implementation process and avoid wasted efforts during the final step of implementation.

Because the actual functional testing is planned to take more than 40 hours, the project manager breaks the testing into smaller subtasks for tracking purposes. In this case, each functional test subtask is a subset of the test cases. In this manner, if the first subset of the test cases has not been completed within the planned time, the project manager knows that functional testing is behind schedule instead of waiting 3 weeks to learn that there is a problem.

MHS and Technology Hardware Procurement. Figure 10-18 shows the major tasks under the second major step of phase 3, hardware procurement. The project manager has broken this step into a major task for each different piece of hardware that is to be procured in conjunction with the project. The list of items shown in this example project plan is fairly typical of most warehouse automation projects. The list for any specific project, of course, may not need some of the items or may include others depending upon the actual requirements of that project. It should be noted, however, that this plan does include items that are frequently overlooked and should be considered, for example, an air compressor for conveyor air and terminals, telephones, and printers to support the new warehouse processes.

As was the case with the system software, many of the procurement tasks contain similar lists of subtasks. Figure 10-19 shows an expansion of a typical procurement task, the pallet conveyor. It can be seen that the design is finished by completing drawings, reviewing the drawings for completeness and correctness, and obtaining final approvals. The conveyor is then ordered, manufactured, shipped, and received at the warehouse. It should be noted that the duration for the manufacture conveyor activity is considerably longer than the standard maximum of 40 hours. Whenever possible, the project manager obtains the vendors' internal schedule for manufacturing each piece of equipment and adds these subtasks to the project plan. This allows easier monitoring of each vendor's progress and helps avoid unpleasant schedule surprises. It is unusual, however, for a hardware vendor to supply a schedule, and it is difficult for the project manager to break such a task into smaller steps without outside assistance. When this occurs, the project manager must take care to track the task closely by contacting the vendor frequently to determine if the shipping date is still valid. If the hardware delivery is critical to the project's schedule, the project manager sends a team member to visit the vendor periodically, to see the hardware being manufactured literally and to verify that the schedule will still be met.

One task in the hardware procurement in Fig. 10-18, location bar code labels, does have a slightly different list of subtasks. These are shown in Fig. 10-20. Before labels can be ordered, the content of the labels must be determined. This involves devising a naming convention for all of the locations in the facility. Then, a location map of the facility is drawn that shows the bar code identification for every location. Based on this, a list of the labels to be printed is drawn up. Meanwhile, label vendors are consulted on the best label stock to be used for the various types of locations and material on which the labels are to be mounted. Samples of the label stock are obtained and

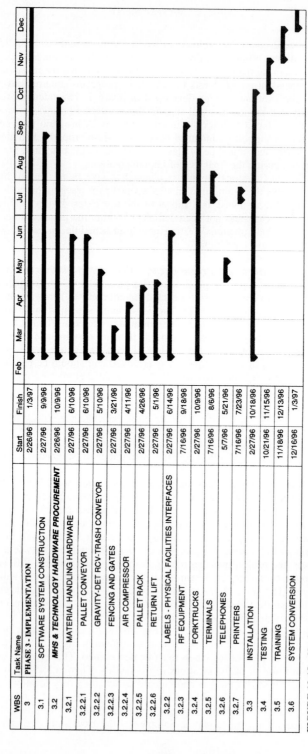

WBS	Task Name	Start	Finish	Feb	Mar	Apr	May	Jun	Jul	Aug	Sep	Oct	Nov	Dec
3	PHASE 3 - IMPLEMENTATION	2/26/96	1/3/97											
3.1	SOFTWARE SYSTEM CONSTRUCTION	2/27/96	9/9/96											
3.2	MHS & TECHNOLOGY HARDWARE PROCUREMENT	2/26/96	10/9/96											
3.2.1	MATERIAL HANDLING HARDWARE	2/27/96	6/10/96											
3.2.2.1	PALLET CONVEYOR	2/27/96	6/10/96											
3.2.2.2	GRAVITY-DET RCV-TRASH CONVEYOR	2/27/96	5/10/96											
3.2.2.3	FENCING AND GATES	2/27/96	3/21/96											
3.2.2.4	AIR COMPRESSOR	2/27/96	4/11/96											
3.2.2.5	PALLET RACK	2/27/96	4/26/96											
3.2.2.6	RETURN LIFT	2/27/96	5/1/96											
3.2.2	LABELS - PHYSICAL FACILITIES INTERFACES	2/27/96	6/14/96											
3.2.3	RF EQUIPMENT	7/16/96	9/18/96											
3.2.4	FORKTRUCKS	2/27/96	10/9/96											
3.2.5	TERMINALS	7/16/96	8/6/96											
3.2.6	TELEPHONES	5/7/96	5/21/96											
3.2.7	PRINTERS	7/16/96	7/23/96											
3.3	INSTALLATION	2/27/96	10/18/96											
3.4	TESTING	10/21/96	11/15/96											
3.5	TRAINING	11/18/96	12/13/96											
3.6	SYSTEM CONVERSION	12/16/96	1/3/97											

FIGURE 10-18   Hardware procurement plan.

10-36

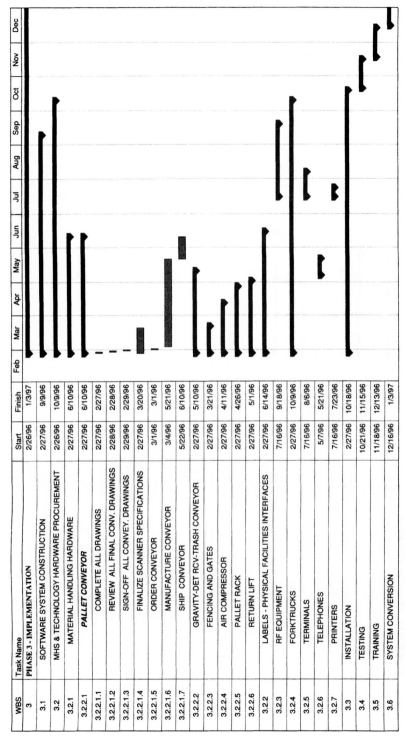

WBS	Task Name	Start	Finish
3	PHASE 3 - IMPLEMENTATION	2/26/96	1/3/97
3.1	SOFTWARE SYSTEM CONSTRUCTION	2/27/96	9/9/96
3.2	MHS & TECHNOLOGY HARDWARE PROCUREMENT	2/26/96	10/9/96
3.2.1	MATERIAL HANDLING HARDWARE	2/27/96	6/10/96
3.2.2.1	PALLET CONVEYOR	2/27/96	6/10/96
3.2.2.1.1	COMPLETE ALL DRAWINGS	2/27/96	2/27/96
3.2.2.1.2	REVIEW ALL FINAL CONV. DRAWINGS	2/28/96	2/28/96
3.2.2.1.3	SIGN-OFF ALL CONVEY. DRAWINGS	2/29/96	2/29/96
3.2.2.1.4	FINALIZE SCANNER SPECIFICATIONS	2/27/96	3/20/96
3.2.2.1.5	ORDER CONVEYOR	3/1/96	3/1/96
3.2.2.1.6	MANUFACTURE CONVEYOR	3/4/96	5/21/96
3.2.2.1.7	SHIP CONVEYOR	5/22/96	6/10/96
3.2.2.2	GRAVITY-DET RCV-TRASH CONVEYOR	2/27/96	5/10/96
3.2.2.3	FENCING AND GATES	2/27/96	3/21/96
3.2.2.4	AIR COMPRESSOR	2/27/96	4/11/96
3.2.2.5	PALLET RACK	2/27/96	4/26/96
3.2.2.6	RETURN LIFT	2/27/96	5/1/96
3.2.2	LABELS - PHYSICAL FACILITIES INTERFACES	2/27/96	6/14/96
3.2.3	RF EQUIPMENT	7/16/96	9/18/96
3.2.4	FORKTRUCKS	2/27/96	10/9/96
3.2.5	TERMINALS	7/16/96	8/6/96
3.2.6	TELEPHONES	5/7/96	5/21/96
3.2.7	PRINTERS	7/16/96	7/23/96
3.3	INSTALLATION	2/27/96	10/18/96
3.4	TESTING	10/21/96	11/15/96
3.5	TRAINING	11/18/96	12/13/96
3.6	SYSTEM CONVERSION	12/16/96	1/3/97

FIGURE 10-19 Pallet conveyor procurement plan.

WBS	Task Name	Start	Finish	Feb	Mar	Apr	May	Jun	Jul	Aug	Sep	Oct	Nov	Dec
3	PHASE 3 - IMPLEMENTATION	2/26/96	1/3/97											
3.1	SOFTWARE SYSTEM CONSTRUCTION	2/27/96	9/9/96											
3.2	MHS & TECHNOLOGY HARDWARE PROCUREMENT	2/26/96	10/9/96											
3.2.1	MATERIAL HANDLING HARDWARE	2/27/96	6/10/96											
3.2.2.1	PALLET CONVEYOR	2/27/96	6/10/96											
3.2.2.2	GRAVITY-DET RCV-TRASH CONVEYOR	2/27/96	5/10/96											
3.2.2.3	FENCING AND GATES	2/27/96	3/21/96											
3.2.2.4	AIR COMPRESSOR	2/27/96	4/11/96											
3.2.2.5	PALLET RACK	2/27/96	4/26/96											
3.2.2.6	RETURN LIFT	2/27/96	5/1/96											
3.2.2	*LABELS - PHYSICAL FACILITIES INTERFACES*	2/27/96	6/14/96											
3.2.3.1	DESIGN LABELS, GARMENT TAG	2/27/96	2/27/96											
3.2.3.2	DESIGN LOCATION NAMING CONVENTION	2/27/96	3/4/96											
3.2.3.3	CREATE FACILITY MAP	3/5/96	4/3/96											
3.2.3.4	DETERMINE LABEL PRINT SEQUENCE	4/4/96	4/30/96											
3.2.3.5	RECOMMEND LABEL VENDOR	5/1/96	5/1/96											
3.2.3.6	GET SAMPLE LABEL	5/2/96	5/2/96											
3.2.3.7	ORDER LABEL SOFTWARE	5/3/96	5/9/96											
3.2.3.8	ORDER LABEL STOCK	5/3/96	5/9/96											
3.2.3.9	LOAD LOCATION FILES TO AS/400	4/4/96	4/10/96											
3.2.3.10	SHIP LABEL STOCK	5/10/96	6/14/96											
3.2.3	RF EQUIPMENT	7/16/96	9/18/96											
3.2.4	FORKTRUCKS	2/27/96	10/9/96											
3.2.5	TERMINALS	7/16/96	8/6/96											
3.2.6	TELEPHONES	5/7/96	5/21/96											
3.2.7	PRINTERS	7/16/96	7/23/96											
3.3	INSTALLATION	2/27/96	10/18/96											
3.4	TESTING	10/21/96	11/15/96											
3.5	TRAINING	11/18/96	12/13/96											
3.6	SYSTEM CONVERSION	12/16/96	1/3/97											

FIGURE 10-20 Bar code label procurement plan.

tested to be sure that the labels are durable, adhere well, and can be scanned reliably. Only when an acceptable label stock has been selected and a comprehensive label specification has been written is a vendor selected and labels ordered. (For more on labels and printing, see Chap. 3.)

Installation.   Once the hardware and software have been built, they are installed in the facility. Figure 10-21 shows the installation portion of the project plan. Note that not all of the installations of various items occur at the same time. For example, the air compressor must be operational before the conveyor can be tested, so its installation is planned to be completed before the planned start of conveyor testing. The project manager takes into account the order in which equipment must be installed and coordinates the start dates of the various installation activities accordingly.

The actual installation of the warehouse management system (WMS) software is a relatively minor activity in terms of time as the verification of the software is done in the testing step. However, it is important that the WMS be installed properly. Two separate sets of data files are established for the WMS. One set of files is designated the production system. These files are used for actual facility operations once the automated warehouse is put into operation. The second set of files is designated the test system. These files are used for testing of the various systems and for training of facility personnel. The production system is used for actual data entry, such as the entering of location information into the system. Periodically, the data in the production system is used to refresh the test system. Testing and training is never done on the production system, only on the test system. This minimizes the risk of an error or mistake corrupting the production data. To minimize confusion, the sign-ons for most of the facility personnel are set up to use the test system. Just prior to cut-over, these sign-ons are modified to use the production system.

The program files for the WMS are also divided into three different sets. One set is designated the production set. These are programs that have been tested and are in use in normal operations. A second set is designated the test set. These are programs that are under test prior to being approved for use in production operations. These programs are set up to only operate against data in the test system. The third set is designated preproduction. These are programs that have been recently approved for production use but are undergoing a "probation" period. Any program that was replaced by a new program is kept unused in the production set. If an unforeseen problem arises with one of the new programs, it can then be easily removed from use without significant disruptions in operations. A single designated coordinator is given the authority to place programs into the test set and to promote programs from test to preproduction and from preproduction to production.

Figure 10-22 shows an expansion of some of the hardware installation tasks in the example project plan. The subtasks shown are typical of those for equipment installation. Prior to the receipt of each piece of equipment, the associated electrical power and compressed air requirements have been identified and the needed power and air have been installed. Also, prior to receipt, an area has been identified and cleared to stage the components of the equipment temporarily as it is being installed. This is frequently overlooked, but components cannot be expected to be stored in the same spot where workers are installing equipment. A separate area must be found where temporary storage will not interfere with installation or with ongoing facility operations. Tasks are designated to install sprinklers that are required by new platforms or racking. Testing of each item is included in the plan, along with the creation of a punch list of any work that was not properly done. Once the testing is complete and the punch list items have been corrected, the equipment can be declared operational. There are, however, some tasks remaining to complete the installation. The vendor must deliver

WBS	Task Name	Start	Finish
3	PHASE 3 - IMPLEMENTATION	2/26/96	1/3/97
3.1	SOFTWARE SYSTEM CONSTRUCTION	2/27/96	9/9/96
3.2	MHS & TECHNOLOGY HARDWARE PROCUREMENT	2/26/96	10/9/96
3.3	*INSTALLATION*	2/27/96	10/18/96
3.3.1	MAJOR PROJECT DEPENDENCIES	2/27/96	2/27/96
3.3.1.1	BUILDING & SITE PREP	2/27/96	2/27/96
3.3.1.1.1	BUILDING OCCUPANCY	2/27/96	2/27/96
3.3.2	MATERIAL HANDLING HARDWARE	3/1/96	9/19/96
3.3.2.1	PALLET CONVEYOR	3/1/96	7/15/96
3.3.2.2	GRAVITY-DET RCV-TRASH CONVEYOR	3/7/96	7/11/96
3.3.2.3	FENCING AND GATES	3/22/96	4/10/96
3.3.2.4	AIR COMPRESSOR	4/12/96	4/19/96
3.3.2.5	PALLET RACK	3/11/96	9/19/96
3.3.2.6	RETURN LIFT	3/14/96	5/9/96
3.3.3	LABELS - PHYSICAL FACILITIES INTERFACES	4/4/96	10/17/96
3.3.4	RF EQUIPMENT	8/15/96	9/30/96
3.3.5	FORKTRUCKS	3/1/96	10/18/96
3.3.6	TERMINALS	8/6/96	9/5/96
3.3.7	TELEPHONES	5/21/96	9/19/96
3.3.8	PRINTERS	7/23/96	9/19/96
3.3.9	WAREHOUSE MANAGEMENT SYSTEM	9/9/96	9/10/96
3.4	TESTING	10/21/96	11/15/96
3.5	TRAINING	11/18/96	12/13/96
3.6	SYSTEM CONVERSION	12/16/96	1/3/97

FIGURE 10-21 Installation plan.

WBS	Task Name	Start	Finish	Mar	Apr	May	Jun	Jul	Aug	Sep
3.3.2.1	PALLET CONVEYOR	3/1/96	7/15/96							
3.3.2.1.1	RECEIVE CONVEYOR	5/28/96	6/14/96							
3.3.2.1.2	DELIVER CONV DESCR OPERATION	5/28/96	5/28/96							
3.3.2.1.3	IDENTIFY POWER & AIR REQUIREMENTS	3/1/96	3/21/96							
3.3.2.1.4	INSTALL POWER & AIR	3/22/96	4/9/96							
3.3.2.1.5	IDENTIFY STAGING SPACE & AREA	3/1/96	3/1/96							
3.3.2.1.6	STAGE CONVEYOR	5/28/96	6/14/96							
3.3.2.1.7	RECEIVE PLATFORM	5/28/96	5/28/96							
3.3.2.1.8	INSTALL PLATFORM	5/28/96	5/30/96							
3.3.2.1.9	INSTALL SPRINKLER UNDER PLTFRM	5/31/96	6/6/96							
3.3.2.1.10	INSTALL CONVEYOR	5/28/96	7/1/96							
3.3.2.1.11	HOST INTERFACE TEST	7/2/96	7/3/96							
3.3.2.1.12	CREATE WALL OPENING	5/20/96	5/26/96							
3.3.2.1.13	IDENT & STAGE TEST INV, PALLETS & TOTES	6/25/96	7/1/96							
3.3.2.1.14	TEST CONVEYOR (HARDWARE & SOFTWARE)	7/2/96	7/4/96							
3.3.2.1.15	LOAD & TEST BACK-UP PC	7/2/96	7/4/96							
3.3.2.1.16	CREATE PUNCH LIST	7/5/96	7/9/96							
3.3.2.1.17	COMPLETE PUNCH LIST	7/5/96	7/10/96							
3.3.2.1.18	CONVEYOR ACCEPTANCE	7/11/96	7/11/96							
3.3.2.1.19	SYSTEM OPERATIONAL	7/12/96	7/12/96							
3.3.2.1.20	WARRANTY (1 YEAR) BEGINS	7/15/96	7/15/96							
3.3.2.1.21	DELIVER CONVEYOR OPERATING PROCEDURES MANUALS	7/2/96	7/2/96							
3.3.2.1.22	DELIVER CONVEYOR "AS BUILT" DRAWINGS	7/2/96	7/2/96							
3.3.2.1.23	DELIVER CONVEYOR SPARE PARTS LIST	5/24/96	6/13/96							
3.3.2.1.24	ORDER SPARE PARTS	6/14/96	6/20/96							
3.3.2.1.25	ORDER BACK-UP PC FOR PLC	5/24/96	5/24/96							
3.3.2.1.26	RECEIVE BACK-UP PC	5/27/96	5/27/96							

FIGURE 10-22 Example hardware installation tasks.

The chart header months: Mar | Apr | May | Jun | Jul | Aug | Sep

WBS	Task Name	Start	Finish
3.3.2.2	GRAVITY-DET RCV-TRASH CONVEYOR	3/7/96	7/11/96
3.3.2.2.1	RECEIVE CONVEYOR	5/13/96	5/13/96
3.3.2.2.2	IDENTIFY POWER REQUIREMENTS	3/7/96	3/27/96
3.3.2.2.3	INSTALL POWER	3/28/96	4/11/96
3.3.2.2.4	IDENTIFY STAGING SPACE & AREA	3/7/96	3/7/96
3.3.2.2.5	STAGE CONVEYOR	5/13/96	5/24/96
3.3.2.2.6	INSTALL GRAV DTL RCV CONVEYOR	5/14/96	6/3/96
3.3.2.2.7	RELOCATE 5 RCV MODULES	5/2/96	5/10/96
3.3.2.2.8	SHORTEN ALL RCV MODULES	5/14/96	5/22/96
3.3.2.2.9	INSTALL NEW TRASH CONVEYOR	5/14/96	6/17/96
3.3.2.2.10	RELOCATE 1 TRASH CONVEYOR	6/4/96	6/19/96
3.3.2.2.11	MOVE TRASH COMPACTOR	5/14/96	6/14/96
3.3.2.2.12	TEST CONVEYOR	6/20/96	6/21/96
3.3.2.2.13	CREATE PUNCH LIST	6/24/96	6/24/96
3.3.2.2.14	COMPLETE PUNCH LIST	6/25/96	6/26/96
3.3.2.2.15	CONVEYOR ACCEPTANCE	6/27/96	6/27/96
3.3.2.2.16	WARRANTY (1 YEAR) BEGINS	6/28/96	6/28/96
3.3.2.2.17	DELIVER CONVEYOR OPERATING PROCEDURES MANUALS	6/27/96	6/27/96
3.3.2.2.18	DELIVER CONVEYOR "AS BUILT" DRAWINGS	7/11/96	7/11/96
3.3.2.2.19	DELIVER CONVEYOR SPARE PARTS LIST	5/6/96	5/24/96
3.3.2.2.20	ORDER SPARE PARTS	5/27/96	5/31/96
3.3.2.3	FENCING AND GATES	3/22/96	4/10/96
3.3.2.3.1	RECEIVE FENCE & GATES	3/22/96	3/22/96
3.3.2.3.2	INSTALL FENCE & GATES	3/25/96	4/4/96
3.3.2.3.3	TEST GATES	4/5/96	4/9/96
3.3.2.3.4	GATES OPERATIONAL	4/10/96	4/10/96
3.3.2.4	AIR COMPRESSOR	4/12/96	4/19/96
3.3.2.4.1	RECEIVE COMPRESSOR	4/12/96	4/12/96
3.3.2.4.2	INSTALL COMPRESSOR	4/15/96	4/16/96
3.3.2.4.3	TEST COMPRESSOR & AIR SYSTEM	4/17/96	4/18/96
3.3.2.4.4	COMPRESSOR & AIR SYSTEM OPERATIONAL	4/19/96	4/19/96

FIGURE 10-22 (Continued)

the final version of the operating principles and the "as-built" drawings. A list of spare parts for the equipment is drawn up and the parts procured. Any needed backup control computers are procured, and a list of personnel to be trained on the maintenance and operation of the equipment is prepared.

Figure 10-23 shows the tasks involved in the installation of bar code labels to the various locations in the facility. Most warehouse automation projects require the bar code labeling of all the locations where inventory is stored. Note that in this example project the installation of labels takes a considerable amount of time. This is not unusual and is fairly typical of most automation projects. Good project management requires planning this activity well in advance so that sufficient time and resources will be available to allow completing this task prior to the loading of the WMS inventory files.

Testing.   The installation plan includes tasks for the testing of the automation hardware and software. The details of how to test and the subtasks under testing are covered in Chap. 11. A comprehensive discussion of how to plan for testing is also presented in Chap. 11.

Training.   Once the installation of hardware and software is complete, the facility's personnel must be trained in the use and maintenance of these systems. The details of the subtasks under training are covered in Chap. 12.

System Conversion.   Once the automation system is ready to be brought into operation, the cut-over phase begins. This is a major topic on its own and is discussed in detail in Chap. 13.

PROJECT IMPLEMENTATION ACTIVITIES

Once the project plan has been set up, the project manager is the person responsible for keeping the project on schedule and within budget.

Schedule and Budget Tracking

During the weekly team meetings, the project manager notes any deviations from the project plan. If an activity is falling behind schedule, the project manager takes action immediately without waiting for it to fall further behind. A number of actions are possible: additional personnel can be assigned to the activity; the personnel already assigned can be asked to work overtime; if the person assigned does not seem to have the needed skills, the task can be assigned to someone who does; or other activities can be rescheduled to allow extra time to complete the late activity. The project manager evaluates each situation and takes the appropriate action. Knowing which action to take and when to take it is one of the project manager's most important skills and a skill that can only be learned through experience in managing projects.

Updating and Reviewing the Project Plan

The project manager continually monitors the project plan. As the project proceeds, more details are defined about future tasks. These details are entered into the project

WBS	Task Name	Start	Finish	Mar	Apr	May	Jun	Jul	Aug	Sep
3.3.3	LABELS - PHYSICAL FACILITIES INTERFACES	4/4/96	10/17/96							
3.3.3.1	RECEIVE LABEL STOCK	6/17/96	7/24/96							
3.3.3.2	INSPECT AND TEST LABELS	7/25/96	9/16/96							
3.3.3.3	TRAINING ON LABEL SOFTWARE	5/10/96	5/10/96							
3.3.3.4	PAINT FLOOR LOCS - EXISTING BLDG	4/4/96	4/10/96							
3.3.3.5	PAINT FLOOR LOCS - NEW BLDG	4/11/96	4/17/96							
3.3.3.6	PRINT AND APPLY LABELS	4/11/96	10/17/96							
3.3.3.6.1	EXISTING TOTES & PALLETS	5/13/96	5/30/96							
3.3.3.6.2	PICK LOCATIONS	6/11/96	6/19/96							
3.3.3.6.3	FLOOR STAGING, EXISTING BLDG	6/25/96	7/3/96							
3.3.3.6.4	RACK LOCATIONS	7/12/96	7/24/96							
3.3.3.6.5	FLOOR STAGING, NEW BLDG	4/11/96	4/22/96							
3.3.3.6.6	NEW PALLETS	8/1/96	8/30/96							
3.3.3.6.7	INSPECT LABEL INSTALLATION	9/2/96	10/17/96							

FIGURE 10-23 Example label installation tasks.

plan as they are learned. Some plans are found to be incorrect or additional work may be required, so change requests are processed to make the appropriate changes to the plan. The project plan is a living document that continually changes throughout the life of the project. The project manager keeps the plan up to date to reflect the current state of the project.

POSTPROJECT ACTIVITIES

At the conclusion of the project, the project manager gathers the key team members for a "post mortem" meeting. At this meeting the team discusses the project and compares the results against the project's goals. Areas for additional improvement and ways to achieve that improvement are identified. These may be areas in the warehouse or distribution center that need improvement or may include other areas in the company. Also, the team discusses the project process and ways in which the project could have been improved. The change control list is reviewed to determine if any changes were due to oversights or errors that could have been prevented by modifications to the project process. A final report documenting attainment or nonattainment of the project goals, recommendations for further operations improvement, and project process suggestions is created and presented to management. This is vital should additional upgrades be contemplated by the company and if other distribution centers have yet to be automated.

BIBLIOGRAPHY

Charland, Thomas C., *Project Management—Advanced Techniques Handbook,* Management Control Institute, Centreville, VA, 1995.

Davis, Edward W., *Project Management with CPM, PERT & Precedence Diagramming,* Blitz, Middleton, WI, 1986.

Gilbreath, Robert D., *Winning at Project Management: What Works, What Fails and Why,* Wiley, New York, 1995.

Harvard Business Review Staff, *Project Management,* Harvard Business School Press, Cambridge, MA, 1991.

Hastings, Colin, Geddes, Michael, and Briner, Wendy, *Project Leadership,* Ashgate, Brookfield, VT, 1993.

Kerzner, Harold, *Project Management: A Systems Approach to Planning, Scheduling and Controlling,* Van Nostrand Reinhold, New York, 1989.

Kharbanda, O.P., and Pinto, Jeffrey, *Leading Your Team to Success,* Van Nostrand Reinhold, New York, 1995.

Knudson, Joan, and Bitz, Ira, *Project Management: How to Plan and Manage Successful Projects,* AMACOM, New York, 1991.

Lewis, James P., *Project Planning, Scheduling and Control: A Hands-on Guide to Bringing Projects in on Time and on Budget,* Probus, Chicago, 1995.

Miller, Dennis P., *Visual Project Planning and Scheduling: A Personal Approach to Project Management,* Fifteenth Street, Boca Raton, FL, 1994.

Obeng, Eddie, *The Project Manager's Secret Handbook: Secrets of Successful Change,* State Mutual Book & Periodical Service, New York, 1994.

Pinto, Jeffery K., *Successful Information System Implementation: The Human Side,* Project Management Institute Communications Office, Webster, NC, 1994.

Viveralli, James, *Systems Development and Project Management Study Guide,* Insurance Data Management Association, New York, 1990.

Whitten, Neal, *Managing Software Development Project: Formula for Success,* Wiley, New York, 1990.

Wideman, R. Max, *A Framework for Project and Program Management Integration,* Vol. 1, Project Management Institute Communications Office, Webster, NC, 1991.

CHAPTER 11
SOFTWARE TESTING

INTRODUCTION

Testing is necessary to ensure that all functions required to support the distribution center's business processes perform according to the design specifications. When the developed system becomes the new operating system for the DC, a thorough testing process ensures the individual components work not only by themselves but also together. The testing process begins as soon as the first piece of software is developed and continues until the system is ready to "go live." Each test is developed and executed to provide the quality assurance necessary to reduce risk in the both the hardware and system development process.

The need for testing is not limited to projects with new software development. Testing is also vital for projects where packaged software is used to ensure supporting hardware is installed correctly and interfaces to existing systems perform consistently and accurately. Technology equipment such as RF terminals and bar code readers must be tested to guarantee that each unit interfaces reliably and efficiently with the warehouse management system (WMS). In addition, material handling equipment requires proper testing to ensure reliability and proper timing within the WMS functions. For example, a sortation conveyor system used to divert cartons to shipping lanes is tested to validate the carton and lane assignment records that are downloaded from the WMS to the conveyor control system prior to each carton physically traveling by the shipping lane identification scanner. It is important to test this system under peak volume conditions to make sure the system can support the volume of data that must be reliably communicated between the WMS and conveyor.

In order to ensure maintainability of the system after implementation and especially after the original project team has moved onto other projects or responsibilities, it is important to test all system documentation thoroughly for completeness and accuracy. As the final test, the complete system, including all documented operating procedures, must be validated by the end users as the accepted process within the new automated distribution center. The key distribution center automation components that require testing include:

- Hardware
- Software
- Host-system interfaces
- Material handling equipment and interfaces
- Technology equipment
- Documentation

FIGURE 11-1   Why test?

In addition to identifying what to test, a testing methodology is integrated into the process to determine how to test. A thorough testing methodology is similar to that of constructing a building. If the building foundation is not very strong, then no matter how well constructed, the building will be weak. If the foundation of testing is not performed well, later stages are more prone to more complex and unnecessary problems. If building materials are of high quality but are put together poorly, a structure will not stand. Likewise, even though a subelement is tested thoroughly, the resulting function or system can be defective due to a problem or error in the sharing of data, in the response time or system performance, or in the communication with other functions (see Fig. 11-1).

OVERVIEW

A basic tenet of testing is that all solution pieces are first tested individually and then in bigger and bigger building blocks. Concentration is first focused on small software pieces or units where problems are easy to detect and simple to fix. As each subsequent block of complexity is added, attention is focused on how the already tested pieces relate to each other and to the planned design. By testing in this hierarchy with the progressive introduction of "live" data, problems are recognized sooner and resolved faster. Timely problem resolution then leads to reduced costs.

There are five major testing areas (see Fig. 11-2). Each of these tests builds upon the accomplishments of the previous steps in order to construct a solid testing methodology:

Unit test.   Tests on an individual unit, program, or feature in an isolated environment for the purpose of validating the unit against the internal design specifications (see Chap. 9).

Functional test.   Validation that a group of units that perform a certain function works as defined in the external design specifications (see Chap. 9).

Integration test.   Testing a complete and integrated set of functions that make up a business process, including both hardware and software interfaces.

System test.   Executing an entire system in a logical order in a simulated "live" environment, addressing aspects such as multiwarehouse operator usage, peak transaction volume, and data management.

Acceptance test.   End users evaluate the complete system, including the software, hardware, and operational processes.

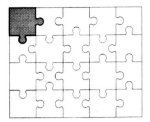

Unit Test

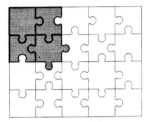

Functional Test

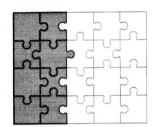

Integration Test

**System and
Acceptance Test**

FIGURE 11-2 The phases of testing.

Each testing building block is discussed in more detail in this chapter. In addition, the various components to be addressed when developing the overall testing methodology are presented, including the test plan, schedule, and test scripts. Each of these components gives structure to the testing steps, which in turn ensures completeness and accuracy in the testing process. Finally, various tools used to assist the testing process are discussed at the close of this chapter.

TESTING METHODOLOGIES

A formal testing methodology helps ensure that the project undergoes a successful test. To ensure uniform compliance, standard procedures are used to answer questions such as who creates the testing documentation, what criterion defines a successful test, and how testing is represented and tracked on a project plan. By following a structured methodology, consistency and understanding are shared by all parties involved. The following are various components to be considered when defining the testing methodology.

Testing Economics

The first step in developing the testing methodology involves examining the economics of testing. The planning standard for customized software testing is to add 40 to 50 percent to the estimated programming duration. As a general rule of thumb, using traditional coding techniques, there are approximately 4 to 8 errors per 100 lines of code that can be found during testing. This can be significantly reduced if CASE or object-oriented approaches are used since these techniques have a significant portion of the unit testing built into the coding process. Owing to the amount of resources and time required for proper testing, there is a tradeoff between risk and cost that must be reviewed and resolved before testing begins. It might not be financially justifiable to test an inquiry function for the same amount of time as a function that controls financial information. The amount of time that is expended on a function test should have a direct relationship with the amount of importance and risk in that function. If risk is not weighed against system function, overtested peripheral pieces and undertested central features result. It is acceptable and preferrable to define a test by the function's relevance to the system as a whole.

Test Scripts

Test scripts are valuable tools in the testing process because they provide a road map for the items to be tested by providing a precise definition of the input data to a function and the expected results. These road maps ensure uniformity and consistency across the testing process and provide the documentation required for successful test completion accountability. Test scripts also provide a vehicle for problem investigation and isolation and allow orderly and efficient regression tests. Regression testing is required when a function fails a test script or a change is made to the function after initial testing. Such regression testing requires that all parts of the test script be executed again to ensure that the change has not affected any other results or created any new problems (see Fig. 11-3).

Test scripts are used in each testing phase, from unit test through acceptance test. Depending on the test phase, they are created using the internal and/or external design documentation as well as any process flows created during the design process.

```
┌─────────────────────────────────────────────────────────────────────────────┐
│                         Warehouse Automation Project                          │
│                          Functional Test Script                               │
│ ─────────────────────────────────────────────────────────────────────────── │
│ Test Case ID:        ORD010R                 Date:        3/3/94              │
│                                                                               │
│ Function Name:            Pick Planning             Tester:  James Smith      │
│                                                                               │
│ Program Name(s):     Wave Creation        Program Number(s):    ORD001R       │
│                      Wave Release                               ORD002R        │
│                      Order Maintenance                          ORD005R        │
│                      Pick Maintenance                           PCK003R        │
└─────────────────────────────────────────────────────────────────────────────┘
```

Function Description: This function allows the user to perform pick planning activities including wave creation/release and order/pick maintenance process.

File Usage: ILCLOC - Read
               ILCPKQ - Read, Update
               ILCUNT - Read, Update
               ILCORD - Read, Update
               ILCORH - Read, Update

PASS/FAIL   **Test Script Processing**

__/__   A.   Verify that the Programmer's unit testing checklist and unit test cases have been
                    completed.

             B.   GENERAL PROCESSING

__/__   -   Create multiple waves for various order types.

__/__   -   Release selected waves for picking.

__/__   -   Perform order maintenance activities (hold/cancel/change quantity) on a release
                    order.

__/__   -   Modify picker information on a pick queue entry.

FIGURE 11-3   Example test script.

Although the test-script format is tailored to meet the testing requirements in each phase, the following information is common to all test scripts:

1. A test-script ID number and name. The test ID number may be the same number as one of the programs in the test. The naming and numbering convention should allow for easy referencing and cataloging.

2. The test date and names of both the test executor and creator.

3. The function or process name and program number(s) being tested (if applicable).

4. The test-script description, i.e., the objective for performing the test script.

5. Any input files used for the test, or any output files generated, and where they are stored.

C. ITEMS BY SCREEN: The following items will be tested on each screen.
Verify the flow from screen to screen.

Screen 1 (Wave Creation):

_/_   From the pick planning menu, select option 1 to retreive this screen.

_/_   Type a "1" in the order selection field, order type____ in "from" field and order type ___ in "to" field.

_/_   Press enter. Verify screen 2 is shown with all orders sorted by order type for the wave.

_/_   Select orders for the wave by keying "1"next to the number.

_/_   Press F8 to create the wave.

_/_   Press F5 to verify wave creation.

_/_   Press F9 to display the "Wave Release" screen (screen 2).

D. SPECIFIC DATA REQUIRED: Record the data necessary to execute the test script.

Customer orders for the following:
1. More than one cutomer; ____, ____, ____.
2. More than one order type; ____, ____, ____, ____.
3. More then one wave; ____, ____, ____, ____, ____.
4. More than one picker; ____, ____, ____.

E. TEST CASES

_/_   1. Verify that multiple waves can be created for various order types.
_/_   Select option 1 from the Pick Planning menu via menu security.
_/_   Key the desired order type ranges in the "from" and "to" fields.
_/_   Press enter. Verify all orders for the wave and that they are sorted by order type.
_/_   Select wave orders by keying "1"and pressing F8.
_/_   Press F5 to verify wave creation.
_/_   Repeat this process using order type ___ and ___.

Expected Results: All order types in the desired range should appear on the screen sorted by order type.
Waves will be created by selecting the order and pressing F8.
Actual Results: _____

_____

_____

_____ Tested by: _____

Date: __/__/__ PTNs _____

FIGURE 11-3   *(Continued)*

6. A list of each test case, specifying the input data and expected outcome(s).

7. A place to record actual output or results.

8. A place to mark sucessful or unsuccessful execution of a test condition.

9. Any problem tracking number (PTN) resulting from the test. The test-script ID may be also entered into an on-line PTN system for cross-reference purposes.

An important consideration when developing test scripts is determining what constitutes an adequate test. Given the constraints on resources and time and the economics of testing, it is not feasible to test a function using every possible input condition. One technique commonly used is known as "error guessing," where based on intuition and experience, test-script creators try to guess what the coders and designers may have overlooked. This is an area where consultants and software or hardware suppliers can provide valuable insight into the testing process. Based on their experiences with previous inplementations, they should be knowledgeable about which input criteria are important to test to promote system reliability. This includes the development of test scripts for assumptions not made explicit in the design specifications, including invalid and unexpected input conditions. For example, a good test-script creator attempts to anticipate situations in the receiving process that may not be explicitly defined in the design document, such as when multiple receiving operators are unloading a single truck. Here a test script is required to verify that multiple end users can efficiently process receipts to the same purchase order.

The project team members approve the test scripts prior to the start of the specific test, checking for test-script completeness and accuracy of the input and expected output conditions. After the test script has been created and approved, the test is executed. A good test script is created in a way that allows anyone to execute the script, not only the creator of that script. In general, coders should not be solely responsible for testing their own programs since finding errors in one's own work is typically difficult to do. In addition, errors may occur due to the coder's misunderstanding of the design specifications. Breakdowns and mistakes during the communication and translation of user requirements are a primary cause for many software errors. The responsible party for the creation and execution of each test depends on the type of testing to be done:

- Unit testing is scripted by the individual responsible for the technical creation of that unit, but preferably tested by another programmer; ideally, the programmer responsible for the unit that calls this unit. For example, the coder responsible for the receipt program is a good candidate to test the receiving adjustment unit.

- Functional testing is both scripted and performed by the project tester who has been involved with and understands the external design. This is also a good stage to have users on the project team participate in the testing.

- Integration testing is usually scripted by the project tester but performed both by the tester and the project leader(s), since the scope is typically expanded to include both hardware, software, and any technology or material handling equipment used for the process being tested.

- System testing utilizes integration scripts tied together by a project leader and executed by the tester and the project leader(s).

- Acceptance testing is both scripted and performed by the warehouse users of the WMS. If a module does not do what the end user reasonably expects it to do, then an error is present.

Regardless of who is performing the test, every test must be signed off as complete. Typically the project leader is responsible for the sign-off on all tests. In order to provide sign-off (approval of a successful test), a set of criteria is required as the basis for the sign-off decision. The criteria used to withhold approval must be established ahead of time, be clearly documented, and be applicable to all types of testing and situations. Since each test script has pass or fail criteria, the project leader bases his or her approval on the documented results, without having to recreate the test.

Once sign-off is accomplished, the equipment or program is ready to move to the next level of testing. The project leader or tester is notified that the test is complete in order to update the project schedule. Then, the successfully passed test script is logged in a project file containing all test scripts and sign-offs.

TESTING SCHEDULE

When the overall master schedule is being created at the start of the project, adequate time should be allotted to testing in order to achieve success. Often, however, when a project gets behind schedule, testing is viewed as an area that can absorb the project overruns since it is one of the last steps in the process. However, reduction in the testing cycle decreases the ability of the testing phase to catch errors, resulting in expensive rework and poor reception by warehouse operators after implementation.

Many testing activities can be scheduled in parallel with other testing and programming tasks (see Fig. 11-4). Thought given to special schedule sequencing when developing the project schedule can improve both testing efficiency and results. When integration testing begins, it is important to schedule the functional tests in a logical order. For example, if the interfaces between the host systems and the distribution center are tested first, there is a mechanism that allows the project team to utilize real data for the remaining tests. Having real data in the test environment not only saves time, it makes all subsequent tests more accurate. The following gives guidelines for scheduling the different testing stages:

- The unit-test stage starts when the first program is complete and finishes when the last program is successfully unit-tested. Since an individual's module and/or equipment is being tested, given adequate resources, it is possible to conduct much of this testing in parallel.

- The functional-test stage starts any time after the unit test of the first program is completed, up until midway through the software development effort, and finishes after the last piece of development is completed.

- The integration-test stage starts in the middle of functional testing and ends sometime after the last functional test is successfully completed. Integration testing does not need to wait until all functional testing is complete since the integration of subsystems can be first tested independently, and then larger and larger portions of the system can be integrated and tested as the functional tests are completed.

- The system-test stage starts when all integration tests are successfully completed since the entire system, from receiving through shipping, and all the required interfaces are being tested.

- The acceptance-test stage starts when all integration tests are successfully completed. Additional risk is added if production begins before the acceptance test is successfully completed.

UNIT TEST

The unit test, the first building block in the testing process, is defined as the execution of a test on an individual unit, program, or feature in an isolated environment. The unit test's main purpose is to compare the function of a unit or program to the internal

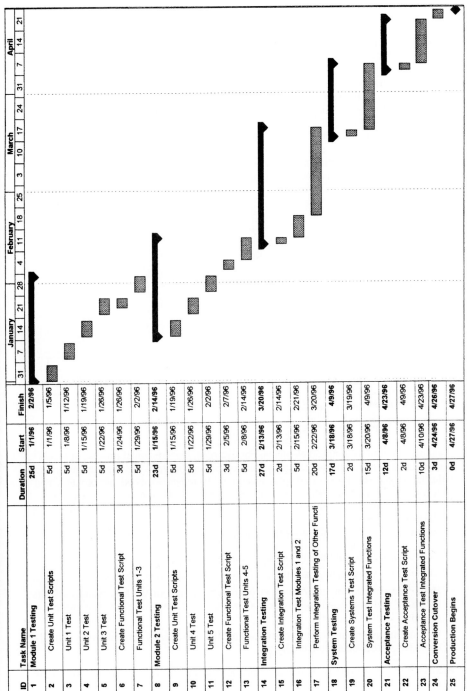

ID	Task Name	Duration	Start	Finish
1	Module 1 Testing	25d	1/1/96	2/2/96
2	Create Unit Test Scripts	5d	1/1/96	1/5/96
3	Unit 1 Test	5d	1/8/96	1/12/96
4	Unit 2 Test	5d	1/15/96	1/19/96
5	Unit 3 Test	5d	1/22/96	1/26/96
6	Create Functional Test Script	3d	1/24/96	1/26/96
7	Functional Test Units 1-3	5d	1/29/96	2/2/96
8	Module 2 Testing	23d	1/15/96	2/14/96
9	Create Unit Test Scripts	5d	1/15/96	1/19/96
10	Unit 4 Test	5d	1/22/96	1/26/96
11	Unit 5 Test	5d	1/29/96	2/2/96
12	Create Functional Test Script	3d	2/5/96	2/7/96
13	Functional Test Units 4-5	5d	2/8/96	2/14/96
14	Integration Testing	27d	2/13/96	3/20/96
15	Create Integration Test Script	2d	2/13/96	2/14/96
16	Integration Test Modules 1 and 2	5d	2/15/96	2/21/96
17	Perform Integration Testing of Other Functi	20d	2/22/96	3/20/96
18	System Testing	17d	3/18/96	4/9/96
19	Create Systems Test Script	2d	3/18/96	3/19/96
20	System Test Integrated Functions	15d	3/20/96	4/9/96
21	Acceptance Testing	12d	4/8/96	4/23/96
22	Create Acceptance Test Script	2d	4/8/96	4/9/96
23	Acceptance Test Integrated Functions	10d	4/10/96	4/23/96
24	Conversion Cutover	3d	4/24/96	4/26/96
25	Production Begins	0d	4/27/96	4/27/96

FIGURE 11-4 Proposed testing schedule.

design that defines the unit (see Chap. 9). The goal is to show that the unit meets the design specifications and to identify areas where it deviates from these specifications. This type of testing is done for customized or new programs and not necessarily for package functions.

Unit testing is logic-driven, that is, the focus is on testing that the program logic executes as the programmer intended; therefore, programmers versus users are involved in this testing phase. A proper unit test helps improve software quality by ensuring that agreed-to standards are adhered to, such as the definition of function keys and the method for entering and saving data. Another objective is to determine that this unit, when inserted into a function or system, will do its job as a "black box," by accepting specific inputs and producing specified outputs (see Fig. 11-5).

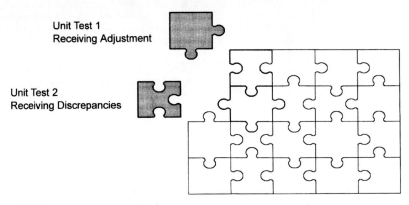

Unit Test 1
Receiving Adjustment

Unit Test 2
Receiving Discrepancies

FIGURE 11-5   Unit-test example.

To determine what pieces of software need to be tested as a unit, each process is broken down into its simplest elements. For example, the functional process of receiving might have one unit that performs the receipt, another that performs the receiving adjustment, and yet another to display receiving discrepancies. Units are broken down to a more finite element when software engineering tools (CASE) are used to construct the system since these tools often perform a great degree of unit testing as the application is written.

The major advantages achieved from the unit test include:

- Determine that the unit meets its individual specifications
- Aid in test management as attention is focused on small units of work
- Help ease the task of pinpointing errors since the unit of work is manageable in both size and scope; when an error is found, it is known to exist in a particular module
- Present the opportunity to test multiple units that make up a complete system simultaneously, rather than sequentially

Examples of typical questions that are to be asked during a receiving adjustment unit test include:

1. Do the command and function keys displayed on the receiving adjustment screen perform correctly?

2. Do the screens flow and navigate in the correct order?

3. Does the purchase-order file get updated correctly?

4. Does the receiving adjustment screen layout match its design specification? Do mandatory fields require input?

5. If an invalid purchase-order number is input, does the program display the message "Invalid PO"?

FUNCTIONAL TEST

The functional test follows and builds upon the unit test. Functional tests are required if design changes are made within a software package or a new function has been created. The functional-test process is defined as the execution of a collective set of units to validate that a particular process or function works correctly from the warehouse operator's perspective. This test differs from unit testing since the functional test validates the external design or customer-acceptance criteria versus the internal design. For this reason, a person other than the unit's programmer or technician performs the test.

A proper functional test improves the system quality beyond that of unit testing due to increased focus on scope. The functional test focuses on a specific group of units that perform a certain function. The main focus in this testing phase is to ensure that a function executes properly from beginning to end, across multiple units. The project tester approaches this test in a less technical way and with more of an "end user" viewpoint using the external design as a road map. In addition, the tester verifies consistency between different units within the function.

Functional testing is not to be a repeat of unit testing; rather, it builds upon the unit test (see Fig. 11-6). An example of a functional process is receiving. To perform the functional process of receiving, the system must successfully be able to perform all units that comprise receiving in an integrated and consistent manner:

- Receive against a purchase order

- Verify SKU number by its description

- View all receiving discrepancies

- Perform a receiving adjustment on that specific receipt

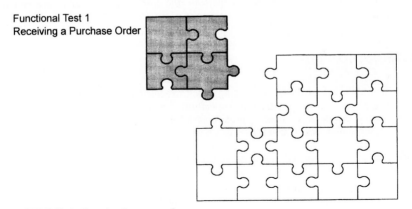

Functional Test 1
Receiving a Purchase Order

FIGURE 11-6　Functional-test example.

Note that the units involved in a functional test can all work correctly, but the functional test can still fail if the function does not do what the user reasonably expects it to do. The error then needs to be traced to flaws in either the internal design or external design. Since the external design document has been previously approved by the user team, any additions or changes should again be reviewed by the team for concurrence.

Benefits derived from the functional test include:

- Ensures consistency between the warehouse operator specifications and the functional requirements of the warehouse processes
- Aids test and project management by focusing and planning based on particular functions
- Verifies that the unit test was performed properly
- Assists in defect localization

Furthermore, many errors and problems appear early by testing the hardware interfaces to the software during this stage of testing. By enabling communication between the equipment and the software early on in the process, the hardware and its interfaces are tested throughout the subsequent testing stages, ensuring full and redundant testing. Testing that is hard to capture by a test script happens naturally when the actual equipment is utilized in the testing process.

Examples of typical questions that are to be asked during a receiving functional test include:

1. Did receiving, SKU look-up, receiving discrepancies, and receiving adjustments processes get successfully unit-tested?
2. When the user went from receiving discrepancies to receiving adjustments activities, did the receiving information display correctly on the adjustment screen?
3. Is the receipt information being translated into receiving adjustments in a real-time manner?
4. Did the purchase-order file, receiving file, and receiving adjustment file get updated correctly?
5. When two warehouse operators tried to perform a receiving adjustment on the same purchase-order and item combination at the same time, did the second operator receive the message "PO in Use"?
6. Do the command keys and options perform the correct function and have consistent use?

INTEGRATION TEST

The integration test is the next building block executed after the functional test has been completed and is the first test to perform if a software package installation is being used. The integration-testing process is defined as the execution of a specific set of functions and processes that make up a business process. Integration testing encompasses various elements out of the unit and functional tests, but its primary objective is to join all functions and elements that make up a business process and test them together, including hardware and interfaces. Both the project tester(s) and the project leader(s) take part in the execution of this test.

Business processes can cross over different systems, use different material handling equipment, or have multiple warehouse operators performing the same task at

Integration Test 1
Processing a Purchase Order
from Creation to Final Receipt

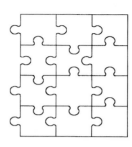

FIGURE 11-7   Integration-test example.

the same time. Many different types of tests must be performed during this stage to validate an entire process and all of its elements. In addition, the system is stressed with simulated volume transactions in specific areas (i.e., interfaces and file updates).

An example of a business process is the processing of a purchase order from creation to final receipt (see Fig. 11-7). To perform this business process, the system must successfully be able to:

• Transfer the purchase-order information from the host system to the distribution center system

• Allow the warehouse operator to perform receiving activity against that purchase order using RF equipment

• Allow the warehouse operator to perform purchase-order inquiry using a CRT terminal if a problem exists

• Interface to a conveyor to be able to identify the received product and divert the product to the necessary zone in the distribution center

• Gather data within receiving and successfully transfer it from the distribution center system back to the purchasing system where it properly updates host data bases

Examples of typical questions that are to be asked during a full-cycle receiving integration test include:

1. Did the receiving interfaces, receiving function, purchase-order inquiry, and conveyor sortation functions each pass their functional tests?

2. Were the purchase orders created on January 1, 1995 by the purchasing system received by the distribution center in their entirety?

3. After receiving against one of the sent purchase orders, was that receiving transaction immediately updated in the system so that it was visible through purchase-order inquiry?

4. Can the purchase-order interface be running to the host system while warehouse operators are receiving against open purchase orders?

5. Did the sortation system divert the carton to the appropriate conveyor for the put-away operation?

6. Was the response time in the receiving function acceptable when handling an entire day's worth of transactions in one batch?

SYSTEM TEST

The system-testing process is defined as the execution of the entire system in a logical order, while addressing multiple warehouse operator usage, peak transaction volume, and live-data aspects. Testers should test against the original customer requirements and customer documentation, while adopting the customer's point of view. System-level tests are the final tests of the software and hardware before the project team releases the system to the users. The project leader(s), project tester(s), and distribution center employees participate during this testing phase.

This test is similar to the integration test but is generally performed at the distribution center, using real conditions that are closely monitored. The primary objective of the system test is to perform a test that will mimic "live production" as close as possible. If receiving is normally performed by 10 warehouse operators, then 10 warehouse operators are used to test receiving. The test uses real data that has been isolated in the test environment. The normal day-to-day issues and situations that arise are tested along with any unusual scenarios that occur as a result of simulating the live environment.

During system testing, all the system components are tested including the hardware, software, host interfaces, material handling equipment and interfaces, technology equipment, and documentation (see Fig. 11-8). Technical issues that are addressed during system testing include:

- Testing for unauthorized security access or manipulation
- Recovery from hardware and software failures and validation of audit trails
- Checking for undesirable interactions when multitasking
- Ensuring compatibility of interfaces to other software systems and various hardware configurations
- Stress-testing complete system function and performance under peak transaction volumes

The system test touches all distribution center activity. To successfully run the DC, the system must successfully perform all of the functions designated in the specifications. These include, but are not limited to, the following functions:

- Receiving
- Put-away

System Test
Warehouse Management
System Full-Cycle Test

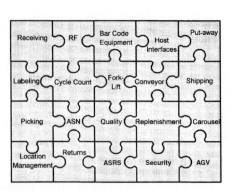

FIGURE 11-8   System-test example.

- Replenishment
- Quality
- Cycle counting
- Order processing
- Picking
- Shipping

Examples of typical questions that are to be asked during a system test include:

1. After picking is performed, are the correct replenishments generated for the locations in need of additional quantity?
2. If a location is being cycle counted by one warehouse operator, can another warehouse operator remove inventory from that same location or can a function allocate inventory to an order from that location?
3. After enough activity to simulate a peak business day is completed, did the end-of-day programs finish in the allotted time frame?
4. If a purchase order is not received in full at the distribution center, did the host purchasing system create a new purchase order to satisfy the demand?

ACCEPTANCE TEST

The acceptance test is the "top floor" in the testing process. The acceptance-testing process is defined as the execution of the entire system in a logical controlled order, comparing actual results with predefined results. The primary objective of the acceptance test is to let the functional users evaluate the software, hardware, and operational processes that have been previously approved by the functional users during internal and external design. This includes accuracy, functionality, and usefulness in the normal business operations. The functional users have total responsibility for creating and executing the acceptance test.

The acceptance test is a very important stage in testing because it allows all business processes to be viewed and judged by the functional users in a controlled and risk-free environment prior to implementation (see Fig. 11-9). Feedback is provided to the project team so that areas of concern can be addressed before the implementation. The acceptance test will focus on:

- Usability
- Convenience and practicality
- Screen and report layouts and help menus
- Performance, throughput, and response time

TESTING TOOLS

Software testing for a warehouse automation project is a significant task. A large portion of project time is used to perform a solid and dependable test. Much of this time can be attributed to performing repetitive tests on the same software or hardware as changes are

---

Warehouse Automation Project
Acceptance Test Script

Test Case ID: Date: **Page 1 of 1**

System Module: Cycle Count   Tester:

GOALS:
Verify that valid cycle count entries are created by end of day and on-demand processing. Cycle count
discrepancies are flagged and corrections can be made. Codes file entries are operational.

SETUP:

1. Codes File
    a. Number of empty locations to count - 4
    b. Cycle count sequence number - 1
    c. Lock location on error - Y
    d. All other values for cycle counting - use default values

2. Locations
    a. At least 5 locations with zero quantity and blank cycle count date.
    b. At least 5 locations with on-hand quantity greater than zero, blank cycle count date, and velocity
        code A.

3. Other
    a. Cycle count queue is empty.
    b. Journaling is started.

Step No.	Actions to Be Performed	Desired Results	Pass/Fail	Severity
1	Select end of day from main menu	End of day menu should appear.		
2	Select "Submit End of Day Process"	Message that job has been submitted to batch queue.		
3	Return to main menu. Select cycle count menu. Start journaling.	Message that journaling has been submitted.		
4	Select "Maintain Cycle Count Queue"	CYC016R.SFLFMT screen is displayed showing all locations selected for cycle counting.		
5	Delete records that are not wanted (i.e., drop zones). Exit from program.	Records should disappear from screen.		

Pass/Fail: Date:

Signature: PTNs:

FIGURE 11-9   Example acceptance-test script.

made and fixed problems retested. As a result, there is a growing use of automated tools
to aid in the testing effort and help ease the burden of repetitive testing.

Automated tools can be grouped into a few categories:

- *Instrumentation.*   Involves the insertion of additional code into software to record
  the execution of statements, branches, and modules. Instrumentation tools can be
  developed in house or bought commercially.

- *Comparators.* Involves putting the expected output of a function into a file and then running the test case. The actual output is put into another file and the comparator then detects discrepancies between the two results.

- *Keystroke recorders.* Involves a software package that records the input and output to a test case and compares the results of that input to the actual output generated by the software during the test.

Automated testing tools have both strengths and weaknesses when compared to conventional testing. As a general rule, automated testing requires more setup time and initial expenditures than manual testing. However, if a test is required multiple times, then automated testing might be the lesser expense over time. It is important to examine each project to determine when a testing tool is the right choice as the test vehicle for a project.

Positive attributes of automated testing:

- Clear directive
- Harder to brush over results
- Uniformity
- Repeatability
- Simulate system loading

Negative attributes of automated testing:

- Longer cycle for setup
- No guarantees
- Possible false sense of security as missing paths can go undetected

CHAPTER 12
TRAINING

INTRODUCTION

Although a warehouse system may be automated, it still must be operated by capable and qualified individuals. Frequently, however, the installation of an automated warehouse system is not met with open arms by many warehouse employees due to the fear of change, potential threat in the loss of jobs, giving up the "paper security blanket," and losing the job security of knowing where every item is stocked. These concerns must be addressed by management prior to conversion to the automated system.

The best tool to minimize these fears and concerns is to train warehouse personnel properly by providing them with the adequate means and time to learn the new processes and procedures. A complete training program begins early in the implementation process and returns long-term benefits to the company. It is important, however, not to underestimate the amount of training required to achieve a successful warehouse conversion.

Spiral and repetitive training sessions are used to ensure that provisions are made for the full range of learning abilities that may exist across the work force. Any change to a complex process such as warehousing and distribution will require repetitive exposure to the concepts to allow personnel to understand the new system fully. Keep in mind that the current system, which everyone understands, did not get created overnight and it takes time to understand the new system's operation. System training sessions typically yield only 20 to 30 percent retention due to the degree of change being imposed on current thinking. Multiple training sessions (repetition) with increasing level of detail (spiral) are required to ensure maximum productivity during the conversion period during which the WMS is put into production. Training sessions are structured such that the quantity of new information is limited so as to not overwhelm personnel, yet providing a plan that is consistent with cost constraints and the warehouse implementation schedule.

First, as soon as the design phase for the automated warehouse is complete, planning for training of warehouse personnel should begin. Initial system training is accomplished by warehouse management reviewing the benefits that the system will bring to various warehouse departments. While it is not necessary, nor desirable, to quantify the financial benefits associated with the automated warehouse system, qualifying the need for change through reviewing the business requirements for reduced cycle times and better customer service assists warehouse personnel in understanding management's decision for change. The scheduled conversion date for the new warehouse system should be reviewed at this time so that everyone understands management's expectations. Improved understanding by all warehouse personnel improves system acceptance and enthusiasm while reducing rumors on how individual jobs will be affected. Also, obtaining early acceptance often results in garnering invaluable feedback that can improve the system's design and reduce possible oversights. Initial training should be reinforced with periodic status meetings to keep everyone informed with the project's progression while providing a valuable feedback mechanism to ensure management is answering questions as they arise.

The next training phase is the high-level system review. This training phase focuses on the cultural shock associated with a computerized warehouse management system (WMS). Warehouse personnel who are not experienced with computers are often intimidated when required to type on a keyboard and/or lack an understanding of what the computer does. To reduce this cultural impact, this training phase emphasizes basic computer terms and introduces the concepts of bar codes, bar code scanning, radio frequency terminals, and mechanized equipment. New material handling (MH) equipment training is given to specific maintenance personnel at this time to allow them to study the new equipment's requirements and modify their operating procedures. In addition, this training phase explains the WMS's scope and how various warehouse departments utilize and interact with the system and provides initial introduction on how the WMS tracks inventory and directs work activity.

The final "formal" training session involves hands-on training, where all warehouse employees (managers, supervisors, and operators) have their first opportunity to run the system. Hands-on training can be accomplished in two modes. First, if it is management's desire that warehouse operators be able to perform all functions, then a classroom environment is used to train large groups. This form of training utilizes teams to work together to assist one another with understanding concepts and gives employees an appreciation of the effect their jobs have on others. Second, "on the floor" training is used to focus specific functional training on warehouse operators who will be performing the specific process. This form of training allows operators to see the system interacting with the physical warehouse and inventory.

Training materials and procedures are investments that provide benefits beyond the initial automated warehouse implementation. New employees require training after the new system has been put into production. Also, training frequently must be conducted at other facilities if the automated warehouse is implemented throughout the company. A thorough training program improves employee morale, minimizes conversion problems, and establishes good communication channels with warehouse personnel.

OVERVIEW

While individual training sessions can be viewed as a series of simple processes, advance planning is required to maximize employee understanding while minimizing cost to the project. Repetitious training is conducted to allow warehouse operators multiple opportunities to understand and adapt the new automated warehouse to their individual responsibilities. Repetition, however, is not conducting the same training session multiple times. Modern education has revealed the need for information to be related through multiple means to provide a learning opportunity for a wide variety of individuals. This chapter presents structured training methodologies with proven results. These methodologies give each employee multiple medias through which to learn the new automated warehouse system.

Training is broken down into two groups: WMS and MH. The primary focus of this chapter is on training the warehouse operator on the WMS. WMS training begins after the system is designed and continues through the conversion period. Warehouse managers, supervisors, and the system design team are all involved in conducting WMS training sessions. MH training is required for warehouse supervisors and maintenance personnel. This training is typically provided by the material handling equipment suppliers through standard offerings and is often required when operators must use new mechanized equipment such as pick-to-light systems, carousels, and turret trucks. However, MH training must be done prior to system conversion to allow sufficient time to modify maintenance procedures and reconfigure maintenance tool bins and supplies.

WAREHOUSE MANAGEMENT SYSTEM TRAINING

The primary lesson learned from other warehouse managers who have installed a fully automated warehouse is not to skimp on training due to tight schedules and budget constraints. Limiting training to on-the-job experience during the conversion period is inadequate and will add to the project costs in the long run. The work of unqualified operators results in false system problems due to manual work workarounds that then propagate into additional problems, which artificially extend the required system conversion period.

First, warehouse operators are given a full understanding of why a WMS is required and how it will affect their jobs. *Business assessment* training is conducted by warehouse managers and supervisors to justify the business requirements for a new warehouse system. Next, *basic* training is conducted to reduce the culture shock that the new warehouse system might have on the employees. Modern technologies with their complex concepts impose a challenge for personnel who have become accustomed to a comfortable environment. Basic training also previews how the system will work with the warehouse operator in each warehouse department. The final training session is hands-on, where each warehouse operator works with the WMS and MH equipment to perform warehouse tasks. Hands-on training is either conducted in a classroom or on the floor, depending on the number of employees being trained and the amount of cross-functional training desired (see Fig. 12-1).

Business Assessment Training

Business assessment training is named in accordance with the time frame that the training is conducted. While the WMS is being prepared for implementation, business

FIGURE 12-1 WMS training.

assessment training is used to justify why the WMS is required to the warehouse opera-tors. Warehouse operators are fully aware that changes will be imposed within the warehouse from the conducting of interviews through the WMS design session and increased exposure to people from outside the warehouse operation. However, the full WMS impact to the warehouse is not fully understood until after the WMS has been designed. Business assessment training provides management the communication vehi-cle to answer the uncertainties that may have developed among warehouse personnel.

Business assessment training is a series of meetings held with warehouse operators beyond normal departmental gatherings. First, warehouse management provides quali-tative reasons why the WMS is required. Qualitative reasons for change are included in the project justifications (see Chaps. 1 and 4), which include the following examples:

Customer order demand. Customer demand is shifting toward smaller orders with more frequent deliveries. Smaller orders cause the need to reconfigure the warehouse stocking methodology. Increased shipments require more handling with shorter time frames that cannot be supported by the current system without a sig-nificant increase in resource and overtime.

Improved inventory accuracy. Current inventory accuracy problems pass addi-tional costs to the customer that make product prices noncompetitive with those of other companies. Surplus inventory must be carried within the warehouse to ensure customer demands can be met in a timely fashion. Warehouse operations must sup-ply the tools to reduce these costs to stay competitive.

More complex customer service. Customers are requiring advanced shipping information and compliance labeling to streamline their operations. The WMS assists the warehouse operators in providing these services that cannot be efficient-ly and effectively provided by the current warehouse system.

The above sample justifications for a WMS are restricted to qualitative conversa-tions and do not present the quantitative justification numbers used within the return on investment (ROI) analysis. It is important for the warehouse operator not to feel that his or her job is being eliminated. Business assessment training provides ware-house management the opportunity to unite employees based on a combined success and the need for cooperation by all employees and introduces the foreseen changes in individual responsibilities.

Business assessment training is the management's and employee's forum to express concerns over the future. This training phase is not a one-time occurrence. Business assessment sessions are a series of ongoing meetings convened periodically while the WMS is being developed that reinforce when and why the WMS is being pursued. Beyond opening up the communication channels between management and employees, business assessment sessions provide employee input that can affect the WMS design where operating requirements may have been overlooked.

Basic Training

Basic training is conducted 5 to 7 weeks prior to the conversion period (see the section on planning, to follow). Basic training allows warehouse operators to become com-fortable with the new technologies that they will be asked to use within the new WMS. Basic training is the one-time formal training conducted in a classroom envi-ronment that is available off hours to employees to gain extra hands-on experience from a laboratory environment.

Typical basic training sessions are conducted in a series of classroom meetings with a total duration of approximately 40 hours. Basic training is designed to give individual employees an opportunity to understand and use new computer equipment without hands-on running of the WMS in a warehouse environment. Examples of basic training include the following information technologies:

Computer literacy. Many warehouse employees do not understand the basics of how a computer works. Some people view it as a "thinking machine" that will eventually perform their job. The physical computer is shown to the warehouse operators, and components of the computer such as disk drives, tape drives, backup power supplies, and communication modems and ports are explained to expand their understanding of what makes a computer system. In addition, learning the fundamentals of how a program is loaded into the computer's memory to work with individuals while they are physically performing their jobs helps to remove the mystery surrounding the "black box" in the corner.

Keyboard entry. Most warehouse operators do not have keyboard training. The task of hunting for the correct key to hit can be burdensome and a roadblock to using the WMS. While the warehouse operator is not expected to become an expert typist, understanding where keys are and what special function keys do can save several seconds while interacting with the WMS to perform their job. In addition, RF terminals typically have a compact keyboard and require several keystrokes to perform a task that takes only a single keystroke on a CRT or PC keyboard. RF terminal keyboards are frequently more difficult to read and the key locations differ dramatically from those on a CRT or PC keyboard (see Fig. 12-2).

RF communication. Most warehouse operators are familiar with the concept of radio frequency communication thanks to the proliferation of cellular phones in the everyday environment. The RF system is illustrated to all employees so that they fully understand how computer information is passed to their mobile terminal. In addition, operators are taught how to handle terminals, what happens when battery power or radio communication is lost, and how frequently batteries should be changed (see Fig. 12-3).

Bar code literacy. Like radio communication, bar coding is familiar to most warehouse operators from exposure in everyday life thanks to UPC labeling in stores. In addition, most warehouses are performing bar code compliance labeling even with a manual warehouse operating environment. However, most operators do not understand the fundamentals of bar code technology and how the white and dark spaces are translated into information (see Chap. 3). In addition, warehouse operators who understand what constitutes a quality bar code can help identify poor bar codes and problems before they proliferate throughout the distribution channel.

Bar code scanning. After understanding how a bar code works, most warehouse operators benefit from hands-on practice of scanning bar codes. Some bar code input devices, such as wands, are more dependent on operator training that others, such as moving beam scanners. Bar code scanning practice without interacting with the WMS helps the employee understand when there is a bar code input problem or a problem with the system during WMS conversion. In addition, warehouse operators are trained how to handle and maintain their scanning devices to provide long-term performance.

Bar code printer operation. The introduction of the WMS will bring with it the need to have warehouse operators use bar code printers. Operators will appreciate

FIGURE 12-2 Examples of RF reduced size or function keyboards, in truck-mount (top) and hand-held (bottom) setups. (*Courtesy of Teklogix, Inc.*)

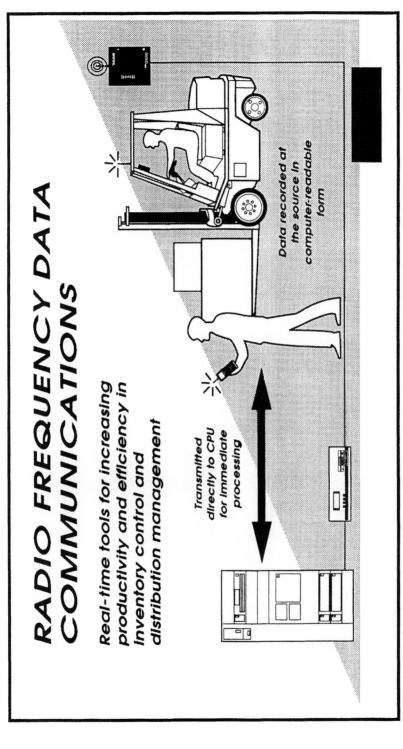

FIGURE 12-3 Example of RF communication within a warehouse environment. (*Courtesy of Teklogix, Inc.*)

the need for quality bar codes after understanding bar code fundamentals described above. Bar code printers, especially thermal printers, require regular ribbon replacement and maintenance to ensure high-quality printing. Operators are trained how to replace ribbons or toner cartridges, add label stock, and know when the printer is on-line and ready to print. This spreads the responsibility of a single maintenance person to keep printers up and running to the warehouse operators who will be interacting with the printers throughout their everyday tasks (see Fig. 12-4).

FIGURE 12-4   Typical bar code label printer used in warehouse operations. (*Courtesy of Zebra, Inc.*)

In addition to the computer and peripheral training above, basic training for the warehouse operators includes the introduction to the WMS and how it interfaces with everyday warehouse activities. The fundamental concept is to provide a functional overview that assists the warehouse operator in understanding the full WMS scope in a classroom using a traditional presentation media.

The basis for the functional overview is the external design or user's manual supplied with the WMS (either internally created or from the vendor). However, the format of these documents is not traditionally suitable for classroom training of warehouse operators. The user's manual or external design is used to extract key concepts that are reformatted into a functional overview training guide.

The functional overview training guide is a presentation document that is used with standard audiovisual equipment (such as a slide projector or an overhead projector). The presentation is created such that it teaches the general WMS concepts without the detailed system rules or hands-on practice. While the creation of a functional overview presentation may seem extravagant, this training guide allows a warehouse operator to understand the entire WMS functionality and is used repeatedly to train new employees and to support training at other facilities implementing a WMS.

One effective form of this type of presentation is a "comic book" format. A comic book format relates the physical action to the corresponding system entry screen and explains what action is expected from the operator. Figures 12-5 and 12-6 illustrate the concept behind a comic book functional overview training guide.

Figure 12-5 explains basic terms as related to the WMS. For instance, a location can be a selective rack, a dock door, a drop zone, or a fork truck. A movable unit is any container that holds inventory such as a tote, shipping carton, or pallet. As appropriate, physical examples, such as location and movable unit labels, should be used to emphasize the concepts discussed in the presentation.

Figure 12-6 shows an example of picking pieces through a pick-and-pack operation. Store orders for less than a full-carton quantity of an item are picked into a tote from carton flow rack and shelving locations. Multiple stores are picked into individual totes by a picker as they pass through the piece pick zone. The picker initiates the process by scanning an empty tote for each store being picked. After all totes have been scanned, the picker is directed to pick items for each assigned store until the tote is full or the store is complete. The comic book overview allows the employee to associate screens with actions and the physical environment without the pressure of using the system. Comic book training shows the routine operation but does not explore the wide variety of exception handling that is part of hands-on training.

Finally, basic training contains specialized systems operator training sessions to instruct the warehouse's systems operator on how to install the WMS and how to perform regular maintenance. This training comes directly from the WMS installation and maintenance guide that is provided from the WMS vendor or programming group. The WMS installation and maintenance guide provides a detailed listing of the files and fields that must be initialized before the system can be put into production. This guide also provides instructions on how to submit system background jobs that are required (such as host interfaces) and how and when to initiate system maintenance functions (such as tape backup and reorganizing disk space). In addition, the computer supplier provides training on regular maintenance procedures and how to manage the system's error log. A warehouse systems operator should be trained on how to manage both the warehouse management and computer systems even if the internal IS department is to provide periodic system maintenance.

Hands-On Training

Hands-on training is the final phase of training that is provided on the WMS. This final training session should result in 80 to 90 percent system comprehension on routine functions and 40 to 50 percent system comprehension on exception handling, assuming completion of prior training phases. Routine functions are defined as everyday system use that does not require problem handling. Warehouse operators leave training with a thorough understanding of basic system interactions. It should be anticipated that the conversion period will also serve as an informal training session, allowing warehouse operators to understand the WMS's process and functional capabilities better.

Location Types

Storage locations are locations that hold items on a semipermanent basis. Case, shelving, flow racks, and selective racks locations fall into this category.

Generic locations are locations that hold items while they are in transit from one location to another. The receiving and shipping docks, drop zones, fork trucks, and conveyor are examples of this type of location.

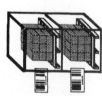

Movable Units

The WMS utilizes a concept called the movable unit. A movable unit is any container that is capable of holding inventory and can be moved from one location to another. Examples of movable units are totes and pallets. The WMS tracks the inventory being moved in each of these movable units so that the system will always know where in the DC an item is located. The movable unit concept also provides the user with the exact location of an item from receiving to storage, to a container on the conveyor, to the dock, and even to a truck. The movable unit is identified with a bar code label that can be scanned at any time to provide information regarding what items are on the unit and what location the unit is in.

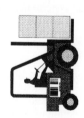

FIGURE 12-5   Basic terms associated with the WMS.

Pick and Pack

Pick and pack is used for filling store orders with less than case quantities. Start picking by selecting the pick and pack function on the RF terminal.

```
WMS Main Menu
Enter Option ___ 1 or Fast Path ID ___

1 Full Pallet Picking        FPP
2 Full Case Picking          FCP
3 Pick and Pack Picking      PPP
4 Location Inquiry           LIQ
```

Retrieve empty totes at the start of the picking area and scan each tote into the WMS. The WMS will assign a store order to each tote scanned. When all totes are scanned, press the enter key to begin picking.

```
                    PCK006R PMTLCC
Scan tote to
have store order      F1=Help

Assigned

Scan Unit Number
_____
```

The system the first location from which to pick. Scan the location address label to indicate to the system you are at the correct location.

```
Go to location      PCK006R PCKFMT
60-001-01-3          F1= Help

                     F3=Exit
Scan loc _____    F8= Cycle Count

or Enter Ref ID ____
```

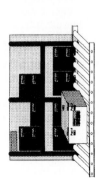

FIGURE 12-6   Comic book format example for pick-and-pack picking.

Hands-on training is started no later than 4 to 5 weeks prior to the planned conversion date and is typically performed within a 3- to 4-week period when installing a full-function (receiving through shipping) WMS (see the section on planning below). Acceptance testing for a function should be complete prior to conducting hands-on training to minimized trainee confusion with system errors that may arise. Hands-on training can be conducted in two formats: classroom or on-the-floor.

Classroom Training. Hands-on classroom training is typically used for training large groups (greater than five warehouse operators) in a single class and/or to provide training across several functional areas. A single classroom training session is focused on a specific functional area such as receiving, put-away, picking, or shipping. Each functional area training session is conducted within a 1- to 2-day period for all warehouse operators in which several training scripts are used to simulate various operating conditions. A warehouse operator receives cross-functional training by attending multiple hands-on sessions.

Classroom hands-on training has the advantage of allowing warehouse operators to work in small teams while working through several training-script scenarios. Small teams allow operators to help one another understand and relate to the system while providing an operator the opportunity to observe other operators using the system. Typically, a warehouse operator is more open to learning and discussing everyday problems with other warehouse operators than if taught solely by a trainer.

Classroom hands-on training also has the benefit of repetition. Training is conducted by creating data sets that each operator (or team) can work within. Training data is created by using acceptance test data or by capturing "live" data downloaded from the host system. For instance, a set of purchase orders and customer orders are downloaded from the host system. A subset of warehouse data for locations, items, and other master files is also loaded into the training data set. After the training script is performed for receiving, the data is saved into a new data set that can be used for put-away training. This process is repeated for all remaining functions, and data sets are restored during training sessions to allow repetition of a function.

Classroom hands-on training, however, has the disadvantage of not allowing the physical movement of "real" inventory and containers through the warehouse. The absence of the physical environment sometimes acts as a roadblock to a warehouse operator connecting the WMS's informational direction with everyday activities and MH interaction.

Every attempt should be made to simulate the physical environment for classroom training. For instance, the system input devices used in the warehouse should be available in the classroom. If possible, do not perform a function designed for RF operation on a CRT. In addition, trainees are required to scan bar codes and move miniature (or simulated) products and containers based on the system's direction. For instance, a set of shelving can hold cartons that simulate pallet racking and pallets to allow the operator to perform an operation physically that reinforces confidence in the system. An alternative to physical shelving is to draw a schematic of storage locations on a white board (chalkboard), use enlarged photographs, or use large-scale engineering renderings. These simulated locations are labeled with bar code addresses to allow bar code license plate labels for simulated inventory to be physically placed in the locations.

Furthermore, the hurdle of associating information movement with physical movement is usually overcome by conducting "what-if" scenarios with the class after warehouse operators have had the opportunity to work through the training script. What-if scenarios are nothing more than verbally repeating sections of the training script to reassure operators that they can apply what they have learned without directly reading a script. In addition, classroom training sessions should be videotaped to provide operators the opportunity to review system functions with which they are uncomfortable.

On-the-Floor Training.   On-the-floor training is typically used for training small groups (less than five warehouse operators per trainer) on a specific function. On-the-floor training is usually done to expedite WMS start-up procedures by limiting the amount of cross-functional and exception training provided to multiple warehouse operators. This type of training can be accomplished within the same time duration as classroom hands-on training (above), but individual training sessions are typically limited to 2 to 4 hours.

The same training data sets described for classroom training are used in on-the-floor training. Instead of a classroom environment, however, on-the-floor training is accomplished by partitioning an area of the warehouse (including racking, fork trucks, and so forth) for training sessions. A slow-moving or off-season product is used to allow movement of inventory within the training area.

On-the-floor training has the advantage of allowing warehouse operators to train within the physical environment in which they will be working. The warehouse environment helps to enforce operating procedures and allows additional practice in scanning bar codes at a distance. It also improves overall integration training through the use of the WMS with the various MH equipment required to complete a task or function.

On-the-floor training, however, has the disadvantage of limiting the number of people who can train within a session. Effective training of large groups in a single session is limited due to the need to "chase" the warehouse operator who is actively interacting with the WMS. Also, repetition training is more difficult because inventory must be physically reset at the beginning of each training session.

Hands-on training is the most important form of training that is conducted, yet is the most difficult to perform adequately. It is important to focus hands-on training on the functions that are complex in tomorrow's environment not today's. For instance, picking within the current warehouse system is usually the most complex function requiring the highest skill operators to handle the paper environment and to follow procedures to ensure quality service to the customer. In addition, picking in the current system includes determining where inventory is located and proper staging into the shipping area or trailer. However, picking is typically a much simpler operation within the automated warehouse environment. The picker is simply required to remove the correct amount of inventory and build a shipping container (i.e., pallet) or place inventory onto an automatic transport (i.e., conveyor) to shipping. However, replenishment may require additional operator training within the WMS to ensure that the operator knows procedures for preparing inventory to put into a primary pick (fixed SKU) location along with how to handle problems if inventory or locations appear incorrect.

As described above, classroom and on-the-floor hands-on training both have their advantages and disadvantages. The benefits of the two types of training can be achieved by utilizing both within the hands-on phase. For instance, the classroom is used to provide an overview and demonstration of a function. In this portion, exception handling and operating procedures are emphasized to large groups of operators. On-the-floor training is then used to repeat the classroom demonstration on an individual basis with the warehouse operator performing the training-script instructions.

Both forms of hands-on training require training scripts. A training script is written to focus on a specific task within a function. Multiple training scripts are usually required to provide a complete functional area (receiving, put-away, and so forth) training session. For example, a training script for a multiple-item put-away activity is provided in Fig. 12-7. The script focuses on the single task of identifying the totes for which to pick, the system directing the operator to a location, and the operator confirming the quantity picked. Other picking training scripts are required to teach the operator how to pick multiple stores at a time and how to handle exceptions such as placing a location on cycle count when the item or quantity is incorrect or dropping a

Pick-and-pack training script (1 of 5): A single store order

Instruction number	User task	Data	System response
1	• Select the option for pick-and-pack picking • Press Enter		• Screen 1 displays to allow scanning of totes to pick into
2	• Scan the movable unit label for the first empty tote for which to pick • Press Enter	Unit = XXX0001, where XXX = your initials	• The system ensures the tote is valid and not assigned, and will assign a group pick for a store to the tote • Screen 1 is redisplayed
3	• Note: only one store is being picked • Press Enter		• The system displays screen 2 indicating the first pick location for the store
4	• Scan the location address label	Location = 60001013	• The system validates that the correct pick locations is entered • Screen 3 is displayed showing the item's description and quantity to pick
5	• Verify the item's description • Enter the quantity picked • Scan the tote picked into • Press Enter	Quantity to pick = 5; unit = XXX0001	• The system validates the quantity picked and the correct tote movable unit label was scanned • Screen 2 is displayed with the next pick location
6	• Scan the location address label	Location = 60003021	• The system validates that the correct pick location is entered • Screen 3 is displayed showing the item's description and quantity to pick
7	• Verify the item's description • Enter the quantity picked • Scan the tote picked into • Press Enter	Quantity to pick = 10; unit = XXX0001	• The system validates the quantity picked and the correct tote movable unit label was scanned • No more picks remain • Screen 4 is displayed to allow the power conveyor (take-away) to be scanned
8	• Scan the conveyor location label; push tote onto take-away conveyor • Press Enter	Location = CONVYR	• The system validates the power conveyor was scanned • Screen 1 is displayed to allow a new tote or store to be picked —END OF SCRIPT—

FIGURE 12-7   Sample script for hands-on training.

tote prior to completing picking for a store. The training scripts also incorporate standard operating procedures that the operator is required to follow such as removing empty cartons from the pick location or placing a store bar code label on a tote.

Training scripts are written using generalized data. For instance, if an operator is required to create a purchase order for receiving, the purchase order number is created with the trainee's initials as a prefix to avoid conflict with other operators that are training simultaneously. In addition, data sheets for items, carriers, customers, and vendors used within the training scripts should be provided to clarify training data. Training data sheets should include printed bills of lading, packing lists, and purchase orders if required by the function.

The final step in the training process is to access each employee's level of system understanding. Training is ineffective if a person is unable to perform his or her job with minimal guidance. An employee should be given the opportunity for additional training if he or she is not able to perform assignments adequately. Guidelines for accessing an employee's new skills and system comprehension are as follows:

1. *The ability to use the language of the discipline.* New terminology will be introduced with the WMS. A good indicator of concept comprehension is the fluent use of new terms and proper reference and application to jobs.

2. *The ability to perform in unanticipated situations.* Exception-handling functions are the most complicated system capabilities to teach and learn. The difficulty comes from the ability to create all possible exception occurrences adequately. An employee who understands the system will be able to apply it to problems he or she has experienced and to anticipate situations. Trainees should be encouraged to ask questions during training that reflect problems they have experienced in the past.

3. *The ability to show, explain, or teach the skill to another person.* Employees are encouraged to work in teams during the hands-on training. The ability of an employee to teach other team members after the initial concept has been presented is a good indicator of comprehension level.

SYSTEM TRAINING PLANNING

Coordination and timing of warehouse employee training require careful planning. If training is conducted too far in advance of system conversion, many concepts and valuable knowledge may be lost. Likewise, if inadequate time is allotted to training, system conversion to the new WMS can extend beyond the planned period and potentially impact customer service and increase overall project costs.

When the various training sessions should be conducted is dependent on two elements: planned conversion date and completion of system acceptance testing. First, the planned conversion date is established using the guidelines presented in Chap. 13. Training should be completed no less than one week and no more than 2 weeks prior to the conversion date.

Next, a start date is established for hands-on training. The start date for hands-on training is established using the table shown in Fig. 12-8 with the following guidelines:

1. *Establish the training topics.* Training topics are identified by the functions that are incorporated within the WMS. It may be necessary to create subtopics for the training plan to manage the training for large functional areas (such as receiving) or to provide supervisor and management training for functions not required for warehouse operators (such as inquiries, master-file maintenance, productivity tracking, and so forth).

Training topic	Optimum number of trainees	Total employees to train	Number of sessions required	Duration of a session	Day of week to conduct training
Receiving	6	9	2	6 hours	Mon., Wed.
Put-away	4	7	2	2 hours	Mon.

FIGURE 12-8   Hands-on training plan estimation table.

2. *Estimate the optimum number of trainees per session.*   The number of employees per session is dependent on the type of hands-on training being conducted and should follow the guidelines provided for in the classroom and on-the-floor discussions above. Keep in mind that the optimum number of people per session is also dependent on the complexity and exception processing required to perform a task within the new system.

3. *Establish the number of employees that require training on each topic.*   The number of employees to be trained for a functional area is dependent on the extent of cross-functional training desired prior to conversion. Cross-functional training can also be divided into phases such as phase 1, for receiving, cycle count, and put-away operations, and phase 2, for picking, replenishment, and shipping operations.

4. *Calculate the number of sessions required.*   Divide the number of employees per topic by the optimum number per session. This results in the number of sessions per topic. The project plan should also reflect at least one make-up session per topic for those employees requiring additional training to perform independently or for employees to make up for sessions that they could not attend.

5. *Plan the days to conduct training.*   Consideration must be given to the timing of training sessions relative to a warehouse operator's normal working schedule. Training sessions can be held before, during, or after a regular workday or can be held on a weekend day. Also, plan hands-on training so as to minimize the extra burden on the employees. This will improve the trainees' attitude. Training before or after normal hours is likely to increase an employee's burden by lengthening an already full day. Likewise, training during regular hours puts an extra burden on employees if their responsibilities are not reduced during training periods. While training on "off" days, such as a weekend day, may be initially resisted by employees, this option typically provides the most positive training results. Employees are typically more relaxed, are better rested, and have a better mental attitude toward learning.

6. *Coordinate the training sessions.*   With the number of training sessions and weekly time frames established, hands-on training sessions can be added to the project plan. Remember that personal schedules along with completion of other project tasks, such as acceptance testing, may affect the tentative training planned up to this point.

The project plan must also account for time to create training scripts and data. The amount of time required to create this training package is dependent on the source of data (use of test data or live data), the type of training (classroom, on-the-floor, or combined), the number of people available to create training scripts and data, and the extent of formal training required on each topic. Typically, 6 to 8 weeks should be planned for at least two people to create a training package if starting from scratch.

The project plan should not only reflect the hands-on training sessions, but should identify when basic and business assessment training is to be conducted. The Gantt

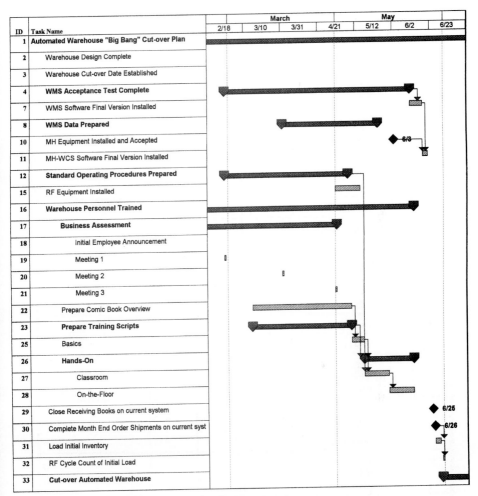

FIGURE 12-9 Sample project plan for WMS training tasks (16 through 28).

chart shown in Fig. 12-9 reflects the estimated durations and project dependencies for a WMS training tasks.

When purchasing a WMS package, it is important to determine the level of training the vendor supplies. Figure 12-10 provides a list of training categories that should be evaluated when deciding on a WMS package. Vendors are evaluated on the extent of training provided and whether it meets the company's expectations. In addition to the topics reviewed earlier in this chapter, the evaluation matrix addresses whether the vendor provides training to all employees or training to the company's training staff. The latter method is commonly called "train the trainers," in which the vendors provide training for a specified number of people who are responsible for training of warehouse operators.

Vendor Training Evaluation	Evaluation Criteria C=Completely Meets Requirement P=Partially Meets Requirement N=Not Provided			
Req. #	Description	Vendor A	Vendor B	Vendor C	Comments
1	Basics Training				
1a	Keyboard				
1b	Printers				
1c	Scanning				
1d	Bar Codes				
1e	Comic Book Overview				
2	Hands-on				
2a	Classroom (On-site or Vendor Location)				
2b	On-the-Floor				
2c	Training Scripts				
2d	Data Preparation (Provided or Existing)				
2e	Train the Trainers				
2f	Train All Employees				
2g	Video Tapes (Provided/ Allowed)				
3	Cutover Assistance				
3a	On-site				
3b	Remote				

FIGURE 12-10   Evaluation matrix for vendor-supplied WMS training.

MATERIAL HANDLING TRAINING

Automation of a warehouse often contains new material handling (MH) equipment to improve product handling. New MH equipment can range in sophistication from a turret truck replacing current counterbalance fork trucks to automatic storage and retrieval systems (ASRS) replacing current manual storage and retrieval. Training on new equipment must be provided to personnel who will maintain and work with this equipment.

Training for new MH equipment is typically provided by MH suppliers as part of their contract. Special or extended training beyond a supplier's standard offering should be explicitly requested in the contract with the MH supplier. The purpose of this section is to highlight the training topics that should be provided by the MH supplier. The three main areas of MH training that will be reviewed include operations, maintenance, and software support.

Operations

Operations training teaches warehouse operators how to safely and efficiently operate the MH equipment. Operations training is typically provided on-site by the MH supplier through a packaged training program. Operations training sessions should be videotaped for training that is outside the initial offering from the MH supplier. These training tapes can be used for review sessions and new hire education. The following are typical deliverables for an operations training package:

- *Review of terminology associated with the equipment.* New MH equipment will have new terms and functions. Every operator is made familiar with the MH equipment's purpose and common terms used in everyday operation.

- *Use of each operator control.* Each operator is trained on every control device for a piece of MH equipment regardless of its simplicity or familiarity for experienced operators.

- *Review of safety features and practices.* Operator safety is of primary importance while using a piece of MH equipment. Training to avoid unsafe operation or situations can avoid most problem situations.

- *What to do in emergency situations.* Each operator is made aware of possible emergency situations and is taught procedures and guidelines on how to react.

- *Troubleshooting procedures for the operator.* Troubleshooting addresses error recovery procedures that can be conducted by an operator. This type of training can limit the MH operation downtime if the operator can diagnose and resolve a problem without a supervisor's or maintenance person's involvement.

- *Features and function of the MH equipment.* Functional and operational training reviews the capabilities of a system and provides hands-on training for each operator and operation certification by the supplier (if applicable).

Maintenance

Maintenance of MH handling equipment is commonly done "in-house" by the warehouse's maintenance department. Electronic technicians and mechanics should be trained on the new equipment's regular maintenance requirements and trained to perform maintenance procedures. A preventive maintenance program is imperative to avoid unnecessary breakdowns and to provide maximum equipment utilization. The following maintenance topics should be provided by the MH equipment supplier:

- *Mechanical system detailed review.* Mechanical system training provides mechanics with knowledge of all mechanical system components and their capacity limits.

- *Electrical system requirements and detailed review.* Electrical system training provides the electricians with the equipment power and signal system requirements and safety practices.

- *Control system points and detailed review.*   Control system training provides both mechanical and electrical maintenance personnel with a working knowledge of all controls and the control system block diagrams.

- *Preventive maintenance and schedules.*   This training provides mechanical and electrical personnel the ability to conduct preventive maintenance procedures. Current warehouse maintenance plans should be updated to reflect these new requirements to ensure full-time equipment availability.

- *Parts removal and replacement.*   This training provides electrical and mechanical personnel with certification to identify, repair, adjust, and replace all devices for the MH equipment in a safe manner.

- *Troubleshooting and diagnostic procedures.*   This training provides electrical and mechanical personnel with system tests that can be conducted to identify a problem's source.

After mechanical and electrical system training is complete, tool and parts bins should be updated to allow proper and timely maintenance. In addition, the MH equipment supplier can supply videotapes to assist maintenance personnel on how to perform specific tasks. The MH should provide multiple copies of the following documents:

- Training handouts and graphical aids
- Electrical manual
- Mechanical manual
- Electrical and mechanical logic diagrams and charts
- Recommended spare parts list
- Manufacturer's component catalog

Software Support

Warehouse supervisors and system support and maintenance personnel should be trained on how to interface with warehouse control systems (WCSs) that may be required for the MH equipments operations. Pick-to-light, ASRS, carousel, and conveyor sortation systems are common MH equipment types that require a WCS. WCS training is provided by the MH supplier or WCS vendor and is customized according to the WCS implementation characteristics. The project plan should reflect MH training tasks, incorporate the supplier's recommendations, and coordinate with the system's planned conversion date. The following is a list of typical WCS training requirements:

- *WCS power-up and power-down procedures.*   Warehouse personnel are taught how to safely power-up and power-down the WCS to minimize loss of data and program problems.

- *WCS start and stop procedures.*   Warehouse personnel are taught the program commands for loading and ending the WCS software for regular maintenance and are shown the problems that might arise.

- *Computer program loading and backup requirements.*   Warehouse personnel are trained on how to load the WCS software and are made aware of program and data backup requirements. It is common to provide a modem attachment to the WCS computer to allow vendor dial-up to provide regular software updates and troubleshooting assistance.

- *WCS console operation (if applicable).* Warehouse personnel must be competent in understanding the WCS console display. Console displays vary significantly in ease of use. Warehouse personnel are shown the timing of information updates and how to use the information to diagnose problems.

- *Operation of peripheral equipment.* Warehouse personnel are taught how peripheral equipment, such as in-line scanners or scales, is interfaced with the WCS and how to troubleshoot this equipment. Regular maintenance and calibration procedures are also demonstrated to warehouse personnel.

- *How to use operator interfaces.* Operator interfaces are common in MH equipment that must interface with a WCS to control picking. All warehouse operators are taught procedures on how to interact with the WCS control points.

BIBLIOGRAPHY

Caine, Renate, and Caine, G., "Making Connections: Teaching and the Human Brain," Association for Supervision and Curriculum Development, May 1995, p. 156.

Uerkwitz, Janell, *Indiana Mathematics Teacher*, Vol. 6, No. 1, 1991, pp. 18–20.

CHAPTER 13
SYSTEM CONVERSION

INTRODUCTION

Long-anticipated benefits of an automated warehouse are realized following a well-planned and successful system conversion. The system conversion marks the beginning of increased productivity and better customer service within the warehouse. This phase of the automated warehouse implementation process is initiated after the warehouse management system (WMS) and material handling (MH) equipment have undergone rigorous testing, problems have been resolved, and operator training has been completed. Planning for this phase, however, begins during the design phase where the "go live," or production start-up, milestone for the automated warehouse is established.

Conversion of the automated warehouse from the current warehouse system is critical to the acceptance of the new processes within the company and benefits realized by the customers. Careful planning and execution are necessary to minimize business risks, obtain acceptance from warehouse personnel and related departments, and ensure that customers receive uninterrupted service. Since the automated warehouse effects are far reaching, the conversion process should not be expected to be a minimal effort. The conversion process may range from 3 to 9 months before full benefits are realized (see Fig. 13-1).

Numerous conversion methodologies exist. The best scenario for a company depends on its business environment. The "big bang" conversion methodology installs all warehouse operations from receiving through shipping, including modified host-system applications and interfaces, simultaneously. This plan is most widely used for implementing the automated warehouse at additional facilities or when the current system is manual and/or the installation period is during a low shipping volume cycle. A "phased" conversion methodology allows isolation of specific warehouse operations that are switched to new processes and procedures and limits the risk associated with initial system installation. Although the selection of the correct phased conversion plan depends on the automated warehouse's scope, a typical phased functional grouping includes receiving through put-away (inbound) and order processing through shipping (outbound) operations. A phased conversion methodology is commonly used when high shipping volumes are expected over the installation cycle and/or where customer order fulfillment is complex and thereby requires new automated processes.

Further isolation of warehouse operations is obtained by a divisional or customer account conversion methodology that is coupled to the big bang or phased conversion methodologies. Conversion of a division or customer within the warehouse offers a more controlled environment and further risk reduction to company profits. Finally, MH equipment offers another conversion option if its use is not tightly coupled with the new WMS. This conversion option is more easily utilized if the MH equipment is not dependent on information controlled by a warehouse control system (WCS) that is

Old New

FIGURE 13-1 Conversion from the old warehouse system to the new automated system.

likewise independent of information from the WMS (i.e., new fork trucks, conveyor transport with no sortation, new racking, and so forth).

As the conversion plan reduces risk to the business, the conversion period for the entire warehouse typically grows longer. The additional installation time must be weighed against the business risk to select the conversion plan that is right for the warehouse. A positive start-up of the automated warehouse is a reflection of the solid planning and commitment by all employees and builds the platform for future implementations at other facilities.

OVERVIEW

This chapter presents commonly used methodologies for managing the conversion process from an existing warehouse operation to an automated warehouse. First, a target date is established for *when* the automated warehouse will be put into production to run the warehouse operation. The conversion date is selected after careful consideration of the business environment, risks, and the overall automated warehouse complexity.

Next, *how to* convert is considered. Many conversion methodologies exist, which include, but are not limited to, the big bang, phased function, divisional product line, and material handling equipment. Each of these conversion options can be performed in parallel (simultaneously) with the current system to further reduce risk to the business.

As the conversion date draws near, plans are put in place to ensure that preparation activities are complete and an implementation team is available. Standard operating procedures, training, inventory preparation, and so forth are completed before warehouse operations are converted to the automated system.

Finally, implementation of the automated warehouse is executed. This stage requires a predefined goal to establish when the new system is deemed operational and accepted. Procedures are put into place to manage problems as they arise. Changes to the warehouse system and operating procedures are carefully managed so that their impact on the implementation is minimized. Successful conversion and acceptance of the new automated warehouse is a team effort reflecting the enthusiasm and participation of warehouse operators, warehouse management, and outside departments.

PLANNING

Proper planning for the warehouse system conversion, or implementation, phase is important to the project's success. Even if every portion of the automation project is completed on time and within budget, these successes are overshadowed if severe problems are encountered when switching to the new system. Poor conversion planning can cause poor acceptance by warehouse personnel, prevent or delay customer shipping and invoicing, and cause severe overruns in scheduled time and budget.

The decision on how the automated warehouse is to be converted is made early in the project either during the external design process or shortly after it is completed. Special considerations may be necessary during the design phase to allow a smooth transition from the current operating procedures and system. For instance, it is common practice to run the current system and automated warehouse system in parallel for a short period of time as part of the conversion plan. To assist in the parallel test, special inventory and shipping reports are necessary to allow easy comparison of the two systems. In addition, it is often necessary to have termination programs, such as the assisted picking program discussed in Chap. 5, that permit shipment and billing of customer orders through the current system while interfacing with the new WMS to maintain its inventory and location integrity.

There is not a single conversion plan that is perfect for all warehouses or distribution centers. However, the "best" conversion plan must, at a minimum, take into consideration the following factors:

- *Shipping volume.* If seasonal trends exist where customer shipments peak, then it is best to convert to the automated warehouse during off season when shipping volumes are at their expected low. This helps to limit the impact of problems that might occur during customer order fulfillment. If seasonal trends do not exist, it becomes more important to consider a parallel conversion option. In either case, major account representatives or customers should be contacted prior to the warehouse conversion to advise them of the automation plans and of the benefits that they will derive, and forewarn them of potential delays or errors in shipment or billing.

    Selecting the correct time of month and day of week can further reduce start-up shipping volume errors. If high shipping volume is skewed to the end of month, then conversion of the automated warehouse should be done at the beginning of a month. Furthermore, beginning conversion during a weekend provides workers additional leeway to get orders out on time. For example, picking of customer orders scheduled for Monday and Tuesday shipments can begin on Saturday or Sunday. Shipping volume is thus reduced by allowing the warehouse to spread the initial week of shipments over 7 days instead of the standard 5 days assuming weekend operation is not customary. Be sure to consider, however, that conversion during a weekend has the added expense of overtime and requires cooperation with the order entry department to release orders to the warehouse possibly earlier than usual.

- *Complexity of processes.* The more complex the process, the greater the risk of errors. If simple warehouse processes are employed within the WMS, then less risk is associated with conversion to the new WMS. System complexity issues are usually found in the picking and shipping processes. Most receiving and stocking processes are sufficiently tested prior to conversion and are more easily bypassed or procedurally managed during conversion. Fulfillment of customer orders is where the greatest data variability is encountered and interfacing to other software (warehouse control and host) systems is required.

  For instance, if the automated warehouse does not have significant mechanization, such as conveyor sortation from consolidation picking, or a sophisticated picking subsystem like the pick-to-light system, then manual procedures can be put into place should problems occur during conversion. Straight RF order picking onto customer shipment pallets is more easily bypassed or fixed than sophisticated interfaces and multiple programs interacting to provide picking.

- *Business risk.* The value provided by a warehouse or distribution center is realized only when a product is shipped in a timely manner to the end customer. The consideration of business risk goes beyond conversion during low shipping volumes, described above, and accounts for the ability to utilize another warehouse facility for customer order fulfillment to minimize business exposure. If the warehouse being automated is not the only facility stocking a product line(s), then other facilities may be used as a backup plan for customer order fulfillment if shipping problems occur. In addition, other facilities provide the option of simulating low shipping (or seasonal) volumes by strategically diverting customer order fulfillment (by account or shipping destinations) ahead of the planned conversion period. While this may cause additional shipping charges, it may alleviate the potential for customer dissatisfaction if problems with the implementation occur.

- *Current warehouse system.* If the current warehouse system is manual in nature, then these manual procedures act as a backup plan if WMS problems are encountered. Manual warehouse systems do not have the need for referencing system data that exists in semiautomated (single-function host computer or isolated PC-based-computer locator or shipping system) warehouses. Through standard query facilities, manual pick lists are created to support current manual processes. If a manual system is not available as a backup system, then plans should be put in place to allow conversion back to the current warehouse system should the conversion to the automated system fail.

CONVERSION METHODOLOGIES

Consideration of the above conversion factors leads to a limited set of conversion methodologies that are commonly used in implementing an automated warehouse. The selection of the "best" methodology should be based on minimizing the overall associated project risk while obtaining project completion within an acceptable time frame to the business. The risk of conversion is often overlooked, or assumed to be minimal, in lieu of simply basing methodology selection on the shortest conversion schedule or minimizing "planned" cost. Unforeseen system problems, including factors outside the warehouse's control, can quickly negate the schedule and cost benefits if the best methodology is not selected. The following list of conversion methodologies is presented in decreasing business-risk sequence, but is not intended to be all inclusive and should be modified to meet individual scenarios.

Big Bang

This conversion methodology implements all warehouse operations at the same time (from receiving through shipping and host-system applications). By implementing all WMS and MH systems within a narrow time frame, the warehouse can begin using the newly designed processes sooner. This conversion option is commonly used when the following conditions exist:

1. Low shipping volumes where risk to the business is minimized

2. New MH equipment and processes that require full implementation to be functional

3. Implementation of a fully functional automated warehouse system at additional warehouses

4. WMS implemented as a package with isolated and minimal customizing

The big-bang conversion option should not be used unless adequate integration and acceptance testing have been completed on all WMS and MH components and all problems and required changes have been solved and implemented with regression testing. Problems associated with this type of conversion are frequently not discovered at the problem's source and do not reveal themselves until shipping problems occur. This can make it difficult to determine the cause of a problem, resulting in a prolonged conversion period. Furthermore, risks associated with this conversion option include inaccurate or impeded customer shipments for a period of time, erroneous customer billing due to interface problems with the host system, and reduced acceptance by warehouse operators of the system resulting in nonstandard operating procedures.

Phased Functions

This conversion methodology implements warehouse operations within phases. Each subsequent phase is predicated on the previous phase being fully operational. The number of phases and definition of a phase's functional scope are based on the warehouse's design. For instance, some departments of a warehouse have easily isolated operations such as returns, cycle counting, work-in-process (WIP), ticketing, and so forth. Isolated operations are prime candidates for phased conversion to help simplify the start-up of other more integrated and complicated functions.

For example, a common phased conversion approach is to group all functions that affect inbound merchandise into phase I and all remaining outbound functions into phase II. Inbound-merchandise-related functions include receiving, location management, put-away, cycle count, and inbound interfaces including inventory adjustment. A controlled method of removing inventory, such as assisted picking, is required to relieve inventory if a product needs to be shipped during phase I.

Phased implementation offers the benefit of implementing inbound processes, such as receiving, put-away, and cycle count processes, to gain control over their inventory quickly with reduced risk to the business. In addition, these functions allow operators to become familiar with the operation of new equipment and gain confidence in the system's ability to assist them in doing their job. This conversion methodology is commonly used when the following conditions exist:

1. The WMS has complex picking and shipping functions or equipment, so that the risk of problem occurrence is high. Phased implementation allows the separation of these more complex functions so that problems can be minimized and better contained.

2. Moderate to high customer order volume is anticipated, making it difficult to manage shipping problems.

3. The fully automated warehouse is not available for installation at the same time. This can be caused by customizing requirements where changes need to occur to the WMS's processes or where physical modifications are being made to the warehouse.

4. The automated system is a *first-time* implementation. Live data variance tends to show more varied results when introduced into first-time installations.

Phased conversion offers more control over the start-up period. However, phased conversion typically has a longer plan period due to the requirement of completing one phase before installing another. The plan period can seem heavily skewed in comparison to a big-bang conversion plan, and the actual amount of extra time is dependent on the complexity of problems encountered. In addition, the phased conversion plan typically requires additional programming to allow termination of the WMS's inventory into manual (or current) processes. These programs usually have a limited useful life defined by the conversion period.

Divisional Product Lines

This conversion methodology is used to limit further the scope of the business impact that can occur during WMS implementation. Divisional conversion is used in conjunction with the big-bang and phased conversion plans where warehouse SKU volume and customer order complexity can be limited to a subset of the warehouse's responsibilities. This conversion method can also be implemented for a department or for a specific customer account(s).

The key to a divisional conversion plan is the selection of a division that has product and customer orders that fully exercise a majority of WMS functions. Limited system knowledge and testing are gained if a division is selected for start-up that has overly simplified product handling, such as floor bulk storage and picking or single-line customer orders that are not typical of other divisions within the warehouse. This conversion option is commonly used when the following conditions exist:

1. Warehouse products, or customer accounts, are split logically within division, or departmental, boundaries.

2. A majority of the warehouse operations are utilized within the selected division but significantly limit the product volume over the conversion period.

3. It is a first-time implementation of the automated warehouse system where business impact risk needs to be controlled.

A divisional conversion plan, like a phased conversion plan, typically extends the planned conversion period longer than the big-bang conversion plan. In addition, a division that is selected based on providing the widest product variance but low shipping volumes may not yield the same level of data variance that may be experienced from host systems. This can lead to false security when other division are brought on line.

Material Handling Equipment

New material handling (MH) equipment that is implemented as part of the automated warehouse system can, in some instances, undergo conversion prior to the WMS. As identified in Chap. 9, MH equipment can be classified as independent of or dependent on the warehouse control system (WCS). MH equipment that is WCS independent offers the greatest opportunity for conversion prior to the WMS since it can be operated independent of system control and data. For instance, new fork trucks can be implemented within the current system and operator training and equipment break-in time provided. Likewise, new pallet racking can be implemented within the current warehouse system though its capacity may not be fully utilized.

MH equipment that is WCS dependent requires more planning for cutover in advance of the WMS than does WCS-independent equipment. The feasibility of separate conversion of WCS-dependent MH equipment must be analyzed on an individual basis. For example, a conveyor sortation system is a prime candidate for advanced conversion if the conveyor interface to the WMS is through the use of an intelligent bar code (refer to Chap. 9). It may be possible to utilize a portion of the WMS software to print bar code divert labels manually or manipulate the WCS to accept a simpler version of the intelligent bar code. The benefit of running the conveyor system can be realized even if it simply used for transportation to a predetermined spur rather that sortation to many spurs. However, the advanced conversion of a pick-to-light system is significantly more challenging than a conveyor sortation system due to WMS data dependencies and has less benefit from the advance break-in.

Conversion of MH equipment in advance of the WMS provides an additional debugging and training period for warehouse operators. However, the benefit of this equipment may not be fully realized until the WMS is installed, which may introduce additional MH system stressing. If the MH equipment is heavily dependent on the WMS, operators may adopt a false impression of the equipment's use and benefit if temporary procedures are burdensome.

There are a wide variety of options available in developing a conversion plan. For instance, it is often advisable to use a parallel run time for the WMS with any of the above conversion plans. A parallel run time allows for the current warehouse system to be used as the "system of record" and, thereby, make the ultimate inventory and shipping updates to the host-system applications. This is done by trapping the interface data from the WMS into a reporting mechanism that is compared to current system results. Parallel conversion also requires the warehouse operators to make data entry into both the current and WMS systems. In addition to slowing operators with extra data entry, special reporting programs for the parallel test have a limited useful life. Live data testing resulting from parallel conversion, however, is invaluable in uncovering program problems to secure a highly operational production system.

PLANNING CONSIDERATIONS

After the desired conversion methodology has been selected, a plan of events must be put into place to ensure that proper communication and responsibility assignments are made between vendors, the host system, and warehouse personnel. Inadequate conversion planning, often caused by tight schedules and inadequate staffing, may lead to the short-term success of achieving the conversion target date, but inevitably leads to long-term system failure in achieving the desired results. Numerous tasks must be completed before the WMS is brought into production. This section highlights some considerations that assist in conducting a smooth conversion period. Sample conversion plans are presented at the end of this section.

Audit list has been validated.   Prior to starting the conversion period, an audit of code completion, training completion, data integrity, and other key items listed above should be conducted by the implementation team's project leader. This audit is intended to ensure that all required conversion requirements have been met beyond standard completion reports submitted by the participating groups.

Identify an implementation team.   The implementation team is responsible for the success of the conversion plan. This team consists of core individuals from the design team, warehouse supervisors, and team leaders. The design team refers to project managers and vendor representatives responsible for the initial system design. Supervisors represent the key people responsible for managing the warehouse operations on a daily basis. Team leaders are warehouse operators who are highly respected by their peers and provide guidance and reassurance to their associates while performing warehouse tasks.

Completion of acceptance testing.   (Acceptance testing is described in Chap. 11.) All testing for the functions being converted to production should be complete before use within the warehouse. Testing is considered complete when all problems related to the conversion functions, regardless of severity, have been regression-tested and resolved. It is up to the implementation team's discretion what design changes, if identified, should be implemented before conversion can begin. (Design changes are problems identified during testing that are out of the initial external design's scope.)

Standard operating procedures are in place.   Standard operating procedures (SOPs) are used to define how processes are to be completed by warehouse operators. In addition to providing personnel with basic job and task descriptions, SOPs provide guidelines on how to interact with the WMS and any special activities that are required that are outside the WMS's control. Figure 13-2 illustrates the typical content and format for a SOP describing a pick-and-pack operation using RF terminals. SOPs should be available for each warehouse operation prior to starting the conversion phase.

Personnel have been trained.   Chapter 12, Training, describes the different levels of warehouse operator and system operator training that should be completed. Warehouse operators who have not received adequate training often find ways to work around the WMS to complete their jobs because it is difficult for them to accept changes they do not understand. Even though a warehouse operator has the best of intentions, the effect of these manual workarounds can cause other warehouse operators who are trying to utilize the WMS to become frustrated due to inaccuracies caused by circumventing the system. However, these manual circumventions implemented with a supervisor's approval as a "temporary fix" to a system problem are acceptable when necessary to keep the warehouse operational.

Problem handling procedures have been identified.   Warehouse operators must have a predefined means for reporting problems to supervisors. Supervisors must ensure that problems are reported quickly and accurately, and that any data or circumstances surrounding the problem are recorded. Predefined problem reporting processes ensure that timely fixes to problems are provided to allow the conversion period to be completed on schedule. Problem reporting procedures should include feedback of status and resolution validation to the warehouse operator who reported the problem. Feedback helps the warehouse operator to feel confident that his or her voice was heard.

Manual tracking forms are available.   Tracking forms are used to validate warehouse productivity where host-system data is inadequate or otherwise unavailable. Manual tracking forms should include data captured within the WMS productivity

Pick and Pack Procedures	
(page 1 of 5)	
Overview	Pick and pack case picking is the picking of individual items in less than full carton quantities. Individual items include small-sized items, items requiring additional packaging, and items requiring repack before shipment. Pick and pack locations are located in shelving and flow rack locations in the mezzanine above the west end dock doors. The locations' address range is 60001011 through 69040051.
Procedures	Pick and pack case picking involves removing individual items from a shelf or flow rack location and placing the items into a store tote. The picker uses a hand-held terminal with an attached medium-range scanner. The RF terminal and scanner allow viewing of locations to be picked, the number of stores to pick, the number of items/stores, scanning of the pick location label, and scanning of the store tote after a pick.

General Terms	Location Tag
	    The numeric ID attached to each location bin.
	Store Group
	    The group of stores being picked by the picker (more than one store can be picked at a time).
	Ship Carton
	    After a store is picked, the items will be removed from the pick tote and packed into a ship carton.
	Tote
	    A plastic container used for picking that has a fixed movable unit tag which can be reused after packing.

Prior to Picking	Before the picker can begin picking, a RF terminal and medium-range scanner must be obtained from the picking supervisor. The picker must ensure the equipment is in good condition, the battery is charged, and the scanner is working. The terminal number is written on the sign-out sheet along with the picker's initials.
Beginning the Picking Process	1. Turn on the hand-held RF terminal, and enter your user ID. Press the down arrow and enter your password. Press Enter; the LOCK-H message is displayed while the WMS prepares your picking session. 2. Select the pick and pack option from the RF menu. If this option is not displayed, see your supervisor for assignment.

FIGURE 13-2 Sample standard operating procedure (SOP) for broken-case pick and pack.

files. For instance, during receiving and put-away operations, the warehouse operator records how many movable units and/or items that he or she has processed on a shift. A cycle count operator might record the number of locations counted and discrepancies found. Likewise, picking and shipping operators record the number of movable units processed and the total number of orders filled.

Master-file data integrity has been checked. All master data files, such as location, item, carrier, customer, and so forth, for the production system should be compared to setup data in testing files. This ensures that the production environment setup does not cause unnecessary problems from the onset. All problems, even simple data initialization errors, are seen by warehouse personnel as WMS problems and can result in mishandling of customer shipments.

Page 2 of 5: Pick and Pack Procedures	
Setting Up Totes	3. Walk to the beginning of the pick and pack location, and obtain totes to pick into. At least 3 totes (stores) should be picked every pass (unless less than 3 are available). 4. The WMS prompts for a movable unit tag to be scanned for each tote into the system in order to have an order assigned. After each tote is scanned, the WMS: - Acknowledges a store is assigned, or - Indicates no more stores requiring picking, or - Warns that the tote scanned is invalid (already assigned to a store). Scan another tote until at least 3 stores have been assigned for picking, or no more stores are available. 5. After all totes have been assigned, press Enter without scanning a tote and the WMS will display the first pick location in ascending location sequence according to the stores assigned.
Normal Picking	6. Scan the location label to confirm you are at the correct location. 7. The WMS displays the item description, quantity to pick, and movable unit (tote) to pick into. Confirm the item is correct, enter the quantity picked, and scan the directed tote. 8. The WMS displays the remaining quantity to pick for the same tote from the same location, or for another tote (store) from the same location. Complete picking as in 7. 9. When all picking is complete from a location, the WMS displays the next pick location. Proceed from number 6.
Picking Complete	10. When picking is complete for a tote/store, the WMS requests that the conveyor location be scanned. After scanning the conveyor, push the tote onto the power exit conveyor and confirm complete by pressing the Y key. 11. If a tote is full, press F7 from the item description/quantity screen to push the tote out onto the conveyor. The WMS displays a confirmation screen at which time the conveyor location label is scanned and completion is confirmed by pressing the Y key. The WMS prompts to scan another empty tote if picks remain for the store.
Exception Handling	12. If the displayed item description is not in the confirmed location, press the F8 key to request the cycle count. The WMS determines another location to pick from, and displays the next pick for the stores. 13. If the quantity to pick is not in the location, the quantity available is picked and the tote scanned. When the system prompts for the remaining quantity to be picked, press the F8 key to request a cycle count. 14. If the location is picked empty, the WMS requests that the location be confirmed empty. After ensuring the location is empty, press the Y key. If the location is not empty, press the N key and continue picking.

FIGURE 13-2 *(Continued)*

Start-up inventory has been loaded.   Time must be in the plan for loading of the WMS inventory files unless the warehouse is empty and can be loaded through normal receiving procedures. Loading of warehouse inventory is usually accomplished through a shutdown of the warehouse over an extended weekend. There are two commonly accepted methods for building these inventory files. First, existing location or inventory data can be downloaded from the host system using a specialized interface program. This allows rapid building of the inventory files. However, the data format of the host-system files may not be compatible with the WMS. For example, the host system probably does not track inventory with unique license plate numbers as a WMS does. Also, a physical count of all locations using the RF cycle count module is required after inventory data is downloaded to ensure data accuracy. One of the primary reasons for the new WMS is to give better inventory accuracy, and just accepting the host-system inventory introduces errors from day one and creates a false impression of inaccuracy within the WMS.

A more common approach to loading start-up inventory is to utilize a specialized RF cycle count program as described in Chap. 5. A RF load initial inventory program allows a warehouse operator to scan a location, enter the quantity and item that are in the location, and assign a movable unit ID if necessary. In addition, any locations that have inventory loaded are submitted to the cycle count queue for a count verification to ensure inventory accuracy within the WMS.

Loading of start-up inventory using this RF-based application can usually be conducted within a 3-day period for a large warehouse of 500,000 to 700,000 square feet that is 70 to 80 percent full. This short count period is made possible by populating inventory containers with license plate bar code labels while the current manual system is running the warehouse. This reduces the need to pull containers out of storage locations to apply labels during the initial loading process (which is the time-consuming portion of completing this task).

Host programs are implemented. Host-program modifications are implemented while the warehouse is shut down to audit the inventory and final facility preparation prior to conversion. While the host programs are implemented, host and WMS interfaces are also implemented. The conversion plan should identify the need to enter test data into the host purchase-order and customer order entry systems, transfer the test data through the WMS interfaces, and return test results back to the host system. This ensures that the communication lines are working and results in all host programs having been properly implemented. Item, carrier, customer, vendor, and other common master-file interfaces should have been implemented at least 1 month prior to this shutdown period to allow the warehouse to update the necessary unique warehouse attributes.

Key accounts or customers have been contacted. Contacting key accounts in advance of conversion to the new WMS can allow the customer to give extra attention to product received and create a feedback loop to bypass normal customer complaint channels. The disadvantage of advanced warning is that the customer might reduce the order volume to limit exposure. It is up to the warehouse or distribution manager and sales team to determine if this is a good approach and which accounts should be handled in this way.

Backup plans have been made. All system backup plans and procedures are validated for completeness and understanding by the necessary personnel. All backup plans and procedures should be tested prior to conversion. Backup plans are wide ranging and encompass all warehouse operations.

First, basic computer and RF system redundancy should be tested. Computer backup systems include an uninterrupted power supply (UPS), disk mirroring, duplicate communication controllers and a secondary host-system communication line, and system backup to tape. Computer backup plans may even include a separate off-site processor. Any tape backup should be coupled with off-site storage plans in the event of a fire or other disaster. RF redundancy should include a second terminal controller, overlapping adequate antenna coverage, and at least 5 percent additional RF terminals and scanners (or other technologies used). (Host-system IS departments usually have similar disaster and backup plans in place as described above.)

The WMS software is designed to facilitate backup procedures. For example, if the RF network is unavailable, then paper-driven functions are used to interface with warehouse operations. While the speed and efficiency may decrease with the use of paper, the goal of picking and shipping customer orders can still be achieved. If paper systems are used during a RF system failure, then a cycle count of all inventory and locations should be done after the RF system has been recovered.

Likewise, MH systems should have backup plans. Like tape backup of the computer system, MH backup plans include regular maintenance procedures and stocking of vital parts and accessories. Maintenance procedures are especially important for mechanical equipment such as fork trucks, carousels, and conveyors. The MH backup plan also specifies the procedures for working around each WCS that may exist. For example, supervisors must know how to operate a pick-to-light system in manual mode by utilizing paper pick lists printed from the WMS and inventory look-up within the WCS.

Job descriptions have been identified and updated. Implementing an automated warehouse will dramatically change the roles of most warehouse employees. Prior to the conversion period, all warehouse employees should be informed of their new responsibilities. This not only clarifies what they are expected to do, but fosters additional teamwork required during this start-up phase.

Some roles, such as data-entry clerks, are virtually eliminated with the real-time data tracking of a WMS. Other job descriptions will not only be changed, but new roles created. For example, all warehouse employees are made responsible for ensuring that RF equipment and scanners are properly handled and not abused. Warehouse operators once responsible for stocking of merchandise will have the additional responsibility of picking and replenishment using the new multitasking function. In addition, new case- and piece-picking processes may be defined such that teams are formed for a warehouse zone or functional area (broken-case, non-conveyable, and so forth). Team leaders are usually identified to be responsible for the efficient operation of these areas.

New material handling equipment will likewise change job descriptions. Supervisors or team leaders must be made responsible for monitoring the performance of any WCS systems and related equipment. Maintenance personnel are not only responsible for fork trucks, but have the added responsibility of ensuring that conveyors and other material handling equipment undergo regular maintenance.

Finally, the conversion period requires the identification of team leaders who act as the first line of problem management. Team leaders assist other operators by answering procedural and system questions to ensure quick response to situations that arise. Team leaders have the added responsibility of representing functional areas in weekly review meetings, reporting problems and identifying potential procedural fixes, and identifying process improvement areas.

The actual conversion plan requires very detailed planning to ensure the warehouse is prepared and product volumes are ramped up over a planned period. Figure 13-3 illustrates the conversion portion of a project plan associated with the big-bang methodology.

Tasks 1 through 16 of Fig. 13-3 are preparation steps common to any warehouse implementation plan. The physical conversion portion of the project plan begins with task 32. As previously described under the big-bang conversion methodology, all functions in the automated warehouse are started at the same time (receiving through shipping). However, receipts and shipments should not be planned for full volume capacity on day one. Rather, receipts and shipments should be ramped up to production volume over a period of time. This allows the warehouse to settle into the new processes and

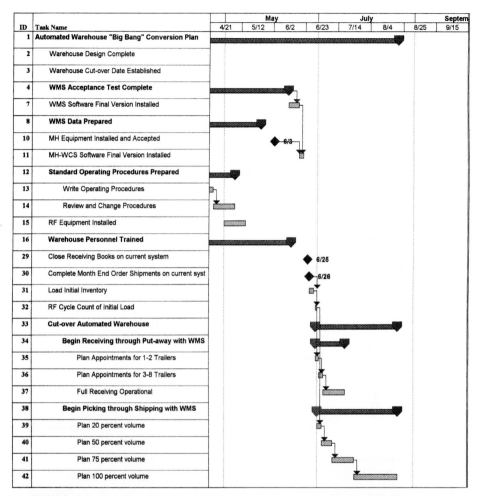

FIGURE 13-3   Example project plan illustrating conversion tasks using the big-bang methodology.

allows adjustments to be made to warehouse processes. In addition, the plan shown can be modified with a parallel plan (described earlier in this chapter) to allow the current system to make updates to host-system inventory and invoicing systems.

A second conversion plan illustration is provided in Fig. 13-4. This plan utilizes the phased-conversion methodology for receiving and shipping operations. Tasks 1 through 30 are the same as the big-bang plan shown above. The phased project plan begins with task 31. The first phase includes receiving through put-away and cycle count operations, which allow the warehouse to gain control over how locations are stocked. The conversion period for phase I (receiving through put-away) is ramped up like the big-bang plan shown above. However, the conversion period for phase I is extended to allow a month-end conversion of phase II from the current shipping system to the automated warehouse. The phased conversion plan also includes a physical count of all locations using the RF cycle count function to ensure inventory accuracy.

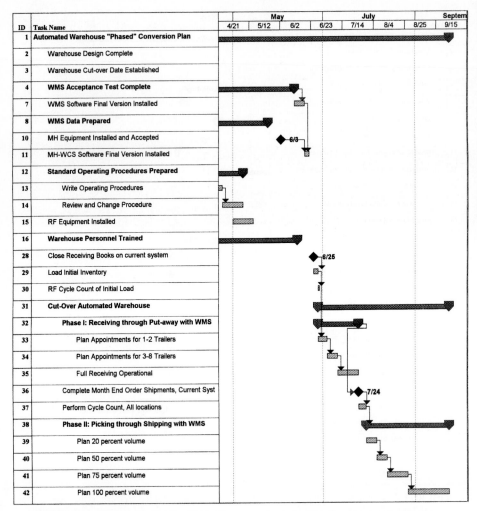

FIGURE 13-4 Example project plan illustrating conversion tasks using the phased methodology.

The phased conversion plan can be modified for additional phases or divisional start-up. In addition, the conversion periods shown can be extended depending on the system's complexity and perceived business risk. Finally, the training task (16) is frequently split into phases to provide training according to the planned operations to be converted, especially phases that extend for more than 1 month.

IMPLEMENTATION

The end of a long and trying project is in sight at this point. The WMS has been designed, MH equipment installed, the entire system has been tested in a "laboratory" environment, and data files have been populated. The final step in the project plan is to convert the WMS into production management of warehouse operations. The execution of this final step requires guidelines for determining when the conversion period is complete along with maintaining the enthusiasm and cooperation of all warehouse personnel.

Completion and Problem Tracking

As part of selecting the method for conversion of the warehouse, completion criteria are established to identify when the conversion period is complete. Completion criteria can take many forms depending on the implementation. One common and effective method of determining conversion completion is by the elimination of high-severity problems and/or implementation of required change requests.

Elimination of high-severity problems during the conversion period is an indication of system stability. The implementation team should agree on the period of time that the WMS should run free of high-severity problems. High-severity problems are those that prevent a function or warehouse operation from being fully functional or prevent accurate receiving or shipment of inventory. (A low-severity problem may be an attribute modification to a screen or a problem that has a workaround that does not significantly impact productivity or accuracy.)

Implementation of change requests. Changes to the design that need to be implemented prior to conversion completion are more subjective and must be evaluated based on the nature of the change and the impact to warehouse operations. Design changes are often required in the new system, and certain impacts to current operations cannot always be foreseen. The wide-ranging effects of the new processes are sometimes difficult to estimate. For example, the designed conveyor throughput may be hampered by the physical orientation of induction points and unanticipated carton volumes. High-velocity item profiles may require rewarehousing of the inventory that was loaded from the old warehouse system. Receiving check-in processes designed to mirror the old warehouse process may result in a gating process to product flow that is actually not necessary with the automated warehouse system. The result is that plans to manage change should be reviewed and the processes and procedures for reporting and prioritization of these changes should be identified.

Daily implementation team meetings. These allow team leaders and supervisors to report on the productivity status, procedural changes, and warehouse functional (system) problems to the warehouse managers and system designers. These meetings are conducted at the beginning of shifts and should be documented with minutes to provide implementation process records.

Productivity status allows the implementation team to identify where receiving and shipping volumes are relative to the design targets. Even if no apparent problems exist with the automated warehouse system, the warehouse may not be achieving the desired throughput that leads to the need to reevaluate procedures. For example, while picking and shipping processes are working as designed without any required changes, throughput may be limited because of how orders or picking waves are being managed from order processing. Procedures for order processing can be refined to designate inventory picking waves to provide better balancing of workload to the warehouse floor. In addition, the warehouse may be shipping all of the orders provided by the host system, but sales and order processing may have diverted order volume away from the warehouse that may be limiting throughput targets.

Daily implementation team meetings allow the team leaders and supervisors to refine procedures based on what seems to be working best. These open forum discussions allow the entire implementation team to review a procedural change and identify if a warehouse system change is required and the potential impact to other functions. Identified procedural changes that are implemented as a permanent process are updated within the SOPs and within the WMS help text and training systems.

Discussion of procedures and productivity results often leads to identification of problems and appropriate change requests. Both system problems and change requests are documented with the data to support problem evaluation and the change request design. Each documented problem and change request is assigned a problem or change severity and priority. The problem or change severity identifies the impact to the business. The higher the severity, the quicker the problem must be fixed. The problem or change priority identifies which problems or changes with the same severity should receive the first resolution. Unlike severity, priority only establishes a ranking and does not usually impact the conversion completion criteria. Periodic review of problem and change request status is included in the implementation team meetings.

Preparation and Climate

Successful completion of the conversion period within the scheduled time frame is not only dependent on a quality design and testing, but is a reflection on the commitment to achieve a singular goal by all warehouse employees. The conversion period will likely involve planned overtime by most if not all employees and the implementation team. In addition, good communication channels between warehouse operators and warehouse management are vital.

Kickoff. The automated warehouse is a significant investment by the company and is a strategic move to provide quality service to customers and to reduce operating expenses. The value of the system to the company should be reflected by a comparable kickoff meeting with all employees involved in the distribution channel. A kickoff meeting with fanfare, T-shirts with a project logo, and a free lunch helps to highlight the company milestone to employees.

Plan for overtime. Conversion to the automated warehouse uncovers unforeseen problems and delays no matter how complete and thorough the conversion plan. Shipping volume targets will not be accomplished within the first day, week, or potentially the first months of operation. The conversion period reflects the need for warehouse operators to adjust to new equipment and processes and the anticipation of problems impacting productivity. Climbing up the learning curve and achieving full system utilization require overtime by the implementation team and at least some warehouse employees. Planning for overtime allows managers and employees to block out personal plans during the conversion period and allows budgeting for overtime costs.

Implementation of a problem reporting hierarchy. The implementation team must be available to work "on the floor" during the conversion period. Portions of the implementation team, namely the system designers, are responsible for fixing problems that arise in a timely fashion. This results in limited and sporadic floor involvement by the designers. A problem reporting hierarchy is established to reduce the burden of answering simple procedural questions from warehouse operators from critical implementation team members. A high percentage of functional problems are simple misinterpretations by a warehouse operator that can be clarified and resolved by team leaders. Warehouse supervisors provide team leaders with clarification of procedures and offer the authority to modify or enhance SOPs, if required, without submitting problem reports. If the warehouse supervisor cannot adequately resolve a reported problem, then proper reporting of the problem and tracking of its occurrence are delivered to the warehouse design representatives for resolution.

Weekly meetings with warehouse personnel conducted at the beginning of each week during the conversion period. Weekly warehouse meetings are conducted by the warehouse manager or supervisor and include members of the implementation team and all warehouse operators. These meetings are designed to inform all employees of the conversion status and present known and fixed problems or changes and to open communication channels to keep morale high. The success of the warehouse system conversion is dependent on the cooperation and commitment of all employees. Communication on warehouse productivity results, awards for employees showing extra effort, and feedback on problems and changes ensure that everyone is properly recognized and involved in the project's success. Finally, these weekly meetings provide a communication channel to keep the performance benchmark moving upward to achieve the warehouse design volumes and allow everyone to focus their efforts on achieving the warehouse performance criteria.

FUTURE CONSIDERATIONS

Successful conversion of an automated warehouse is a significant milestone in a company's business reengineering effort and highlights a thoughtful and quality design. However, no matter how good the design or quality of the implementation, a poor conversion reflects poorly on the entire project and team.

The job of a company's automated warehouse is not complete with the conversion to the new system. Continued attention is required to ensure that the system is returning the anticipated value to the business. What is the inventory accuracy after the first 3 months, 6 months, first year, and so forth? Are the customers receiving accurate shipments? Are customer shipments completed on time? What improvements or changes can be made to further enhance productivity? Are shipping volumes meeting expectations? Questions such as these become a daily inquiry that is facilitated by the new WMS.

Completion of the automated warehouse conversion should be followed by a rigorous documentation effort. It is typical for the project's design and implementation team to be fragmented after successful conversion. The value of the project's experience is lost without good record keeping.

Each team member should submit a "do and don't" list to describe what went right with the project, what went wrong, and any suggestions for improvement. This assists the company and vendors in improving future projects. In addition, the project records should, at a minimum, include the following:

1. Warehouse project plan
2. Design and implementation team's skill inventory and role descriptions
3. All notes and memos itemized sequentially
4. Statements of work, purchase orders, and project bills

5. Hardware and software warranties

6. External and internal designs

7. Training and installation guides

8. System backup plans

9. Maintenance guides for all hardware and software

10. Expected and actual return on investment

Following through the conversion period with the records and audits allows the company to ensure successful automated warehouse implementation to other facilities. These projects have simplified hardware and software design tasks and are accomplished in a fraction of the original project's time frame. Conversion of other facilities to the automated warehouse still requires many of the same project phases including requirements mapping, project planning, personnel training, and conversion.

The success of an initial automated warehouse implementation can breed complacency and overconfidence. Further implementation of the automated warehouse should not be taken for granted. Good project management and attention to details are required to ensure successful implementations. Key personnel, including team leaders and warehouse supervisors from the initial DC, are assigned to an implementation team to help transfer the knowledge and experience gained from the previous implementation. Poor implementations at other warehouses are as expensive to the company as is failure of the original automated warehouse implementation.

INDEX